Conservation Biology

Conservation biology is a flourishing field, but there is still enormous potential for making further use of the science that underpins it. This new series aims to present internationally significant contributions from leading researchers in particularly active areas of conservation biology. It will focus on topics where basic theory is strong and where there are pressing problems for practical conservation. The series will include both single-authored and edited volumes and will adopt a direct and accessible style targetted at interested undergraduates, postgraduates, researchers and university teachers. Books and chapters will be rounded, authoritative accounts of particular areas with the emphasis on review rather than original data papers. The series is the result of a collaboration between the Zoological Society of London and Cambridge University Press. The series editor is Professor Morris Gosling, Professor of Animal Behaviour at the University of Newcastle upon Tyne. The series ethos is that there are unexploited areas of basic science that can help define conservation biology and bring a radical new agenda to the solution of pressing conservation problems.

Published Titles

1. *Conservation in a Changing World*, edited by Georgina Mace, Andrew Balmford & Joshua Ginsberg 0 521 63270 6 (hardcover), 0 521 63445 8 (paperback)
2. *Behaviour and Conservation*, edited by Morris Gosling and William Sutherland 0 521 66230 3 (hardcover), 0 521 66539 6 (paperback)

Behaviour and Conservation

Edited by

L. MORRIS GOSLING
University of Newcastle upon Tyne

AND WILLIAM J. SUTHERLAND
University of East Anglia

THE ZOOLOGICAL
SOCIETY OF LONDON

PUBLISHED BY THE PRESS SYNDICATE OF THE UNIVERSITY OF CAMBRIDGE
The Pitt Building, Trumpington Street, Cambridge, United Kingdom

CAMBRIDGE UNIVERSITY PRESS
The Edinburgh Building, Cambridge CB2 2RU, UK   http://www.cup.cam.ac.uk
40 West 20th Street, New York, NY 10011-4211, USA   http://www.cup.org
10 Stamford Road, Oakleigh, Melbourne 3166, Australia
Rue de Alarcón 13, 28014 Madrid, Spain

First published 2000

Typeset in FFScala 9.75/13 pt [VN]

A catalogue record for this book is available from the British Library

Library of Congress Cataloguing in Publication data

Behaviour and conservation / edited by L. Morris Gosling and William J. Sutherland.
    p.   cm.
   Includes bibliographical references (p.   ).
   ISBN 0 521 66230 3 (hb). – ISBN 0 521 66539 6 (pbk.)
   1. Conservation biology.   2. Animal behaviour.   I. Gosling, L.
Morris, 1943– .   II. Sutherland, William J.
QH75.B45 2000
333.95′ 16–dc21   99-26461 CIP

ISBN 0 521 66230 3 hardback
ISBN 0 521 66539 6 paperback

Transferred to digital printing 2004

Contents

Contributors

MICHEL BAGUETTE
Unité d'Écologie et de Biogéographie, Université Catholique de Louvain, Croix du Sud 4, B-1348 Louvain-la-Neuve, Belgium.

ALLAN J. BAKER
Centre for Biodiversity and Conservation Biology, Royal Ontario Museum, 100 Queen's Park, Toronto, Ontario M5G 2C6, Canada.

PATRICK BATESON
Sub-Department of Animal Behaviour, University of Cambridge, High Street, Madingley, Cambridge CB3 8AA, U.K.

MONIQUE BORGERHOFF MULDER
Department of Anthropology, University of California at Davis, Davis, CA 95616, U.S.A

GRAEME P. BOSWELL
Department of Mathematical Sciences, University of Bath, Bath BA2 7AY, U.K.

ELIZABETH L. BRADSHAW
Veterinary Services, University of Oxford, c/o University Laboratory of Physiology, Parks Road, Oxford OXI 3PS, U.K.

NICK F. BRITTON
Department of Mathematical Sciences, University of Bath, Bath BA2 7AY, U.K.

RICHARD W. G. CALDOW
NERC Institute of Terrestrial Ecology, Furzebrook Research Station, Wareham, Dorset BH20 5AS, U.K.

TIM CARO
Department of Wildlife, Fish & Conservation Biology, University of California at Davis, Davis, CA 95616, U.S.A.

DAVID P. COWAN
Central Science Laboratory, Sand Hutton, York YO4 ILZ, U.K.

SARAH DURANT
Institute of Zoology, Zoological Society of London, Regent's Park, London NWI 4RY, U.K.

SARAH E. A. LE V. DIT DURELL
NERC Institute of Terrestrial Ecology, Furzebrook Research Station, Wareham, Dorset BH20 5AS, U.K.

NIGEL R. FRANKS
Centre for Mathematical Biology & Department of Biology and Biochemistry, University of Bath, Bath BA2 7AY, U.K.

GILLIAN GILBERT
Research Department, Royal Society for the Protection of Birds, The Lodge, Sandy SG19 2DL, U.K.

ELAINE L. GILL
Central Science Laboratory, Sand
Hutton, York YO4 ILZ, U.K.

JENNIFER A. GILL
School of Biological Sciences,
University of East Anglia, Norwich
NR4 7TJ, U.K.

JOSHUA R. GINSBERG
Wildlife Conservation Society, 2300
Southern Boulevard, Bronx, New York
10460-1099, U.S.A.

L. MORRIS GOSLING
Evolution and Behaviour Research
Group, Department of Psychology,
Ridley Building, University of
Newcastle, Newcastle upon Tyne
NE1 7RU, U.K.

JOHN D. GOSS-CUSTARD
NERC Institute of Terrestrial Ecology,
Furzebrook Research Station,
Wareham, Dorset BH20 5AS, U.K.

SIMON JENNINGS
School of Biological Sciences,
University of East Anglia, Norwich
NR4 7TJ, U.K.

OWEN T. LEWIS
Ecology & Evolution, School of Biology,
University of Leeds, Leeds LS2 9JT,
U.K. Current address: NERC Centre for
Population Biology, Imperial College,
Silwood Park, Ascot, Berks SL5 7PY, U.K.

DAVID W. MACDONALD
The Wildlife Conservation Research
Unit, Department of Zoology,
University of Oxford, South Parks Road,
Oxford OX1 3PS, U.K.

RUTH MACE
Department of Anthropology,
University College London, Gower
Street, London WC1E 7BT, U.K.

PETER K. MCGREGOR
Behaviour Group, Zoological Institute,
University of Copenhagen, Tagensvej
16, DK-2200, Copenhagen N, Denmark.

SELWYN MCGRORTY
NERC Institute of Terrestrial Ecology,
Furzebrook Research Station,
Wareham, Dorset BH20 5AS, U.K.

ANDERS PAPE M̌ller
Laboratoire d'Ecologie, CNRS URA 258,
Université Pierre et Marie Curie, Bât. A,
7ème étage, 7 Quai St Bernard, Case
237, F-75252 Paris Cedex 5, France.

KEN J. NORRIS
School of Animal and Microbial
Sciences, University of Reading,
Whiteknights, PO Box 228, Reading,
RG6 6AJ, U.K.

THOMAS M. PEAKE
Behaviour Group, Zoological Institute,
University of Copenhagen, Tagensvej
16, DK-2200 Copenhagen N, Denmark.

RICHARD A. PETTIFOR
Institute of Zoology, Zoological Society
of London, Regent's Park, London NW1
4RY, U.K.

THEUNIS PIERSMA
Netherlands Institute for Sea Research
(NIOZ) and Centre for Ecological and
Evolutionary Studies, University of
Groningen, PO Box 59, 1790 AB Den
Burg, Texel, Netherlands.

JOHN D. REYNOLDS
School of Biological Sciences,
University of East Anglia, Norwich
NR4 7TJ, U.K.

JONATHON C. REYNOLDS
The Game Conservancy, Burgate
Manor, Fordingbridge, Hants SP6 IEF,
U.K.

J. MARCUS ROWCLIFFE
Institute of Zoology, Zoological Society
of London, Regent's Park, London
NW1 4RY, U.K.

LORE RUTTAN
Division of Environmental Sciences and
Policy, University of California at Davis,
Davis, CA 95616, U.S.A.

RICHARD A. STILLMAN
NERC Institute of Terrestrial Ecology,
Furzebrook Research Station,
Wareham, Dorset BH20 5AS, U.K.

WILLIAM J. SUTHERLAND
School of Biological Sciences,
University of East Anglia, Norwich
NR4 7TJ, U.K.

CHRIS D. THOMAS
Ecology & Evolution, School of Biology,
University of Leeds, Leeds LS2 9JT, U.K.

FRANK A. M. TUYTTENS
Department of Mechanisation, Labour
Buildings, Animal Welfare and
Environmental Protection, Van
Gansberghelaan 115, 9820 Merelbeke,
Belgium

MICHAEL P. WALLACE
The San Diego Zoo, PO Box 120551,
San Diego, CA 92112, U.S.A.

ANDREW D. WEST
NERC Institute of Terrestrial Ecology,
Furzebrook Research Station,
Wareham, Dorset BH20 5AS, U.K.

ROSIE WOODROFFE
Department of Biological Sciences,
University of Warwick, Coventry
CV4 7AL, U.K.

PART I

Introduction

Advances in the study of behaviour and their role in conservation

WILLIAM J. SUTHERLAND & L. MORRIS GOSLING

The merging of evolutionary theory and classical ethology produced the field of behavioural ecology, which has flourished over the last 25 years. At the same time, conservation biology has emerged as a loosely knit, but thriving, group of the disciplines that aim to support the conservation of biodiversity. Behavioural ecology has remained principally an area of fundamental research while conservation biology is shaped by increasing concern about accelerating threats to biodiversity and is thus essentially strategic in character. Despite obvious areas of common interest, there has been surprisingly little mixing of behavioural and conservation biology. Most important, there has been very little benefit for conservation from the major advances in the understanding of wild animals that behavioural ecology has produced.

Many conservation projects do in fact include aspects of animal behaviour but usually only in a trivial way. For example, many determine the home range or diet of a species of conservation concern. Typically, these treatments do not refer to the body of theory that is now available and, partly because of this, they are rarely used predictively.

The lack of contact between these areas of theory and application is particularly surprising because many behavioural ecologists have a personal commitment to conservation. A number of field ethologists are deeply involved in conserving their study species and usually to a degree that goes well beyond the selfish motive of ensuring that a population persists until the study is completed. Indeed, some of the sites that are best protected from poaching are those used for behavioural research.

Why then do so few ethologists apply their behavioural skills to answer conservation questions? Or, more generally, why has behavioural ecology not been incorporated into conservation biology, the body of strategic research that is most commonly used to address practical conservation questions? We can think of five main reasons and suspect that all of these sometimes play a role: (1) conservation biology and conservation are not perceived as rigorous subjects and thus are not considered prestigious; (2) there is a cultural separation between behavioural ecology and conservation: most behavioural ecologists work in universities and, because they may not regularly meet practising conservationists who tend to work for state organizations or charities, they have few opportunities to experience real conservation issues; (3) patterns of funding tend to follow and reinforce these cultural divisions: for example, the British Research Councils tend to fund basic research while most funding for conservation projects comes from trusts and charities; (4) there may simply be a historical lag: both behavioural ecology and conservation biology are young disciplines and there is little tradition of combining the two; (5) it is sometimes technically difficult to combine the two subjects. For example, to achieve conservation objectives the role of behaviour is often expressed through population processes and many ethologists may have little expertise in population ecology.

We believe that studies of behaviour and conservation have a great deal to offer each other. This cross-play can happen at a number of levels. For example, the high priority given to conservation helps provide a justification for theoretically based studies of behaviour and this may become increasingly important to justify research spending. Studies of behaviour can also provide essential new insight into intractable conservation problems. Perhaps most important, it can also be argued that an evolutionary understanding of the behaviour of individuals in populations allows us to predict responses under changed conditions with greater confidence than in the case of higher-level processes.

When inviting contributions to this book, we tried to select major developments in behavioural ecology that had important relevance to conservation. These areas are listed in Table 1.1. The table also shows how the chapters relate to the behavioural concepts and their application.

Behavioural ecology and conservation biology have developed largely independently over the last quarter century. Thus, it is surprising that, as shown by the chapters in this book, so many fruitful links have been established between the two fields in the last few years. However, inevitably the occurrence of such links is patchy and there are a number of areas that have rich potential for interesting behavioural research and considerable implications for conservation. We will enlarge on three of these: culture, sensory inputs and the source and consequences of individual behaviour.

Understanding how natural selection and evolution operate provided the key that enabled behavioural ecology to flourish. The understanding of cultural evolution is much less advanced, although there is a theoretical framework for understanding how ideas mutate, are transmitted and are selected for or against (Dawkins, 1976; Cavalli-Svorza & Feldman, 1981; Boyd & Richardson, 1985). We believe that this is a much neglected research area in behaviour. Understanding how culture is transmitted is likely to aid the re-introduction of captive-bred animals (Sutherland, 1998). Many re-introduction attempts have failed because captive bred animals often lack basic survival skills such as the ability to negotiate complex vegetation, recognize and respond to predators, distinguish toxic from palatable food, breed successfully or locate water (Myers *et al.*, 1988; Price, 1989; Beck *et al.*, 1994). Other animals are reintroduced relatively easily after quite simple pre-release procedures (for example, Arabian oryx, *Oryx leucoryx*, Stanley Price, 1989) but we cannot yet predict this variation. Other effects of captivity, such as the ability of animals deprived of extensive social groups for generations to compete for and choose mates, have not been investigated, nor monitored in released populations. Understanding culture transmission will also be essential where it is needed to manipulate the behaviour of animals in protected areas and other sorts of fragmented habitats to help ensure their survival in a changed world. Examples are pioneering attempts to change nesting locations or migration routes (Essen, 1991) to help conserve threatened populations.

There has been relatively little work on the behavioural consequences of the way in which animals sense their environment. Such work will involve the collaboration of physiologists and behavioural ecologists and seems likely to be an exciting research area (Wehner, 1997). Practising conservationists have tricks for encouraging captive and wild animals to breed, care for young, eat novel foods or recognize and avoid predators. However major problems remain, such as behavioural incompatibility in captive-breeding programmes where animals selected for breeding fail to breed; valuable and rare animals, such as clouded leopards, *Neofelis nebulosa*, sometimes fight and may be injured or killed. Two areas of behavioural research have the potential to overcome such problems. The first is the expanding area of signalling theory which promises to provide important generalizations about the way that animals transmit information, and particularly about honest signals of competitive ability and mate quality. Second, it seems likely that detailed understanding of the mechanisms by which animals acquire sensory information may greatly improve the sophistication of

Table 1.1
Developments in behavioural ecology and their implications for conservation

Subject	Consequences for conservation (examples)	Author (chapter number)
Mating systems and sperm competition	Determines effective population size and thus viability	Durant (11)
Sexual selection and status signalling	Costs of display structures reduce mean fitness and thus increase extinction probability	Møller (10)
Sociality and kin selection	Consequences for population ecology and extinction probability	Pettifor et al. (12); Boswell et al. (9)
Life histories	Life-history characteristics and behavioural responses allow an understanding of responses to exploitation	Reynolds & Jennings (14)
	Modelling human reproductive decisions allows realistic prediction of demographic changes	Mace (2)
Evolutionary stable strategies	Can be used to predict density dependence, and thus population size, and the consequences of environmental change	Goss-Custard et al. (5); Gill & Sutherland (4); Pettifor et al. (12)
Social constraints on human resource exploitation	Predicting human impact on habitats and threatened species	Mace (2); Borgerhoff Mulder & Ruttan (3)
Cultural evolution/learning	Pre-release training and behavioural manipulation during re-introduction schemes	Wallace (17)
	Using conditioned taste aversions to protect threatened species from predators	Cowan et al. (16)

Topic	Description	Reference
Communication	Using calling to improve population censuses	McGregor et al. (15)
	Failure to communicate may accelerate population declines	McGregor et al. (15)
Migration	Trading-off fat storage against other physiological requirements	Piersma & Baker (7)
Home range and dispersal	Design of protected areas and predicting local extinction as a result of hunting along edges	Woodruffe & Ginsberg (8), Thomas et al. (6)
	Dispersal can be manipulated to retain individuals that would otherwise leave the area	Thomas et al. (6)
	Reserve shape and habitat structure influence dispersal and the probability of local extinction	Boswell et al. (9)
Foraging behaviour	Understanding habitat choice	Pettifor et al. (12)
Predation and anti-predator behaviour	Predation as a population limiting factor	Caro (13)
	Designing refuges for threatened prey species	Caro (13)
	Using the normal anti-predator behaviour of a species to understand its responses to hunting	Bradshaw & Bateson (19)
Responses to disturbance	Population consequences of disturbance by people	Gill & Sutherland (4)
	Social disruption may lead to increased levels of disease transmission	Tuyttens & Macdonald (18)
	Assessing levels of stress and welfare standards in conservation management and exploitation schemes	Bradshaw & Bateson (19)

captive-breeding techniques. For example, it is now appreciated that the perception of ultraviolet light may be important in determining mate choice in birds (Bennett et al., 1996). Signalling theory and understanding of mechanism will help us to (1) provide the conditions needed for natural patterns of mate choice (for example, glass or plastic screens may reduce ultra violet light and affect mate choice) or (2) manipulate behaviour to achieve practical goals, such as encouraging mating with an animal of low attractiveness (which might otherwise be rejected) to maximize outbreeding in captive-breeding programmes.

Variation in the behaviour of individuals is crucial in explaining demographic processes, such as density dependence, because these are determined by the extent to which density affects the chance of survival or breeding of different individuals (e.g. Sutherland, 1998). For example, individuals may vary in competitive ability and thus in the interference they experience. They may thus differ in the extent to which intake rate declines with density. Individuals may also vary in foraging efficiency or in the intake necessary for survival. Such variation will largely determine which and how many individuals starve at different population levels (Sutherland & Dolman, 1994) and may be related to genotype, environment, parasite load and life history. Understanding the origin and persistence of this variation is thus of fundamental importance in studies of behaviour, in population biology and in practical conservation.

Although the idea of linking behaviour and conservation is very recent, there has been a flurry of interest (Ulfstrand, 1996; Clemmons & Bucholz, 1997; Caro, 1999). The level of enthusiasm was shown by the fact that the joint meeting of the Association for the Study of Animal Behaviour and The Zoological Society of London on which this book was based was one of the best attended of either society. As with most young subjects there is not yet a coherent set of concepts or a core of theory. However, it is pleasing to note a number of repeated themes throughout the book. For example, Woodroffe & Ginsberg (Chapter 8) show that human persecution along the edges of protected areas can explain much of the likelihood of extinction. Thomas et al. (Chapter 6) show that butterflies may also be lost from the edge of habitat blocks by the simple mechanism of wandering off-site. In both cases, critical reserve area can be linked to home range although for the silver-studded blue, *Plebejus argus*, the area is 0.0005 km² while for the African wild dog, *Lycaon pictus*, it is 3606 km²!

ACKNOWLEDGEMENTS

We thank The Zoological Society of London and the Association for the Study of Animal Behaviour for funding the conference that provided a starting point for this book. We are grateful to Unity McDonnell for helping to organize the conference and to Guy Cowlishaw for comments on this chapter.

PART II

Conservation impact of people

2

The evolutionary ecology of human population growth

RUTH MACE

INTRODUCTION

It is the impact of human societies on natural ecosystems that has created the field of conservation biology. Most wild species are in areas where humans either live, or venture into to harvest resources. Humans have always made use of other species and are thought to have been responsible for the extinction of many species, even when living as hunter-gatherers at much lower population densities than today. Over the last 10 000 years, our food has come increasingly from cultivated plants and animals; this has resulted in an expansion in the human population, which has been particularly dramatic in the last 200 years (Figure 2.1). Population growth is at the root of the threat to natural resources. Habitats are being turned from those that are relatively favourable to biodiversity (such as wetlands) to those that are not (such as agricultural monocultures and urban developments). Species are being threatened directly by harvesting, which is becoming more and more technologically advanced. Even when species are not driven extinct, the size of populations and the number of viable sub-populations are being dramatically reduced by these processes.

Population growth inevitably leads to increased demand for resources and an indefinite increase in that demand is not compatible with conserving our resource base. For these reasons, conservationists have often focused on a reduction in human reproductive rate as one of their important objectives. However, the relationship between population and consumption is not always a simple one. Technological advances lie behind both human population growth and the impact each community has on the environment. At present, the populations who consume the most are those from wealthy countries where family sizes tend to be small. Where population growth is rapid, families are typically living frugal lives. An increase in

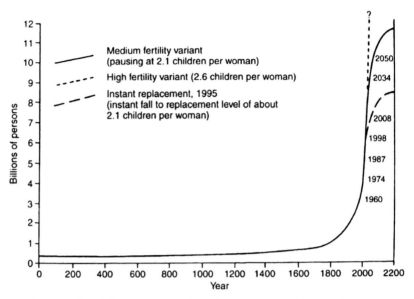

Figure 2.1 Population growth over the last 2000 years, with UN predictions for future trends (source; UN).

standard of living is generally associated with a decline in population growth rate, although the precise relationship between economic development and fertility decline is complex and not fully understood. This demographic transition from large to smaller family sizes started two centuries ago in Europe and North America, much later in Asia and South America and very recently in Africa. Although population growth is still rapid, the rate of growth now appears to be slowing in every country in the world (Cleland, 1995). This offers the prospect, if distant, of an eventual stabilization of the human population. However, our impact on wild living resources is not showing the same trend; there is no evidence that our rate of exploitation of the natural world is decelerating. As economies grow and harvesting becomes more technologically efficient, consumption is fast outstripping the limits that would sustain biodiversity at, or even near, present levels.

Human evolutionary ecology and resource use

Evolutionary and behavioural ecology provide a theoretical framework that predicts relationships between the behaviour and life history of organisms and the resources on which they depend. Thus the field should have much to offer to increase our understanding of the human exploitation of wild species and habitats, and of human reproductive rate. Yet, although these issues have been widely investigated in non-human species, the application of such evolutionary models to human behavioural ecology is a small and relatively new discipline.

Simple models from optimal foraging theory have been successfully used to analyse the behaviour of foraging people. Precise predictions can be made about foraging decisions, such as which prey types a hunter should hunt and which they should ignore, when a forager should move on from one patch to the next (Kaplan & Hill, 1992), what size groups hunters should forage in (Clark & Mangel, 1986) and how the arrival of more efficient hunting technology might influence these decisions (Alvard & Kaplan, 1991). Abbot & Mace (1998) have applied this framework to the harvesting of fuelwood from Lake Malawi National Park, including how women are responding to the risk of being detected and fined by park wardens.

These studies have some important messages for conservationists. There has been a tendency amongst some anthropologists to create an image of the 'ecologically noble savage' (Redford, 1991), whose resource-use decisions are assumed to be based on the motive of conserving that resource. Detailed studies of hunting and gathering decisions that have used the framework of foraging theory have shown that this is not the case (Alvard, 1993). For example, it has been observed that Amazonian hunters typically do not hunt in depleted areas, which are usually areas close to settlements. They usually move on to new areas when the vicinity is becoming depleted of prey. This is consistent with conservation, but is also consistent with optimal foraging models, which predict that hunters should not spend time in less profitable areas (Hames, 1987). However, if hunters are simply maximizing harvesting returns, they should always take prey from the depleted areas opportunistically as they pass through, which is what they have been shown to do (Alvard, 1994). Further, these models, based ultimately on fitness maximization, stress that traditional populations must be seen as groups of individuals, where each person is trying to maximize individual fitness or the fitness of very close kin. Evolutionary models do not predict that such individuals will make sacrifices for the good of the wider group. More likely explanations for the many examples of traditional foragers harvesting resources sustainably are limited technology, low population growth and distance from markets. There are examples of traditional communities co-operating to set rules limiting harvests in places where resource scarcity has become an issue (McCay & Acheson, 1990) but this involves individual fitness maximization as well; the members of a small closed community are likely either to be relatives or to have long-term interactions with each other. Long-term interactions tend to promote co-operation among individuals because actions can be reciprocated.

More complex, but very useful, dynamic optimality models have been developed that can be used to analyse decisions where the state of the decision-maker is important. These employ the technique of stochastic dynamic programming, which is a tool for finding a resource user's optimal strategy in a stochastic environment (Houston *et al.*, 1988; Mangel & Clark, 1988). In human behavioural ecology, Mace (1993) used dynamic optimality to examine which mode of subsistence (farming, herding or agropastoralism) families would be likely to adopt in order to maximize household survival, depending on the resources available to them. Some use of stochastic dynamic models has been made in fisheries science (Walters, 1978; Clark, 1990) and in the analysis of pest-control problems (Jaquette, 1970; Shoemaker, C. A., 1982). The technique has also been applied to terrestrial wildlife management problems (Reed, 1974; Anderson, 1975; Milner-Gulland, 1997) and to conservation decision-making (Possingham, 1996).

Fitness, as in number of descendants, can be used directly as a currency in some problems; Mace (1996b, 1998) has used this approach to understand determinants of human family size in different environments, with a particular view to understanding which environmental variables are likely to be key factors that will motivate people to have small families. I shall describe this model, which is one of very few cases where evolutionary ecology has been used to address the issue of human population growth directly, in more detail below.

Predicting human reproductive decisions using a model from evolutionary ecology

Humans, like most large mammals, invest a great deal in each offspring. Competition for parental investment can be fierce through the long period of childhood. In societies with high fertility, there is evidence that the length of both the preceding and subsequent birth intervals influences the risks of mortality and morbidity of babies and young children (e.g. Alam, 1995; Bohler & Bergstrom, 1995; Madise & Diamond, 1995). There is also evidence that older children compete for food (e.g. Ronsmans, 1995; LeGrand & Phillips, 1996). At maturity, siblings may compete for their parents' heritable resources. There is considerable evidence that wealth is positively correlated with reproductive success in traditional societies. When key resources, like land or livestock, are individually owned, resources inherited by children from their parents can be an important determinant of their future reproductive success (e.g. Low, 1991; Mace, 1996a). Thus evolutionary models of parental investment should help us to understand patterns of resource inheritance. Parents would be expected to allocate resources amongst their children in such a way as to maximize their own long-term reproductive success.

In societies where inherited resources are crucial, children with no prospect of any inheritance may contribute little or nothing to their parents' long-term fitness. The cost of feeding those children may even reduce the potential reproductive success of their siblings by reducing the wealth of the household. Thus decisions about how many children to have and about how many resources to give each of them at the end of the parents' reproductive lives will be related, and both are likely to depend on the wealth of the parents.

A large proportion of the research in human demography over the last few decades has focused on the demographic transition to low fertility. A single, socio-economic correlate of the onset of demographic transition has remained the elusive goal of demographers (Cleland, 1995). Evolutionary anthropologists have, however, taken a broader cost/benefit approach to understanding determinants of family size. On the face of it, the observation of smaller families in the presence of enhanced resources represents a serious challenge to evolutionary theory; although a number of mechanisms by which some quantity/quality trade-off might be occurring have long been familiar to evolutionary ecologists, going back to Lack (1968).

I shall describe a decision-making model that is adaptive in that it optimizes reproductive success and it considers the case where parental investment is important to the future reproductive success of children. I shall use the model to investigate the influence of a number of parameters on optimal fertility and inheritance strategies, to see which are likely to produce outcomes similar to those observed in societies undergoing fertility decline. I investigate mortality risk, risk in the environment (which I model by reducing the risk of drought), and the economic costs of raising children (such as their food requirements). These are three variables generally assumed to be associated with the demographic transition. Development agencies [such as governments and non-governmental organizations (NGOs)] normally concentrate on the first two of these in an effort to increase people's standard of living. The last is often a consequence of an increased standard of living. That such economic changes lead to demographic change, albeit not in consistent ways, is certainly well documented; although a theoretical framework for predicting that change has been largely lacking. I will show here that evolutionary ecology can provide a relevant framework.

Beauchamp (1994), Anderies (1996) and Mace (1996b) have all used dynamic optimality models to analyse family size in humans over a single generation. Beauchamp (1994) and Anderies (1996) use lifetime reproductive success as the currency to be maximized, not taking into account any effect the level of parental investment in each child might have on their future reproductive success once they have survived childhood. In Mace (1996b), I used married children as the currency to be maximized and explored how varying the minimum investment necessary to marry off a child will influence optimal fertility. The model is based on the Gabbra, a group of camel-keeping, nomadic pastoralists in northern Kenya. The results of a simulation of a population following the optimal decision rules arrived at for the Gabbra system, assumed to have started with the average herd size reported for that cohort of women that have completed their fertility, are shown in Figure 2.2. The model is broadly successful at predicting family sizes that are close to those seen in the Gabbra. There is an over-representation of families with 0 or 1 child. Evolutionary models do not generally favour deciding to have no children, or so few that random childhood mortality leaves a person with one or none. However, it is unlikely that these small families are the result of decisions in this case; they are more likely a consequence of involuntary infertility (which is not included in the model). Most women completed their fertility with about six living children (thus on average three living sons), as the model predicted they should.

I demonstrated in Mace (1996b) that the costs of marriage are predicted to have a strong influence on optimal family size. I entered marriage costs as a given; but in fact parental investment in children at marriage also represents a decision by the parents. If we want to understand why family sizes might change as the environment changes, then it is important that the allocation of wealth amongst children is also allowed to change. In the model described below (Mace, 1998), I use a dynamic optimality model to find decisions that maximize number of grandchildren, given an optimal

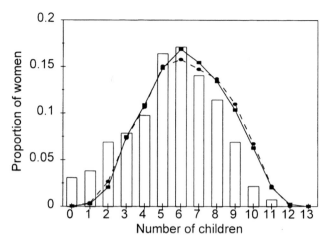

Figure 2.2 The observed distribution of completed family size of Gabbra women over 45 years of age in 1993 (bars, $n = 421$), and the predicted distribution based on women following the decision rules derived by Mace (1996b) assuming families started with initial herd sizes of 10 camels (the mean reported for that cohort); solid and dotted lines are predictions given two different estimates of mortality; see Mace & Sear (1998).

allocation of family wealth between children. The key variable of optimal investment in each child through inheritance (or marriage gifts, which are effectively similar) is derived from the model rather than assumed. I measure the reproductive value of children as the number of grandchildren they are expected to produce, given the wealth allocated to them at the beginning of their reproductive lives.

In pastoralist systems, livestock are the key units of wealth. They provide food directly, through milk and meat, and indirectly, through cash raised from their sale. They can also provide transport, leather and other necessities. They are used for social transactions such as for brideprice and dowry. The food and other yields from a herd relate directly to the size of that herd. Pastoralist herds have often been likened to investment accounts: when pastoralists are given other opportunities to earn money, they frequently put that money straight into increasing herd size. Wealth will rise and fall when the herd grows or shrinks, owing to births and deaths from consumption, sales or wasted mortality (such as in drought). These features of the system characterize many other systems based on the individual ownership of resources; in particular, it encapsulates the processes whereby those with a great deal of wealth will find it easier to maintain and

Table 2.1
Parameter values used to describe herd dynamics in the baseline case

Annual probability of drought	0.2
Variables per one female stock unit*:	
P(birth) − P(death) in non-drought year	0.11
P(birth) − P(death) in drought year	− 0.14
Food yield when living (i.e. milk)	10 units/year
Food yield if sold/slaughtered	20 units
Variables for human food requirements:	
Food requirements (fr) of an adult ($\geqslant$ 15 years)	15 units/year
Food requirements (fr) of a child (< 15 years)	1 age unit/year
Household food requirement (children + two parents)	$fr(ch,t) + 30$
Maximum herd size	45

*1 stock unit approximates to one female camel, although reproductive potential has been enhanced to account for the fact that, for each camel owned, Gabbra will also own a number of faster-reproducing sheep and goats (which are not included explicitly in the model to simplify calculations).

Table 2.2
Parameter values used to describe human demographic parameters in the baseline case

Probability of adult death	0.005 per year
Probability of child death (< 4 years)	0.05 per year
Probability of child death (15 > years $\geqslant$ 4)	0.008 per year
Reproductive life span of an adult	30 years
Economic life span of a household	40 years
Minimum inter-birth interval	2 years
Risk of maternal mortality (death in childbirth)	0.005 per birth

Mortality estimates from Mace & Sear, 1996, 1998.

increase their wealth, whereas those with very low levels of wealth are at greater risk of destitution.

Assumptions in the model of birth rate where resources are inherited

This model is of reproductive decision-making in a subsistence system based on a herd of livestock with dynamics described in Table 2.1, and a human population with the demographic parameters described in Table 2.2. However, in respects other than precise parameter values, the model is general enough to apply to a range of subsistence systems based on individually owned resources.

It is assumed that inherited resources are essential to reproductive success. This assumption is not necessarily true in many societies (external sources of earned income, raiding livestock from neighbouring tribes or extra-pair matings are examples of means by which men could achieve reproductive success without inherited wealth in the Gabbra); but a model where wealth inheritance and reproductive success are strongly related is more likely to give us insights into decisions about resource allocation and its relationship with fertility.

Stochastic dynamic programming is used to find the optimal strategy set of fertility and inheritance rules for maximizing number of grandchildren. The equations of the model used to determine the optimal policy in a range of different environments are described in Box 2.1.

Box 2.1 Model of reproductive decision-making

To model the reproductive decisions of when to have a baby (b) and how many children to give an inheritance (i), several quantities need to be defined. Consider a household in state w,ch. Then

$P(w,w')$ is the probability that a herd of size w will be a herd of size w' after 2 years, based on a binomial probability of each female animal giving birth to another surviving female, and a binomial probability that each animal will die. Details of parameters that influence this probability are given in Table 2.1.

$P(ch,ch')$ is the probability that a household with ch sons, will have ch' sons after 2 years, after having experienced the risk of extrinsic child mortality ($ch' < ch$). The parameters determining the risk of mortality are shown in Table 2.2. (Daughters are not considered as it is the number of sons that receive inheritance and constitute the costs that can be demonstrated as reduced reproduction in the Gabbra (Mace, 1996a,b; Mace & Sear, 1997)).

b is the decision whether to have another baby ($b = 1$) or not ($b = 0$).
$b = 0$ if $t > 15$ (which is after 30 or more years of marriage).
$Ps(b)$ is the probability that the mother survives 2 years, given decision b.

$R(w,ch,T)$ is the reproductive value of a family of ch sons with w units of wealth allocated optimally between them, at the end of the parents' household's life ($T = 20$, which is 40 years after marriage). $R(w,ch,t)$ is the expected reproductive value of a family in state (w,ch)

when t is the number of time steps of reproductive life that have passed. $R(w,0,0)$ is therefore the expected reproductive value of newly-wed parents with wealth w, at the beginning of their reproductive lives. $R(w,ch,T)$ is estimated for the first iteration of the model (to values between 0 and 1), but thereafter can be calculated by

$$R(w,ch,T) = max_i \; R(w/i,0,0) \cdot i \tag{1}$$

where i is the number of sons between which the household wealth will be divided in order to maximize grandchildren, where i is less than or equal to ch.

$R(w,ch,t)$ is the reproductive value of a household at time t in state w,ch and given that they always make the optimal decision b, and it can be calculated from $R(w,ch,t+1)$ by

$$R(w,ch,t) = max_b \; S_{w'ch'}(p(w,w') \cdot P(ch,ch') \cdot (Ps(b) \cdot R(w' - fr(ch,t),ch'$$

$$+ b,t+1) + (1 - Ps(b)) \cdot R(w' - fr(ch,t),ch',T))) \tag{2}$$

where the summation is over all possible values of w' and ch', and $fr(ch,t)$ is the food requirements of a family of ch sons aged t (see Table 2.2 for parameters that influence $fr(ch,t)$). If w falls to zero at any time, the household is considered destitute and fitness, R, is 0.

Initially, $R(w,ch,T)$ is simply estimated (to values where $R(max(w),max(ch),T) = 1$ and $R(0,ch,T) = 0$ and $R(w,0,T) = 0$). $R(w,ch,T-1)$ can then be calculated from equation (2), as can $R(w,ch,T-2)$ and thus, by backward iteration, after 20 time steps $R(w,0,0)$ is calculated and the optimal scheduling of births over the whole reproductive lifespan is also known. When $t = 0$, $R(w,ch,T)$ for the next iteration is recalculated by equation (1), which gives the optimal value of i over all combinations of w and ch. $R(w,ch,T)$ is then normalized (by dividing all values by $R(max(w),max(ch),T)$ and the process is repeated. After a small number of iterations, $R(w,ch,t)$ and the optimal values of i and b for each value of w and ch converge. This is the optimal strategy set of fertility and inheritance rules for maximizing grandparental fitness.
From Mace (1998)

The model is used to predict both the decision rules that parents should follow to maximize their grandparental reproductive success (whether or not they should have a baby and the number of sons between which they should divide their wealth) for each state of wealth and current family size that they may find themselves in. These rules can be used to find the population outcomes of a group of people following those decision rules. The population outcomes are determined by running a simulation of a population following the optimal decision rules, subject to the same environment for which that decision rule is optimal. That environment includes, of course, the stochasticity inherent in the system, which has been taken into account in determining the optimal decision rules. The starting conditions for the simulation were estimated for the first iteration. Thereafter, the distribution of herd sizes at household formation gained by inheritance from parents, that is generated by the model, is used as the starting state of the population. After a small number of iterations, a stable distribution of initial herd sizes is reached.

REPRODUCTIVE DECISION-MAKING IN A RANGE OF ENVIRONMENTS

The model is used to explore how we would expect reproductive decisions, and hence population outcomes, to be influenced by environmental circumstances. I explore these changes by comparing outcomes with a baseline case. The baseline case is that of the Gabbra cohort shown in Figure 2.2, who live in the system just described, characterized by the parameter values given in Tables 2.1 and 2.2. I have shown elsewhere (Mace, 1998) that both the predicted range of completed family sizes and the average allocation of wealth that each new family inherited are broadly in line with that observed in the Gabbra. The aim here is to see which variables would cause these patterns to change. It is in this sense that modelling can be most useful. In the real world, opportunities rarely arise whereby individual variables are altered in a controlled way, such that the effects of such actions can be observed. A modelling framework such as this one allows us to explore which environmental changes might be expected to influence reproductive decisions, and what the response and impact on the population might be.

Figures 2.3–2.6 all show optimal decision rules or stable-state population distributions for four different environments: the baseline case, a high-costs case (where it is assumed that the cost of raising a child is double that of the baseline), a low-drought probability case (where it is assumed

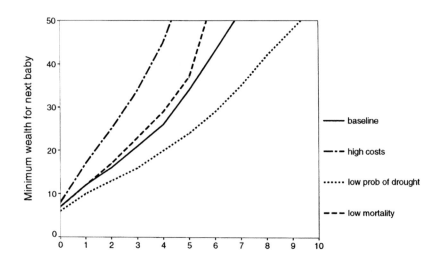

Number of living sons

Figure 2.3 The optimal decision rule of the minimum wealth to have another baby, based on the number of living sons already born, shown for four different environments: the baseline case (line), high costs of children (dashed and dotted line), low probability of drought (dotted line) and low mortality risk (dashed line).

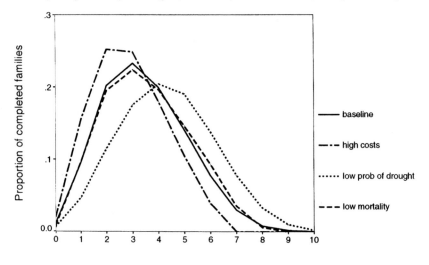

Number of living sons

Figure 2.4 Predicted stable-state distribution of completed family sizes for households after 40 years of marriage, for different environments: the baseline case (line), high costs of children (dashed and dotted line), low probability of drought (dotted line) and low mortality risk (dashed line).

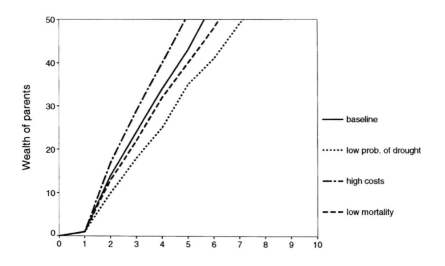

Maximum number of sons to inherit

Figure 2.5 The predicted maximum number of sons between whom the household herd should be divided on inheritance, given the level of parental wealth, for four different environments: the baseline case (line), high costs of children (dashed and dotted line), low probability of drought (dotted line) and low mortality risk (dashed line).

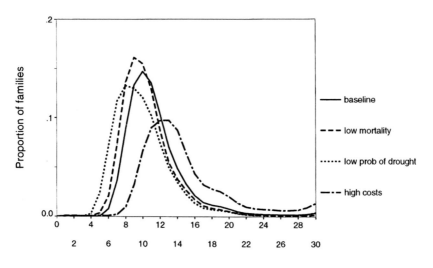

Inherited wealth

Figure 2.6 The predicted stable-state distribution of the wealth of new families at household formation, which is the amount received in inheritance from parents, shown for four environments: the baseline case (line), high costs of children (dashed and dotted line), low probability of drought (dotted line) and low mortality risk (dashed line).

droughts only occur half as often, raising the mean and reducing the variance in the productivity and growth rate of the herd) and a low-mortality case (where all extrinsic mortality risks are an order of magnitude lower than the baseline case).

Figure 2.3 shows the optimal fertility strategy in each of these four cases. The fertility strategy shown is the optimal strategy for a couple married for 20 years, who could thus have between zero and ten children, and have another ten years of reproductive life remaining. The line indicates the optimal number of sons for that level of wealth. Parents in households in the area above the line, who have more children than the optimum for their level of wealth, would reduce their expected grandparental fitness if they had another baby. If a child died or herd size increased, they might move into a region where they were below the line again and thus they should have another baby. Thus decision-making is dynamic, changing with current circumstances. Parents are predicted to quickly 'replace' dead children.

In all cases shown in Figure 2.3, wealth would be expected to be positively correlated with reproductive success. This correlation has been shown in numerous traditional societies, including the Gabbra (Mace, 1996a). The wealth at which it is optimal to have the first child is not especially sensitive to environmental conditions, but thereafter the environmental parameters altered here do have, in some cases, marked effects on the minimum wealth at which to have subsequent children.

Figure 2.4 shows the stable distribution of completed family sizes (living sons) that would result for women who lived throughout their reproductive lives in each of the four environments. Most families are predicted to have many fewer than the maximum possible number of sons, in any of the environments modelled.

Figure 2.5 shows the decision rule for the optimal inheritance of wealth between sons at the end of the parents' economic lives. In all cases, the maximum number of sons to be given an inheritance depends strongly on parental wealth. The wealthier the parents, the more sons they will give herds to. The lines are not curved, indicating that the size of the inherited herds would not increase with parental wealth, unless the number of sons living at that time was below the maximum to be given an inheritance. If the number of sons they have is above that maximum, then no parental fitness advantage is gained by giving the 'extra sons' any inheritance. Primogeniture, in this case, appears to be a response to poverty rather than wealth. I have shown elsewhere that, in the Gabbra, the more elder brothers a man has, the more he is disadvantaged with respect to inherited wealth, his age at marriage and his reproductive success (Mace, 1996a).

Figure 2.6 shows the stable distribution of initial herd sizes (i.e. those herds inherited by sons starting their own households) predicted in each of the four environments. As is evident from Figure 2.5, very few individuals are predicted to be given very large herds, as the division of resources between many sons (if they have them) is a better reproductive strategy for wealthy families.

WHICH EFFECTS ARE MOST LIKELY TO CAUSE FERTILITY DECLINE?

Reducing the risks in the production system

It is clear that simply improving the productivity and reliability of the subsistence base is not predicted to lead to fertility decline; in fact the opposite occurs. The minimum wealth at which it is optimal to have another baby is significantly reduced, for all family sizes, and thus fertility above the level seen in the baseline case is predicted. Completed family sizes are significantly larger, and the inheritance given to each of those children is slightly smaller, such that the distribution of household wealth with which each new generation starts married life is slightly towards the poorer end of the spectrum than in the baseline case. When living conditions are easier, parents can afford more children and do not need to be so generous to each child in order to maximize their grandparental reproductive success.

Reducing the risk of mortality

A reduction in extrinsic human mortality is probably considered one of the most important correlates of fertility decline, although it can either precede or follow it (Cleland, 1995). The case I model here investigates the lowering of mortality by an order of magnitude, which is the difference, in many cases, between the developed and developing world; it is also a level of difference that can nearly be achieved in developing countries when effective medical services are provided in an area that previously had none (e.g. Pison et al., 1993; Weaver & Beckerleg, 1993). The model predicts that reducing mortality will definitely decrease number of births, as decisions are dependent on number of living children. Dead children are replaced by more births whenever possible. This has been shown to occur in numerous demographic studies.

However, what is of interest here is that the effect of a very great decline in mortality on other aspects of reproductive and inheritance strategies is relatively small. Having children that die does waste time and resources, leading to slightly lower wealth and fewer new families each generation. The minimum wealth at which it is optimal to have another baby is higher in the low mortality environment, especially for parents with large families. This compensates, almost perfectly, for those children that would be expected to be lost in the baseline case, such that the stable distributions of completed family sizes are very similar in the baseline and low mortality environments. The maximum number of sons to be given an inheritance, for a given level of wealth, is slightly higher in the low mortality scenario: each son needs slightly less if he will not be wasting resources on raising children that do not survive. However, the difference is not great. This model does assume that most mortality is concentrated amongst very young children and the impact of higher mortality might be greater if it occurred throughout childhood; however, in most societies it is the infants that are by far at the greatest risk of early death.

Demographers have focused on changes in fertility (i.e. rates of giving birth) not least because these are relatively easy to measure. However, it is living children that really matter. Lowering mortality does lower fertility but, according to this model, it is not expected to lower family size greatly or to increase the level of parental investment in each child. These are probably the two most characteristic features of post-transition families, so we cannot look to reduced mortality alone to explain demographic transitions.

High cost of children

Doubling the costs of children causes a marked change in both fertility and inheritance strategies. The minimum wealth at which to have another baby is much higher than in the baseline case. This leads to much smaller completed families, most with only two sons. That the cost of children is expected to influence reproductive rate is a common theme of examinations of reproductive rate in economics and anthropology (albeit with varying theoretical justifications). The framework presented here is clearly based on reproductive success and allows the extent of the effect to be predicted precisely if the relevant parameter values are known.

The number of sons to be given an inheritance, for each level of wealth, is lower than the baseline case and each son is given a larger inheritance. This leads to more couples starting out their married life with larger inherited herd sizes. The average family in the stable population is wealthier.

These are all features of post-transition societies. An increase in costs of children could be associated with sending children to school, either because it costs money to send children to school or because their labour is lost to the household economy when they are at school. Education is frequently cited as a correlate of low fertility, although most demographers concentrate on the educational status of the mother, rather than the costs to her of educating her children, which this model suggests might be more explanatory of her fertility. Why parents feel the need to educate children is beyond the scope of this model, although it does seem unlikely that uneducated children would be successful (either economically or in competition for mates) in societies where all the other children were educated.

Thus, this evolutionary approach makes clear that increasing parental investment in children is likely to be the key to fertility decline. An important contribution to fitness from inherited resources combined with high costs of raising children successfully to adulthood will favour small families; and the model further suggests that these conditions will be associated with a wealthier population in general. The model described here is not explicitly based on a society in a developed country but these are factors that might change as a less-developed society becomes more developed. All parents like to raise children at least to a standard of living approaching their own, and the cost of education and other costs are frequently referred to by parents as deterrents of large family size in modern societies. In developed societies, girls are generally as costly as boys and biased inheritance with respect to sex or birth order is not popular (Judge & Hrdy, 1992). In some cases it is not even legal; for example, primogeniture was outlawed in nineteenth-century France, causing a marked decline in fertility at that time (Johansson, 1987). If it is assumed that all children, rather than just sons, are costly and benefit from inheritance, then models such as those described here could clearly predict very low family sizes as optimal.

It could be argued that the presence of Government-funded support systems may have ameliorated or even removed the costs of large families in some modern societies. One would not necessarily expect our decision rules to keep pace with recent social and political changes of this nature and reach an optimum, although there is plenty of evidence that economic change can have a near-instant effect on birth rates: fertility in East Germany dropped dramatically after unification with West Germany in 1989, when many state benefits relating to children were removed (Conrad *et al.*, 1996). Hoem (1992) documents an increase in the fertility of Swedish women in the 1980s in response to increasingly generous maternity benefits. Thus our reproductive decisions are constantly influenced by our environment, even in modern industrial societies.

Why do richer people have smaller families?

All the cases modelled here do, however, predict a positive correlation between wealth and fertility (as do other evolutionary models, such as Rogers, 1995); yet such a correlation has not been widely demonstrated in post-transition societies. Frequently, area-level or country-level statistics are used to explore comparative relationships. Such comparisons frequently lead to the conclusion that wealth is negatively related to reproductive success. Some have used this to argue that evolutionary models are not appropriate models of contemporary human reproductive behaviour (Vining, 1986). Yet, where this relationship has been investigated in clearly defined, homogeneous populations, such as is frequently the case in anthropological studies (e.g. Irons, 1979; Borgerhoff Mulder, 1987; Mace, 1996a), then the opposite is found.

What is shown here is that, when costs of children are high, levels of inheritance will also be high and the average wealth of families will be high. Thus, if costs of children vary between groups, then the average wealth of those groups is also likely to vary. An important conclusion of this finding is that reproductive decisions concerning parental investment have to be investigated with respect to a very clearly defined group if they are to be understood in ecological terms. Whilst none of the models presented here predict anything but a positive relationship between wealth and reproductive success within a homogeneous group, they do show that if a society actually comprised several different groups, each of which are following different decision rules, then this may not be true. If different people have different fertility policies, perhaps because they have different risks of mortality or different costs of raising children, or different perceptions of these risks and costs (which is not true of a largely homogeneous society such as the nomadic Gabbra, but is true of most modern, conglomerate societies containing peoples of many different origins), then it is likely that those in a higher income bracket might be following a decision rule leading to a smaller family than another family with a lower level of wealth, because they are following a different policy. When looking at correlation between wealth and fertility across society as a whole, fertility and wealth may appear to be de-coupled or even negatively correlated, because the richest families would have the smallest number of children.

A fully evolutionary model of high parental investment does require that increased parental investment is rewarded by increasing the reproductive potential of those children. Education is the most obvious and expensive form of parental investment in modern societies, but is generally considered to have a negative influence on fertility. However, as I stated above, it would be most informative to look within very homogeneous groups to see whether the education they received increased their fertility relative to others in that group. Part of the difficulty in addressing this problem is that there are few studies of modern populations in developed countries that concentrate on strictly homogeneous groups. One rare example is that of Hubback (1957), who charted the reproductive success of English women graduates from Britain's best universities in the mid-twentieth century, who were a most highly educated, and socially and financially privileged group. Those that went to Oxford and Cambridge, which are considered to be England's best universities, were a particularly highly selected group, as only a very small minority of colleges in these universities admitted women up until less than 20 years ago. Family wealth correlated positively with reproductive success. So, incidentally, did the class of degree that these women obtained: those attaining the highest marks in exams had the most children. It is interesting to note that those features traditionally associated with fertility decline, were clearly associated with enhanced fertility within this homogeneous group.

Kaplan et al. (1995) found that ethnicity and birth cohort were much stronger determinants of fertility than was wealth in a large sample of New Mexican men. Children of parents with large families had lower levels of educational achievement and lower incomes as adults, so some features of their reduced level of parental investment had the potential to influence future generations of offspring. Hispanics, originally from Mexico, had higher fertility than Anglos, and Kaplan et al. (1995) argue that some are reducing their fertility over time in order to take advantage of better educational opportunities, which are followed by opportunities to earn higher incomes. However, those with the most children also had the highest number of grandchildren; thus, in this case, there is no evidence that a decision to limit fertility for educational or other benefits is evolutionarily adaptive in the long term.

WHAT ARE THE IMPLICATIONS FOR CONSERVATION AND DEVELOPMENT PROJECTS?

The reduction in the size of families is frequently an explicit aim of development agencies (governments or NGOs) and conservationists. This is usually addressed by providing advice on family planning and making modern contraceptives available. I myself have used this approach when working for an NGO with the Gabbra in northern Kenya. Rural Gabbra women were interested in hearing about contraception but not because they felt any pressing need for it. Infertility was far more likely to be raised as an issue that could benefit from medical intervention. Faced with my own data and my theoretical perspective on it, I should have realized sooner that most of these women who were living a traditional lifestyle, that had probably not changed greatly since the adoption of pastoralism, were probably having roughly the number of children they wanted to have.

It is changes in the economic and environmental circumstances that are more likely to trigger an interest in small families. Development projects are, inadvertently, frequently concentrating on those areas that are likely to move things in the direction of larger, not smaller, families. Improving the productivity and reliability of the subsistence base is generally the main objective of most integrated rural development projects. Other things being equal, that is likely to push fertility up. This can be counter-productive, even causing an overall decline in the average wealth of households. Reducing mortality is the obvious aim of any project with a public health component. That is likely to reduce fertility, i.e. number of births, but this model predicts it will have little effect on the number of living children in a family.

The model described here suggests that it is increasing the cost of children to parents that is likely to reduce family size and, where resources are heritable, this will also improve the general wealth of the population. Increasing security of individual tenure over a resource might mean that inherited resources become more important, which could favour smaller families in some but not all circumstances (Rogers, 1994; Milner-Gulland & Mace, 1998). Increasing the costs of children is predicted to be most effective at reducing family size, but is unlikely ever to be the explicit aim of a development project. Nor is there any real reason why it should be. Increasing the cost of raising children increases hardship in the short term and the population becomes wealthier, partly because poor families fall out of the system. Innovations that reduce the drudgery of life, such as improving water supplies, probably reduce the cost of children and thus would be predicted to push fertility up. However, educational uptake is frequently an aim of governments and NGOs. Putting as much of this cost as possible onto parents is frequently the aim of governments because of a shortage of cash. If parents know that a good education is likely to produce good returns, then they will take it more seriously and consider having fewer children so that they can be well educated. If mortality is reduced, then investing in children at a younger age may also be more likely to occur.

Evolutionary ecology can provide important insights into the factors likely to influence human reproductive rate. However, these insights do not contain any reassuring messages for conservationists. It should be clear that reducing family size, whilst ultimately necessary for the sustainable exploitation of natural resources, cannot, in itself, be considered as a conservation measure. There is neither theoretical nor empirical evidence to suggest that smaller families will be associated with lower or wiser resource use. The factors that lead to fertility decline are the same factors that are associated with high levels of consumption. Populations with higher wealth and higher standards of education are better placed to harvest resources with even more technological efficiency and to market their harvests further afield. In any case, controlling the largely unfettered destruction of biodiversity is too urgent a task to rely on waiting for population stabilization. A whole suite of measures to do with regulation and governance, the provision of sustainable alternatives and the appropriate allocation of costs and benefits of resources are required. These have been discussed at length elsewhere (Milner-Gulland & Mace, 1998). These should be far higher priorities to practical conservationists than population growth which is, anyway, frequently outside their spheres of influence.

3

Grassland conservation and the pastoralist commons

MONIQUE BORGERHOFF MULDER & LORE M. RUTTAN

INTRODUCTION

Conservation commonly refers to the maintenance of biodiversity, in terms of the full set of genetic, species and ecosystem diversity at the natural abundance at which they occur (OTA, 1987). Although this definition is quite widely accepted, it has been more difficult to reach consensus on a satisfactory definition of the human acts that might aptly be called 'conservation'. For most social scientists, conservation occurs when people *intend* to conserve, and they are particularly interested in determining the multiple social and political factors that lead to the successful practice of conservation. In contrast, evolutionary anthropologists (evolutionary ecologists studying human affairs) are cautious about equating conservation with intent and emphasize rather the potentially altruistic, and hence theoretically problematic, aspects of conservation. Their concern then is to understand the conditions under which conservation is individually advantageous.

In this chapter, we take the view that aspects of both positions have merit. Retaining both the assumption of individual self-interest and the methodological rigour that characterize an evolutionary ecological approach, we build a model that incorporates the asymmetries in power and interest that typify most human communities and engage social scientists' attention. More technically, we question two fundamental assumptions within the evolutionary anthropologists' intellectual tool kit. First, that economic efficiency is synonymous with short-term efficiency as predicted by optimal foraging models (and its corollary that economic efficiency is antithetical to conservation). Second, that an act cannot be designated as conservationist unless it is the true cause of the conservation outcome (e.g. Hunn, 1982).

The chapter is structured as follows. First, we give some history to the study of conservation behaviour in humans, highlighting the significant contribution from evolutionary ecology. We then address the question of whether and how conservation acts might be maintained in a population, using the classic example of a pastoral grazing reserve; here we develop a game theory model following closely on the work of institutional economists and behavioural ecologists. Finally, we end with some brief conclusions concerning the implications of our model, and some new ways of thinking about these matters.

DOES THE 'ECOLOGICALLY NOBLE SAVAGE' EXIST?

Living in harmony with nature

Ethnographers and sociocultural anthropologists commonly argue that indigenous peoples live in balance with their environment and, more generally, in harmony with nature. They base this view primarily on the observation that small populations with limited technology subsist on plant and animal species without driving these resources to extinction, and without causing long-term degradation of the environment (e.g. Alcorn, 1989; Posey & Balee, 1989; IWIGIA, 1992), a point first made by Birdsell (1958). In conjunction they note that indigenous peoples exhibit deep reverence for nature (e.g. Nelson, 1982; Kay, 1985; Zann, 1989) and often command an extensive knowledge of their prey species' habits (e.g. Johannes, 1981). Somewhat similar equilibria (be they stable or oscillating) are also observed among non-human predators and their prey. Thirty years ago biologists attributed such systems to the behaviour of predators, speculating that a 'prudent predator' (Slobodkin, 1968) spares the prime-aged reproductive prey in order to secure a sustainable harvest for the future. While this view has become outdated in biology, it still holds sway in anthropology. Indeed most anthropologists maintain that prudence is a *goal* of subsistence hunters. Accordingly, indigenous peoples are attributed the reputation of being natural conservationists (e.g. Durning, 1993) or, in Redford's (1991) more colourful epithet, 'ecologically noble savages'.

As might be expected, evolutionary ecologists have raised theoretical objections to the somewhat romantic view that indigenous peoples are natural conservationists (e.g. Winterhalder, 1977; Smith, 1983; Hames, 1987, 1991; Low, 1996; Beckerman & Valentine, 1996; for a model, see

Winterhalder & Lu, 1997). Individuals cannot be expected to limit present harvests of resources for the purpose of conserving them for future use if this behaviour entails a cost. As resource economists noted long ago, if resources are not owned privately then any restraint in their use is altruistic insofar as the benefits from restraint will be shared by many while the costs are borne individually (Gordon, 1954; Scott, 1955). Such altruism can easily be exploited and hence is unlikely to evolve between unrelated individuals (Williams, 1966).

Definitions and tests of conservation acts

With these theoretical objections in mind, human evolutionary ecologists began conducting field examinations of whether or not the 'ecologically noble savage' exists, with the focus mainly on foraging populations. Hames (1987) had the nice idea of testing for conservation behaviour by pitting its predictions against a set of alternatives derived from optimal foraging theory, which he termed 'efficiency hypotheses'. Here, efficiency refers to returns to the *individual* and not to the group, and therefore differs from economists' use of the term (a usage that side-steps the issue of how the costs of providing efficiency to a group are borne). Efficiency maximization is predicated on two assumptions – that more food enhances individual fertility and survival, and that time spent acquiring food incurs opportunity costs (Kaplan & Hill, 1992). It should be noted, however, that most optimal foraging models (but see Benson & Stephens, 1996) are based on the assumption that exploitation of a resource has no effect on its future abundance. Hence, in testing for efficiency, evolutionary anthropologists predict that foragers should make prey choice decisions that maximize the rate at which resources are taken per unit time spent foraging, *irrespective of the effect this harvesting will have on resource availability in the future.*

While the expectations from optimal foraging theory were already well specified (e.g. Stephens & Krebs, 1986), precise conservation hypotheses were as yet undetermined. Hames (1987, p. 93) captured the essence of the meaning of conservation by emphasizing short-term restraint for long-term benefits, incidentally paralleling political economists' earlier revival of the term 'stinting' (Tate, 1967 in Ciriacy-Wantrup & Bishop, 1975). Thus, evolutionary anthropologists define conservation as the costly sacrifice of immediate rewards in return for delayed benefits with respect to preventing, or at least mitigating future resource depletion, species extinction, and habitat degradation (Alvard, 1995; Smith, 1995). However, because restraint and long-term environmental effects are both hard to identify in field situations, empirical tests of these two hypotheses have tended [but see Alvard (1995) for some exceptions] to view conservation acts simply as those that are not predicted by optimal foraging theory.

Hames' (1987) initiative generated valuable empirical research. Without exception, these studies show that foragers choose the prey that maximizes return rates or efficiency (Hames, 1987; Alvard, 1993, 1994, 1995, 1998; see also Stearman, 1994; Vickers, 1994), seriously weakening the notion that populations living in apparent harmony with their environments necessarily practice a conservation ethic. Furthermore, foragers eschew prey choice decisions that might minimize any impact on the population dynamics of their prey, for example, they do not avoid females and individuals of peak reproductive value (Alvard, 1995). These and other studies have stimulated intriguing speculations by resource economists and evolutionary anthropologists (e.g. Clark, 1973; Hames, 1987, 1991; Rogers, 1991) and many other social scientists (e.g. Ostrom, 1990) on where and when conservation might be observed, in terms of explicit trade-offs between short-term costs and long-term benefits.

Despite these advances in evolutionary anthropologists' ability to identify and predict conservation acts in humans, there is still little consensus on the validity of such analyses (see, for example, Stearman's (1995) and others' comments – Alvard, 1995). Here we focus on one issue – the major simplification that was made (albeit for analytical tractability) in finessing conservation as an *alternative* to efficiency. To differentiate conservation from efficiency, Hames (1987) and others assumed that conservation strategies are economically less efficient for the individual than are optimal foraging strategies because, by definition, the latter maximize rates of return. Yet, rates of return are measured over the period of time that the behaviour in question is employed, usually a period in the order of hours. While this simplification was a practical necessity (with recognized shortcomings, e.g. Alvard, 1995; Hill, 1995), it ignores the fact that in many areas of human endeavour economic efficiency is synonymous with sustained long-term production, a point recognized by Berkes (1987), Bulmer (1982) and others. Indeed a behavioural strategy can be both environmentally friendly and economically rational for the individual. This is particularly true in production systems where producers have more control over the resources they depend on than is the case among foragers. Under such conditions, conservation acts *can be isomorphic with economic efficiency*.

In a nutshell, our point is not that evolutionary anthropologists are unaware of this time-scale problem, nor that they think short-term restraint (conservation) is antithetical to individual self-interest, only that the significance of longer-term outcomes has been undervalued in empirical analyses of when, where and why humans act as conservationists.

Applying game theory to the conservation of the pastoralists' commons

If our goal is to extend the search for an 'ecologically noble savage' beyond the realm of hunting into the realm of food producers, and to identify conservation acts in a more realistic economic and political world, evolutionary anthropology needs to expand its intellectual tool-kit. We look here at the conservation of pastoralists' commons. We retain the lay notion of conservation acts as short-term restraint aimed at maximizing long-term benefits. We do not accept the false corollary that only short-term benefits maximize efficiency. However, because restraint on the part of some individuals opens up the possibility for cheating on the part of others, we pose the problem as a frequency-dependent game. Our approach differs from most other treatments of collective action in that we examine whether restraint can emerge as an evolutionary stable strategy (ESS) when the population is heterogeneous with respect to their wealth and interests. Finally, we speculate on the possibilities for coercion to enter into the game. We use the classic example of a pastoral grazing reserve to show how heterogeneous interests in resource use can generate conservation outcomes. [The pastoral commons was also used by Hardin (1968), albeit inappropriately (see Ciriacy-Wantrup & Bishop, 1975), to illustrate the inevitable nature of the tragedy of the commons.] We believe that our approach retains the strength of evolutionary ecological thinking (development of simple testable models), yet admits greater realism with respect to politics, economics and stratification.

IDENTIFYING CONSERVATION AMONG PASTORALISTS

Grazing reserves and the Barabaig

Despite highly publicized claims that pastoralist communities are responsible for overgrazing and environmental degradation (e.g. Lamprey, 1983), pastoralists are also commonly credited with practising a conservation ethic. Much of the evidence that supports the 'overgrazing hypothesis' is equivocal (Sandford, 1983; Ellis & Swift, 1988; Homewood & Rodgers,

Figure 3.1 An Eyasi Datoga community assembly in 1989, at which crimes and misdemeanors are adjudicated. Game theory can be used to predict the circumstances under which coercion is used by wealthy livestock owners to prevent less wealthy and powerful families from using pastures reserved for dry season grazing (photograph by Monique Borgerhoff Mulder).

1991). Furthermore, there are numerous detailed observations of herdsmen following grazing regimes (e.g. McCabe, 1990; Lane, 1996), observing stocking regulations (e.g. Netting, 1976) and maintaining institutional land-use practices (e.g. Galaty, 1994) that seem to protect grasslands from over-use on an annual scale (between-year range condition is often more dependent on stochastic environmental fluctuations than on grazing pressure). Though an evolutionary ecologist might read these provisional conclusions with some scepticism, there are plenty of data showing that herdsmen do adopt strategies regulating the use of grazing areas in the vicinity of their settlements, and that these can be successful in preserving rangeland quality. The evolutionary ecological puzzle, then, is this: are these strategies individually costly, and if so what mechanisms serve to maintain such conservation strategies in the population?

We take as an example the Barabaig (Figure 3.1), a subsection of the Datoga (Tomikawa, 1979; Borgerhoff Mulder *et al.*, 1989). They keep cattle and smallstock on the semi-arid plains of Hanang District, Tanzania, supplementing their diet with grain, obtained through exchange of livestock or shifting cultivation. Rainfall averages 600 mm/year and periodic

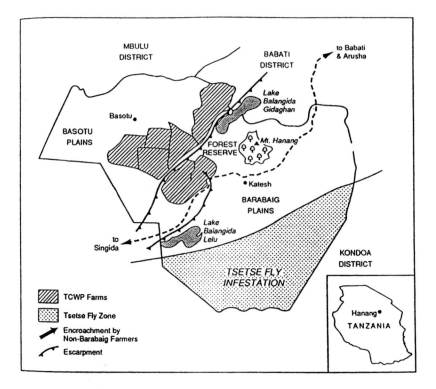

Figure 3.2 Map of Barabaig grazing movements (adapted from Lane, 1996).

droughts occur (Lane, 1996). The dominant vegetation in the area is acacia and commiphora woodland, interspersed with open grassland. In general, as in most East African pastoral groups, grazing is free to all members of the local population, here the entire Barabaig subsection of the Datoga.

Barabaig grazing patterns (Figure 3.2) have been well described by Lane (1996) in the context of a study exposing the devastating social and ecological consequences of land alienation. Most Barabaig maintain permanent homesteads throughout the year on the Barabaig plains, but soon after the rainy season starts (November–December) young men and women take their family herds to the Basotu plains where high-quality forage is available, and set up temporary camps on these wet-season pastures (WSP). Because there are no permanent water sources here, at the end of the rainy season (May–July) the herds are driven back down the Rift escarpment to the Barabaig plains [see Western (1975) for a similar water-limited grazing pattern in East Africa]. Throughout the next 4–8 months grazing becomes increasingly scarce on the dry season reserve (DSR), such that in normal rainfall years animals lose condition, and in years of drought they are forced to enter more densely wooded areas with associated elevated risks of disease and predation. Among the Barabaig, grazing rotation is dictated by water needs (specifically the permanent water of the wells at lake Balangida Lelu on the DSR), by the desire to diversify livestock intake, and the opportunity to capitalize on grass species whose productivity varies seasonally. Most critically for our purpose here, herders need to preserve the grasslands adjacent to the permanent homesteads on the Barabaig plains, and elaborate regulations exist to prevent unseasonal use of these pastures. If these are infringed, all households suffer.

Pastoralists almost universally move their herds around in response to the vagaries of climate (and a variety of other constraints), so as to make the best use of grazing and water resources (e.g. Dyson Hudson & Dyson Hudson, 1969). To facilitate this form of foraging, they operate systems of common land tenure. Lane (1996) describes in detail the extensive rights, obligations and prohibitions associated with the common property system of the Barabaig, as well as the customary rules regulating its use, and the tripartite jural structure entailed in rule enforcement. For instance, a herder who takes his livestock into seasonally protected areas will first be brought to a judicial moot (informal court) and asked to desist from the offence. If he refuses he may be fined, and in more serious cases, cursed. Very similar observations come from other pastoral groups, and indeed have stimulated a whole literature attesting to the conservation ethic among pastoralists (mentioned above).

We model the Barabaig grazing system as an example because it is clearly described and simple to conceptualize (even though it is somewhat unusual in its details). Many intriguing aspects of the system described by Lane (1996) are not preserved in the algebra, in view of keeping the model both simple and of broad generality. For example, we do not include the fact that forage on the Basotu plains (WSP) is of somewhat higher nutritional value than that on the Barabaig plains (DSR). We also defer discussion of encroaching Basotu wheat farms until the end of the chapter. We do however allow differences in herd size among rich and poor herders (as a varying ratio), because this is universal among Datoga (Lane, 1990; Borgerhoff Mulder, 1991) and most other pastoral groups (e.g. Fratkin & Smith, 1994). We explore the dimension of heterogeneity in wealth because the present modelling exercise was stimulated by the idea that richer herders (with presumably a greater investment in the pastoral sector) might be more inclined to favour conservationist strategies.

A game with asymmetrical pay-offs

To keep the analysis simple we imagine there are two types of herders; rich herders, who each own many cattle (m) and poor herders, who each own few (f). The cattle belonging to rich and poor herders are identical insofar as each eats an amount (α) each day. Pay-offs to rich and poor herders are asymmetrical because their value depends on the number of cattle owned by that herder. We model the environmental conditions described above by imagining that the WSP are available for a limited number of days, d, and that the DSR have a fixed area and thus a fixed amount of grass available, A.

The decision we wish to analyse is whether or not a herder should 'co-operate', and move his livestock to the WSP during the rainy season, or whether he should 'defect' and remain in the DSR all year round. By defecting, a herder avoids paying a cost, C, of moving his livestock and setting up a wet-season camp. However, less forage may then be available for his livestock in the dry season. The currency is measured in arbitrary units of livestock survival and productivity, which are an unspecified function of grass intake, how far the livestock need to travel, and the availability of labour within the household.

The matrix (Figure 3.3) summarizes the pay-off to each strategy dependent on the behaviour of the other player. We can see then that if a rich herder co-operates, a poor herder should also co-operate if:

$$(\alpha f d) \cdot f/(f+m) > C \tag{1}$$

A poor herder should co-operate if the cost of going to the WSP is less than the amount they lose during the dry-season, because they themselves 'stole' it.

If a rich herder defects, then a poor herder should co-operate when:

$$\alpha f d - C > (\alpha m d) \cdot f/(f+m) \tag{2}$$

That is, when the amount taken from a distant pasture minus the cost of going there, is greater than what would have been the poor individual's share of what the rich herder is taking from the nearby pasture while the poor individual is away.

If we consider how a rich herder should act when faced with a co-operating or defecting poor herder, we find that the rich individual should base his

	Rich Co-operates	Rich Defects
Poor Co-operates	αmd + A·m/(f+m) − C αfd + A·f/(f+m) − C	αmd + (A — αmd)·m/(f + m) αfd + (A − αmd)·f/(f + m) − C
Poor Defects	αmd + (A − αfd)·m/(f+m) − C αfd + (A − αfd)·f/(f+m)	A·m/(f + m) A·f/(f + m)

Figure 3.3 Pay-off matrix for a game between two herders, one rich and one poor. Pay-offs to the poor herder are in the lower left corner of each cell while pay-offs to the rich herder are in the upper right.

or her decision on the same criteria. However the asymmetrical pay-offs, resulting from differing numbers of livestock owned by the two parties, lead to there being different cut-off points for the two individuals. A defecting herder with few livestock takes less forage from nearby pastures than does a rich defector but the rich herder would normally get a larger share of that 'stolen' forage. Thus, given the same parameters, rich and poor's preferred strategies may not be the same.

We present results for the three variables that show the strongest effects on ESS outcomes, the cost (C) of going to the WSP, the relative wealth differences (or inequality) between wealthy and poor herders [$f/(f + m)$], and the numbers of days the WSP was available (d). Plotting each player's indifference curve we see three ESS spaces (Figure 3.4a). Above both curves the ESS is 'Mutual Defection'; both poor and rich should defect. Below both curves it is to the advantage of both parties to co-operate and there is no temptation to defect. The ESS is thus 'Mutual Co-operation'. In the regions between the curves, the ESS is 'Rich Co-operate Poor Defect' (RCPD); the herder with more animals should always co-operate while the herder with fewer should always defect. The pattern remains similar as the wet season lengthens (Figure 3.4b), and when the cost of moving to the WSP is assessed on a per animal rather than per herder basis (not shown).

Three clear patterns emerge, all of which make intuitive sense and suggest that the model captures some important features of grazing patterns among groups like the Barabaig. First, 'Co-operation' by either herder is precluded when the cost of using the WSP is very high [perhaps these seasonally available pastures are very remote or perhaps construction and

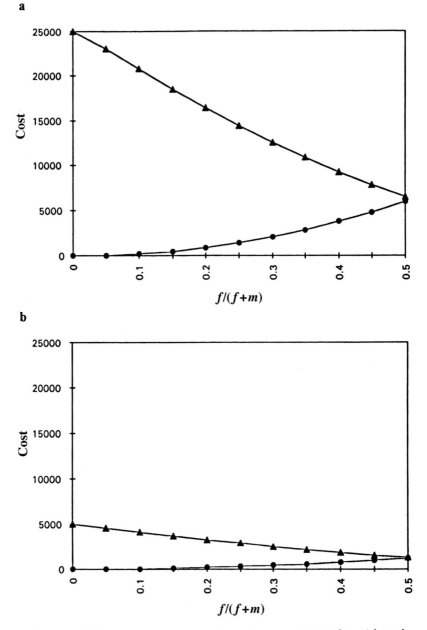

Figure 3.4 ESS regions for a two-person game. Equations (1) (triangles – rich) and (2) (circles – poor) are plotted as a function of cost of moving to the WSP and the poor's proportion of the total number of livestock owned by rich and poor. Units of forage eaten per cattle per day, α, is set equal to 1. In (a) $d = 250$ and in (b), $d = 50$, where d is the number of days the WSP is available.

defence of the wet season camps is too labour-intensive, see Sperling & Galaty (1990), Borgerhoff Mulder & Sellen (1994)]; this is particularly the case when the WSP is only available for a short time. Second, 'Mutual Co-operation' is most likely when costs are low and wealth inequalities are mild, that is when the pay-offs to any given strategy are quite similar for wealthy and poor herders. Third, 'RCPD' (unilateral co-operation on the part of the rich and unilateral defection on the part of the poor) is most common when the costs of using the WSP are moderately high and the inequalities in wealth are sharp, in other words when only some individuals are in a position to benefit from the WSP. Note that this outcome reflects the fact that if a poor herder defects, it still pays a rich herder to leave for the WSP because the value of what a poor herder (with his very few cattle) steals from the DSR in the rainy season does not match what the rich herder (with his many cattle) stands to gain at the WSP.

For the purpose of the present chapter we highlight two general results from this model that are directly relevant to the question of how to identify and account for conservation acts in human communities. First, under many of the conditions we have modelled, respect for dry season reserves emerged as a stable strategy. In other words, conservation occurred for purely selfish reasons, precisely because it contributes to economic efficiency. 'Mutual Co-operation' in this context is somewhat similar to 'By-product Mutualism' (Mesterton-Gibbons & Dugatkin, 1992; Connor, 1995) insofar as it enhances production of the resources to be shared. Second, because of the asymmetrical pay-offs contingent on the differing numbers of livestock owned by the two players, herders have different thresholds for triggering defection, which leads to the result 'Rich Co-operate Poor Defect'. Here again there is no 'problem' in attaining the conservation outcome, it is simply the product of the rich individual's self interest. Analyses of an n-person game indicate that the outcome RCPD still occurs, albeit under a narrower range of conditions (Ruttan & Borgerhoff Mulder, 1999). Note, however, that in the RCPD zone a rich herder could get an *even higher net pay-off* by coercing a reluctant poor herder to leave the DSR in the wet season. What are the opportunities for such coercion?

Some speculations on coercion and policing

Among pastoralists, herd owners who benefit most from conservation of the dry season reserves (men with the largest herds) are commonly those who are the most powerful in their local community, as elders. Though wealth is not a prerequisite to elder status, influential and respected elders are usually rich in livestock. Furthermore numerous ethnographic references point to the role of elders in enforcing grazing regulations and policing infractions through committees and moots (e.g. Lane, 1991). As our analysis shows, given the parameter conditions yielding an RCPD outcome, the rich might benefit from punishing herders who utilize DSR in the wet season, at least if the marginal returns from coercion are positive.

Because we have not explored incentives for punishment in an n-person game with more than one possible punisher, we can only speculate on why Barabaig appear to punish non-cooperators. First, the costs of detecting cheaters are small; given the proximity of the DSR to the permanent homesteads of the Barabaig, cattle cannot be grazed there illegally in the wet season without being spotted. Second, the opportunity costs of bringing offenders to trial may be low given that men with one or more wives and grown sons are not directly involved in the day-to-day herding of their stock. In the Datoga, wealthier men do indeed have more people living in their homestead (Sieff, 1995); furthermore, wealthy pastoralists enjoy more leisure (Fratkin & Smith, 1994). It therefore seems plausible that wealthy old men incur few opportunity costs from sitting on committees and councils, attending trials, and adjudicating moots. Third, when offenders are punished a fine is exacted. Among the Datoga and many other groups, disputes can be resolved with fines of cattle, or with obligations to brew honey beer. In either case, the payment is consumed by the elders on the relevant council(s) (Klima, 1965), often immediately after the ruling is reached; this is so even in disputes where women adjudicate (Klima, 1964). Finally, in the Barabaig (Lane, 1996) and several other groups it is reported that some councils operate in total secrecy, so as to protect members from the possible retribution of those that are punished. Each of these considerations serves to reduce the costs associated with punishment and regulation.

Future directions in modelling

As with any model, a number of simplifications were made. The most critical of these is that we have not considered what happens when the dry season reserves are exhausted before the rains return, in other words when cattle die and where rangeland is irreversibly degraded. One might suspect that cattle loss would be more costly for poor than for rich individuals, on account of their greater proximity to a minimum threshold for survival; the finding that poor herders seem to keep higher *proportions* of their livestock alive during droughts than do rich herders (e.g. Fratkin & Roth, 1990; Herren, 1990) suggests their greater fear of catastrophe. If the poor are more averse to losing cattle, this could lower their threshold for co-operating (in other words make them keener conservationists). Conversely, with respect to rangeland quality, it might be that rich herders discount future pay-offs at a lower rate than poor herders; this notion arises from the observation that in many populations (e.g. Bradburd, 1982; Fratkin & Roth, 1990; Sieff, 1995) poor families are much more likely to drop out of pastoral production than are rich families, or at least to diversify production with honey collection, hunting and cultivation. If the rich do value rangeland quality more than the poor, this should lower *their* threshold for co-operation (and conservation), and in addition provide greater motivation for them to coerce the poor into co-operating. Finally, the cost of utilizing wet season pastures probably varies according to the wealth of the herder, because poorer families often have insufficient labour to maintain two households (Sperling & Galaty, 1990; Borgerhoff Mulder & Sellen, 1994); this would effectively raise a poor herder's threshold for co-operation. Because the Barabaig evidently punish those who do not conserve, our current work (Ruttan & Borgerhoff Mulder, 1999) is investigating the role of punishment (following Clutton-Brock & Parker, 1995).

CONCLUSIONS

General conclusions and implications

Using game theory to explore the consequences of asymmetries in power and pay-offs supplies some much-needed political and economic reality to the study of conflicts over resources within and even between communities (see also Ostrom et al., 1994). Failure to place the study of conservation more squarely within a politico-economic context is one of the main criticisms of earlier evolutionary anthropological studies of conservation in indigenous groups. We should, however, acknowledge that many field workers who study common pool resource systems (notably those of pastoralists) reject the view that individuals in traditional communities can be characterized as selfish and liable to defection. Rather they stress the multiple relationships and interdependencies among social actors that reduce an individual's temptation to cheat. We respond by noting that although predicating models on self-interest may oversimplify issues, it has the power to delineate the potential scope and weaknesses of common property systems under different conditions, and hence is a useful tool (see also Ostrom, 1990).

We end by drawing attention to just two conclusions. First, conservation can be an outcome of economic maximization, or efficiency. Second, asymmetries in pay-offs provide the opportunity for both unilateral co-operation and/or coercion of the weak by the powerful. Regarding the first point, the success of our model in generating conservation outcomes throws into relief the restrictive nature of earlier work that for heuristic reasons opposed conservation and efficiency outcomes. Although it is clear why investigators such as Hames made this practical gambit, now that the search for conservation behaviour is moving beyond the realm of prey choice in foragers, it may be more productive to look for conservation acts *within the framework of efficiency maximization*. One important implication of this approach is that it highlights the fact that not all interactions leading to beneficial outcomes for the group are necessarily structured as a Prisoner's Dilemma Game (or as a Game of Assurance). Another implication is obvious, but more practical: those behavioural acts with conservation outcomes that are also economically efficient will be particularly attractive to potential adopters, and hence of more relevance both to development advisors and community-based conservation activists.

Our second conclusion is that asymmetries in pay-offs contingent on wealth differentials (and other related interests) can generate an array of conservation outcomes. Particularly interesting is the finding that as the degree of asymmetry in pay-offs increases (and the costs of conservation rise) opportunities for unilateral conservation can emerge. From here we argued that if asymmetries in pay-offs map onto asymmetries in power, as they do in many pastoralist communities, dominant individuals are tempted to coerce all individuals into co-operating. The implications of this are the following. First, conservation is most likely to be observed where those with the greatest interest in conservation are those with the most power. Indeed this may be an important reason why pastoralists are often successful in protecting their commons from degradation, at least if these systems are undisturbed by external forces (e.g. Ensminger & Knight, 1997). To take a stark comparison, think of modern industrial societies, where corporate power has far more to gain from environmental abuse than does any single individual – what hope is there in instituting conservationist policies until individuals have equal or greater power than corporate bodies? Second, when field-workers observe conservation outcomes, we

cannot assume that the conservation strategy is indeed necessarily beneficial to all individuals (see also Gadgil & Berkes, 1991); some coercion may
be going on. Third, in a more applied context, empirical studies need to
determine precisely *how* environmentally successful projects obtain their
successes; see for example Gibson & Mark's (1995) evaluation of Zambia's
ADMADE communal resource management project or Metcalfe's (1994)
study of Zimbabwe's CAMPFIRE project, where attempts are made to determine whose interests a project serves.

Our results not only are in line with those of institutional economists
and political scientists, but also provide a broader framework within which
to accommodate their contrasting results. The specific finding that Mutual
Co-operation is most likely to occur when heterogeneity among players is
low (in conjunction with low costs of conservation) is consistent with many
empirical observations on the management of common pool resources (e.g.
Ostrom, 1990; Hanna *et al.*, 1996). At the same time, our finding that
opportunities for unilateral conservation are more likely to emerge as the
degree of asymmetry in pay-offs increases is consistent with the arguments
of other social scientists who propose that a certain amount of heterogeneity facilitates collective action insofar as members of a small 'privileged'
group find it individually advantageous to support the costs of punishment
(Olson, 1967; McKean, 1992; Ruttan, 1998).

Political issues

Studies purporting to show conservation, or its lack, in traditional communities are often viewed as inflammatory with respect to the contentious
political debates surrounding conservation and indigenous affairs (e.g. Alcorn, 1991, 1995; Puri, 1995). However, as Hill (1995) and many others
make quite clear, *conservation performance is not a valid reason for divesting
indigenous people of their land.* Indeed indigenous groups demonstrating
no conservation ethic under traditional conditions readily become conservationists as they adjust their behaviour to the novel, political and economic environment (Kaplan & Kopishke, 1992; Vickers, 1994).
Furthermore, environmental regulations that are based on traditional cus-
tom and sanctioned by community institutions are more likely to be respected than those imposed by external authorities, even where the
regulations themselves are very similar (Stevens, 1997). Hence a dispas-
sionate understanding of where, when and how humans engage in conservation is central to the development of strategies designed to develop

sustainable patterns of resource use, to protect human rights, and to conserve biodiversity.

The implications of the present study concerning the impact of power differentials on conservation outcomes do not, of course, mean that we sanction the old-fashioned view that the poor stand in the way of conservation. In fact our motivation is quite the opposite. Given the current popularity of communal resource management systems among Non-Governmental Organizations, we suspect that such projects need to be closely scrutinized, to see not only *whether they afford biodiversity protection*, but *how and why they work*, and potentially in *whose* best interests (see also IIED, 1994).

Finally, we note that Barabaig are continually losing access to the Basoto wet season pastures because of agricultural incursions that began in the 1960s and accelerated in the 1980s. Herders are now forced to use the Barabaig Plains at times of the year when these reserves should be resting, and to maintain permanent homesteads in zones that were traditionally prohibited from habitation. Accordingly, elders can no longer enforce traditional sanctions. We suspect that under these unfortunate, and all too common (Fratkin, 1997) circumstances, a game theoretical analysis of herding practices based on individually perceived costs and benefits becomes increasingly valuable, precisely *because* the traditional community norms have been undermined (see also Grabowski, 1988). The current land-use crisis in Hanang is rooted ultimately in economic conflict stimulated by national and international development priorities, and can be remedied only through institutional change at the national and regional level.

ACKNOWLEDGEMENTS

Lore Ruttan was supported by a MacArthur Fellowship from the Institute of Global Conflict and Co-operation, University of California. We thank Michael Alvard, Peter Coppolillo, Hilly Kaplan, Peter Richerson and Eric Smith for helpful comments and discussion.

4

Predicting the consequences of human disturbance from behavioural decisions

JENNIFER A. GILL & WILLIAM J. SUTHERLAND

INTRODUCTION

As a result of increases in human populations, their mobility and leisure time, human presence is increasing in many areas that are important for conservation. The restricted nature of the remaining areas of suitable habitat in combination with increasing concerns for species conservation have meant that the issue of whether human presence has adverse effects on wildlife has gained in importance in recent years (Hockin et al., 1992; Hill et al., 1997). A major concern for conservationists is that, in response to human presence, animals may avoid or under-use areas and thus this is equivalent to habitat loss or degradation. Disturbance by humans could therefore be just as damaging as actual habitat loss or degradation, although as the habitat is unchanged, the effects are also potentially reversible.

Animal behaviour is an important tool in the understanding of the response to human disturbance. This chapter will show that an understanding of behavioural responses to disturbance allows disturbance to be described in terms of a trade-off, in the same manner in which animals respond to predation risk (Milinski, 1985; Lima & Dill, 1990). Furthermore, we will show that understanding the consequences of disturbance for species requires an understanding of density-dependence and that a good way of approaching this issue is the use of game theory models based on individual decision-making processes.

DISTURBANCE AS A TRADE-OFF WITH RESOURCE USE

The importance with which the issue of the impact of human disturbance for wildlife is viewed is reflected in the very large number of publications on

Table 4.1

Examples of the consequences of disturbance in the non-breeding season for local bird populations, measured as the proportion of birds displaced from a site by disturbance

Species	Disturbance type	Max. local displacement (γ)	Reference
Pink-footed geese *Anser brachyrhynchus*	< 40 m from road with 20–30 cars/day	− 1.0	Madsen (1985)
Pink-footed geese	200 m from road	− 0.8	Gill *et al.* (1996)
Pink-footed geese	< 100 m from field centre to road	− 1.0	Keller (1991)
Greylag geese *Anser anser*	< 100 m from field centre to road	− 1.0	Keller (1991)
Brent geese *Branta bernicla*	Coloured tape over crops	− 0.85	Vickery & Summers (1992)
Red-breasted geese *Branta ruficollis*	1.5 km from roads	− 0.64	Sutherland & Crockford (1993)
Wigeon *Anas penelope*	Start of angling season	− 0.85	Bell & Austin (1985)
Wigeon	Bait-digging	− 0.93	Townshend & O'Connor (1993)
Teal *Anas crecca*	Start of angling season	− 0.9	Bell & Austin (1985)
Mallard *Anas platyrhynchus*	Increased sailing activity	− 0.81	Batten (1977)
Mallard	Start of angling season	− 0.78	Bell & Austin (1985)
Pochard *Aythya ferina*	Increased sailing activity	− 0.40	Batten (1977)
Tufted duck *Aythya fuligula*	Increased sailing activity	− 0.14	Batten (1977)
Bar-tailed godwit *Limosa lapponica*	Bait-digging	− 0.26	Townshend & O'Connor (1993)
Redshank *Tringa totanus*	Bait-digging	− 0.48	Townshend & O'Connor (1993)
Grey heron *Ardea cinerea*	Repeated deliberate scaring from fish-farms	− 0.85	Draulans & van Vessem (1985)
Bald eagles *Haliaeetus leucocephalus*	Human activity along rivers	− 0.25	Stalmaster & Newman (1978)

the subject; between 1988 and 1997, 308 papers are listed in the Bath Information Data Service with 'human disturbance' as a keyword. Many of these studies have examined one, often heavily disturbed, site and shown that disturbance can result in animals avoiding such sites (Batten, 1977; Burger, 1981; Bélanger & Bédard, 1989; Pfister et al., 1992; Stock, 1993). In other studies, the approach that is adopted is a comparison of the density of animals in disturbed sites with the density in less disturbed or undisturbed sites. This can then provide a measure of the local displacement of animals from a site as a result of human disturbance. Table 4.1 lists studies that have measured the proportion of birds displaced from a site by a given source of disturbance during the non-breeding season.

These approaches, however, often fail to incorporate other factors that may also be determining distribution, for example, the distribution of resources, such as food, nesting sites or roosting sites. The approach that we have taken is to view the effect of disturbance as a trade-off with the use of such resources. In this respect, the approach is identical to the trade-offs between foraging success and predation risk which have been widely used in behavioural ecology (Werner et al., 1983; Milinski, 1985; Lima & Dill, 1990; L'Abée-Lund et al., 1993). This is illustrated in Figure 4.1, in which each bar represents a patch with a given level of resource and the patches are ranked according to the level of disturbance that they experience. In Figure 4.1(a), the animals do not respond to disturbance and if resource availability constrains the animals, the resource in all patches will be used to the same threshold level. In Figure 4.1(b), there is local displacement in response to disturbance such that the proportion of the resource used declines with increasing disturbance. An example of this is the trade-off shown by pink-footed geese, *Anser brachyrhynchus*, in their use of wintering habitat (Gill et al., 1996). This species spends the winter months feeding largely on agricultural land in western Europe, primarily in Britain. The southernmost wintering population, in Norfolk, feeds on the harvested remains of sugar beet, a common crop in this region of Britain. The geese are frequently disturbed from the roads alongside fields and hence they prefer to feed in fields in which they can be far from the roads. This results in a strong negative relationship between the proportion of roots consumed on individual fields and the distance of the fields from roads (Figure 4.2). This trade-off between consuming the roots and avoiding sources of disturbance results in up to 64% of the roots going uneaten.

Pink-footed geese are thus a species which shows a strong behavioural response to disturbance, by avoiding sites close to areas with high human

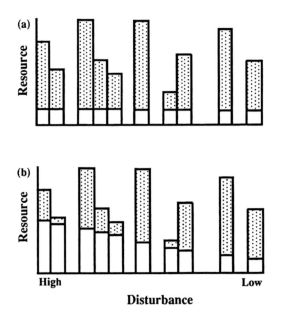

(a)

Resource

(b)

Resource

High   Low

Disturbance

Figure 4.1 The trade-off between disturbance and resource use. Each bar represents a patch ranked according to the level of disturbance. The height of each bar indicates the initial quantity of a given resource in each patch. The dotted area represents the resource used in each patch. In (a) the species is not displaced by disturbance, while in (b) the species is displaced by disturbance and the proportion of the resource used declines accordingly.

presence, and the consequence of this behaviour is that fewer animals are supported on these sites than is possible. Quantifying the effect of disturbance in this way has the advantage of making it possible to predict the consequences for the pink-footed goose population of changes in disturbance levels, thus providing evidence for policy-makers and land managers to employ.

By contrast, black-tailed godwits, *Limosa limosa islandica*, have been studied using the same trade-off approach, but with very different conclusions. This wading bird winters on estuaries in north-west Europe. Estuarine habitats are frequently considered to be amongst the most at threat from the increases in recreational and industrial developments (Davidson & Rothwell, 1993). However, quantifying the impact of disturbance for this species requires an understanding of the other ecological parameters determining the number of birds using a site. The estuaries of south-east England experience high levels of recreational and industrial use. The

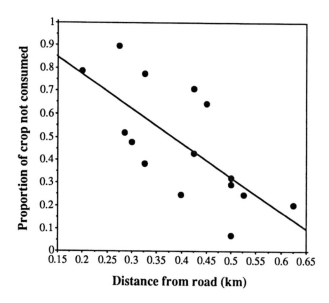

Figure 4.2 The effect of the distance between pink-footed goose flocks at first landing and the nearest road and the proportion of sugar-beet roots that is not consumed by the geese on each field ($r^2 = 0.50$, $n = 15$, $P < 0.003$).

distribution of black-tailed godwits around these estuaries is significantly, positively related to the density of their main prey, estuarine bivalves (J. A. Gill *et al.*, unpublished data). Black-tailed godwits in this region feed largely on specific size classes of the bivalves *Scrobicularia plana, Macoma balthica* and *Mya arenaria*. Studies of the distribution of black-tailed godwits in relation to both the density of these bivalves and the distribution of recreational activities on these estuaries found no effect of recreational activities on the use of the bivalve food source (J. A. Gill *et al.*, unpublished data). These studies took place at a range of spatial scales, from small patches of mudflat up to whole estuaries, and using a range of measures of recreation, from indices of human presence to comparisons of sites with marinas and footpaths, and control sites.

These two cases illustrate the power of viewing the effect of disturbance as a trade-off with resource use. In the case of pink-footed geese, the approach allows predictions of the increase in local population size that may be expected from a decrease in disturbance levels (Gill *et al.*, 1996). In the case of black-tailed godwits, there is a widespread belief that human presence on estuaries in winter has a significant negative impact on wading-bird populations, based largely on the strong flight reactions to human presence that are commonly observed in these species (Burger, 1981; Kirby et al., 1993; Smit & Visser, 1993). However, this approach suggests that even if bird distribution is altered in the short-term by the presence of humans, over an entire winter disturbed sites are not consistently under-used by this species.

Viewing this issue in terms of a trade-off is therefore an effective means of determining the extent to which disturbance restricts habitat use at a local scale. There are a number of situations in which such local constraints may be of conservation concern.

1. If disturbance within a site causes an increase in mortality or reduction in fecundity (often as a consequence of animals being at higher densities elsewhere) such that the species declines as a result.

2. If the species in question is declining or restricted in range for unknown reasons, then reducing disturbance may be a sensible precautionary approach.

3. If the land is deliberately being managed for that species, for example, many nature reserves are designed to attract certain species. Manipulating disturbance may aid this goal, even if it has no benefits for the species as a whole.

4. If the aim is to remove a species from an area. For example, the problems caused by many pest species can be alleviated by manipulating disturbance such that the animals are dissuaded from using areas where they may cause problems (Vickery & Summers, 1992; Gill, 1996).

There may be many cases in which disturbance alters distribution on a local scale and yet none of the above criteria are met. It is clearly important for conservationists to be able to distinguish these cases from those with which they should be concerned. There is thus a need to be able to identify local effects of disturbance (here termed 'local displacement') and the consequences of these local effects for the status of a species as a whole.

SPECIES-LEVEL CONSEQUENCES OF DISTURBANCE

Sutherland (1998b) describes the approach that can be used to calculate the consequences of disturbance for a species. This approach describes the change in the total population size (ΔN) that will result from a given level of disturbance within one site:

$$\Delta N = L\, M\, \gamma\, d' \, / \, (b' + d') \qquad (1)$$

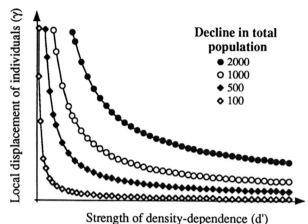

Figure 4.3 The effect of the strength of density-dependent mortality on the total population decline resulting from the local displacement caused by disturbance at one site.

where L is the area affected by the disturbance, M is the density within the site prior to disturbance, b' is the strength of per capita breeding output, d' is the per capita density-dependent mortality and γ is the proportional change in the number of animals in the site as a result of disturbance (see Table 4.1 for examples). The effect on the entire population of disturbance at one site can be seen in Figure 4.3 (for simplicity, b' is assumed to be constant). This figure describes the relationship between the strength of density-dependent mortality (d') and the local displacement caused by increasing levels of disturbance (γ). The four lines show the interactions between displacement and density-dependent mortality that would result in the four different declines in total population size. Thus, very high levels of disturbance may result in large declines in number within the disturbed site, but the effect on the total population size may be negligible if density-dependence is very weak. By contrast, a very slight level of disturbance can have significant effects on total population size if density-dependence is strong.

Thus, if the aim of conservationists is to determine the consequences of different forms and levels of disturbance for entire species, it is necessary to know not only the local impact of disturbance but also the strength of density-dependence both in the breeding and the non-breeding season. Field values of the strength of density-dependence are extremely difficult to measure. The only migratory species for which values exist for both the breeding and the wintering grounds is the oystercatcher, *Haematopus ostralegus*, and these can be used to illustrate the approach. For this species, game theory models of the behaviour of individuals have been used in studies of the breeding biology to give an estimate of $b' = 0.00005$ (Ens *et al.*, 1992; Sutherland, 1996b) and in studies of winter foraging ecology, giving an estimate of $d' = 0.00011$ (Goss-Custard *et al.*, 1995a,d; Sutherland, 1996b). Thus, from equation (1), the density-dependent ratio $[d' / (b' + d')]$ is 0.69. This means that for oystercatchers, any value of disturbance that displaces x birds from a site will result in a population decline of $0.69x$. Clearly, these values are dependent on the estimates of b' and d' being good estimates of the strength of density-dependence operating throughout the species' range.

Achieving good estimates of the strength of density-dependence in the breeding and the non-breeding season is therefore critical to predicting the impact of disturbance for species. However, very few such estimates exist in the literature at present.

MOVEMENT IN RESPONSE TO DISTURBANCE

Whilst the impact of disturbance for species is thus a complex issue to address, estimates of local displacement in response to disturbance (e.g. Table 4.1) do highlight some interesting differences in the response of species to human presence. It has previously been assumed that species which, in the non-breeding season, show strong local displacement in response to disturbance (e.g. pink-footed geese) are likely to be those that are most susceptible to disturbance (Burger, 1981; Tuite *et al.*, 1984; Bell & Austin, 1985) and should be given protection accordingly. However, there are clearly many other factors involved in determining the shape of the behavioural response shown by different species, for example, resource availability, competitor density or individual quality. The combination of all these parameters will define the strength of the density-dependent survival rate, which in turn determines the cost to an individual of moving in response to disturbance. The effect of these costs on the response to disturbance can be illustrated using the technique described in Figure 4.4; in an undisturbed site (site A), survival rate will depend on the density of animals within that site. Disturbance is not likely to result in direct mortality, however, its effects can be viewed as a perceived risk of mortality. The result of

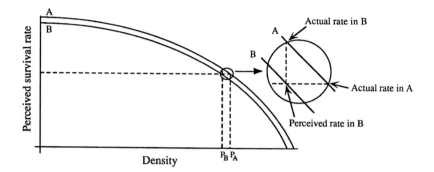

Figure 4.4 The decline in the perceived survival rate with increasing population density for two sites: an undisturbed site (A) in which the perceived survival rate is the actual density-dependent survival rate and a disturbed site (B) in which the perceived survival rate is the actual rate minus the perceived risk caused by the disturbance. The inset shows how such relatively small changes can be viewed as having linear survival rates.

this risk will be a perceived survival rate, which will result from both the actual survival rate and the perceived reduction in survival as a result of disturbance. The perceived survival rate of the disturbed site (B) is thus lower than the actual rate by an amount proportional to the level of disturbance in B (Figure 4.4). Thus, disturbing site B should cause the animals to redistribute between the two sites such that the perceived survival rate in the two sites is equal. Note, however, that the actual survival rate in the disturbed site is in fact higher than in the undisturbed site, because of the lower density present in site B.

The interactive nature of density-dependence and the displacement caused by disturbance is illustrated in Figure 4.5. This shows that a given distribution of animals between sites A and B can be achieved with a relatively large difference in disturbance between the two sites and strong density-dependence (Figure 4.5(a)). However, Figure 4.5(b) shows that exactly the same distribution could result from very little disturbance at B in conjunction with weak density-dependence. This is because with weak density-dependence, the costs of moving to a less disturbed but populated site are relatively small, such that animals should move in response to very little disturbance. By contrast, strong density-dependence will tend to counter the perceived risk from high levels of disturbance; animals will remain in the heavily disturbed site to avoid the high costs of competition in the undisturbed but heavily populated site.

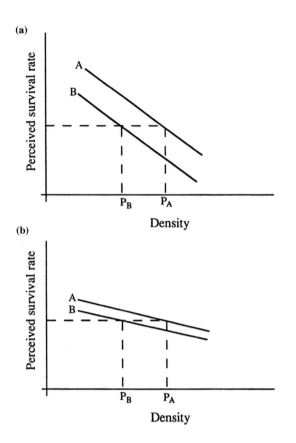

Figure 4.5 The distribution of animals between sites A (undisturbed) and B (disturbed) when (a) B is heavily disturbed (and thus has a much lower perceived survival rate than A) and the survival rate is strongly density-dependent and (b) B is only slightly disturbed and survival rate is weakly density-dependent. Under both sets of very different conditions the distribution of animals (P_A and P_B) is identical.

This approach demonstrates the relationship (Figure 4.6) between the slope of the density-dependent survival rate (d'), the level of disturbance (S) and the decline in the number of animals at a site as a result of that disturbance (γ) (assuming that the area of site A is large enough to accommodate the animals displaced from site B). This relationship is such that γ is inversely related to d', and the exact position of this relationship is determined by the level of disturbance (S) (Figure 4.7). This figure shows that species with high levels of displacement at very low levels of disturbance (those generally considered as highly susceptible to disturbance) must have

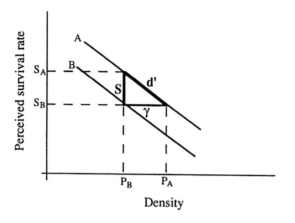

Figure 4.6 An illustration of the relationship between S, the decrease in perceived survival rate caused by disturbance ($S_A - S_B$); γ, the resulting local displacement of animals ($P_A - P_B$); and d', the strength of density-dependent survival ($P_A - P_B$ / $S_A - S_B$).

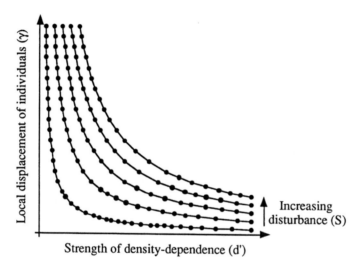

Figure 4.7 The inverse relationship between the strength of density-dependent survival (d') and the local displacement caused by disturbance (γ) for different levels of disturbance altering the perceived survival rate (S).

relatively weak density-dependent survival rates. This means that the costs of moving are small and hence animals will move in response to very little disturbance. The species that are not displaced by very high levels of disturbance (and are thus generally assumed to be unaffected by disturbance)

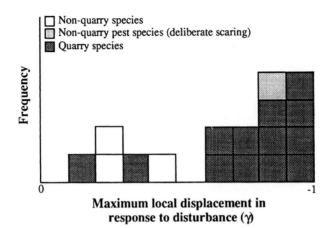

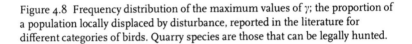

Maximum local displacement in response to disturbance (γ)

Figure 4.8 Frequency distribution of the maximum values of γ; the proportion of a population locally displaced by disturbance, reported in the literature for different categories of birds. Quarry species are those that can be legally hunted.

are constrained by strong density-dependence. Hence, the change in the number of animals inhabiting a site as a result of disturbance is dependent both upon the level of disturbance and the strength of the density-dependent survival rate. Local displacement is not, therefore, a good predictor of which species are likely to be adversely affected by disturbance.

There are, however, some interesting differences between species in the extent to which they are displaced by disturbance. Figure 4.8 shows the frequency of the estimates of the proportion of animals displaced by disturbance presented in Table 4.1. There appears to be a greater number of cases where disturbance has a severe effect (> 50% of animals displaced) than cases where the effect was relatively small. This is quite likely to be the result of under-reporting of cases where the effect of disturbance was small and hence there were few conservation or management implications. Although this is not a phylogenetic analysis, another striking fact in Figure 4.8 is the distinction between quarry (species which can be legally hunted) and non-quarry species. The only instance in which a non-quarry species exhibited a strong response to disturbance was a study of grey herons, *Ardea cinerea*, which were deliberately scared from fish-farms where they were considered a serious commercial problem. Thus, it seems that one significant aspect of the importance of disturbance for animal species may be whether or not human presence presents a direct risk of mortality. In Britain and many other countries in north-west Europe, the hunting of waterfowl is quite strictly controlled and restricted to common and numerous species that are not under any conservation threat. Thus, although quarry species may be the most responsive to disturbance, from a species perspective they are likely to be amongst the least at threat from disturbance. An alternative explanation for the different levels of displacement recorded in different species is the different habitats that they utilize. Many of the geese and ducks that show strong local displacement in response to disturbance also tend to use grassland or arable habitats. By contrast, the species which showed little displacement tended to occur on estuaries or inland reservoirs. It is therefore also possible that the costs of moving between reservoirs or estuaries may be prohibitive whereas the costs of moving to a nearby field may be small and that the levels of displacement simply reflect these costs. In reality, the decisions determining whether an animal moves in response to disturbance are likely to be based upon a suite of factors such as these.

CONCLUSIONS

In conclusion, the impact of human disturbance on wildlife is a subject of considerable conservation concern at the moment (Hockin *et al.*, 1992; Davidson & Rothwell, 1993). Fundamental to an understanding of the impact of disturbance is an understanding of the decision-making processes of individuals. This is because the displacement of animals from disturbed sites has consequences for the density-dependent processes determining individual success. Thus the pattern of distribution adopted by individuals in response to disturbance will depend both on the risk associated with disturbance and the density-dependent consequences of switching to an alternative site.

On a local scale, the impact of disturbance can be viewed as a trade-off with resource use in the same way as predation risk has been shown to alter habitat selection and use. This approach can be used both to identify whether disturbance is constraining the number of animals using a site and to predict the changes in numbers locally that would result from a change in disturbance.

The consequences of disturbance at one site for the entire population can be calculated with knowledge of the proportion of animals leaving the site as a result of disturbance and the strengths of density-dependent breeding output and density-dependent mortality. There are very few field estimates of the strength of density dependence, however, game theory models can be used to derive these estimates. This process has been carried out for one species, the oystercatcher (Ens *et al.*, 1992; Goss-Custard *et al.*, 1995a,b; Sutherland, 1996b), in which the behavioural strategies adopted by individuals are fundamental to determining levels of density-dependence. Clearly, quantifying the importance of disturbance and other factors which can result in the degradation or loss of habitat, will depend on a greater understanding of these density-dependent processes.

5

The role of behavioural models in predicting the ecological impact of harvesting

JOHN D. GOSS-CUSTARD, RICHARD A. STILLMAN, ANDREW D. WEST, SELWYN MCGRORTY,
SARAH E. A. LE V. DIT DURELL & RICHARD W. G. CALDOW

INTRODUCTION

Much attention has been given to devising ways of predicting the effects of habitat loss on vertebrate populations. The immediate effect is to increase density in the remaining areas (Goss-Custard, 1977). If the birth or death rate is already density-dependent, or becomes so as a consequence of habitat loss, equilibrium population size will decrease (Goss-Custard, 1993; Sutherland, 1996a). The magnitude of the decrease following habitat loss in either the breeding or the non-breeding regions depends on whether the precise form of the density-dependent functions in one region are changed and also on how they interact with the density-dependent functions in the other to determine population size (Goss-Custard et al., 1996a; Sutherland, 1996a; Goss-Custard & Sutherland, 1997). Individuals-based models founded on behavioural ecological principles are useful here because they can predict the form of density-dependent functions after habitat loss and these can then be inserted into demographic models for population prediction (Goss-Custard et al., 1995e).

A similar framework can be used to understand and predict how harvesting a natural food supply affects 'co-dependent' vertebrate populations that, directly or indirectly, are influenced by the abundance of target and non-target food species. For animals that consume the harvested species, harvesting can affect the abundance, size distribution and species composition of their food in some or all patches. As harvesters may also disturb them, they may also be forced to feed in less favourable patches where, because of elevated conspecific densities, competition may intensify. Behaviour-based models can be used to predict the changed intake rates of animals forced by harvesting to alter their diet and/or to redistribute themselves over resource patches of varying quality. Importantly, they provide a reliable basis for prediction because the responses to harvesting made by animals in the model are based on decision principles, such as rate maximization, that are unlikely to alter in the new environments even if the particular choices made by animals, and thus their chances of surviving and of reproducing, do change (Goss-Custard, 1996; Goss-Custard & Sutherland, 1997).

However, there may also be indirect consequences of harvesting which behaviour-based models do not consider. For example, as the target species is depleted, competitor non-target species may increase in abundance, thus changing the food supply of vertebrates that do not consume the target species themselves. The harvesting process itself may change the abiotic environment, as when dredging disturbs marine sediments. This might affect both target and non-target species and any other animals that are affected by their abundances. These knock-on ecosystem impacts of harvesting are not considered by behaviour-based models, yet could be very important. Clearly, behavioural models that can predict some immediate and direct impacts of harvesting must somehow be combined with ecological models that predict the wider ecosystem effects. Using fishing as the example, this article briefly reviews cases where harvesting may affect co-dependent vertebrates. It then discusses how behavioural models can contribute to both understanding and predicting any impact before briefly noting ways in which the wider ecosystem consequences can be predicted.

HARVESTING AND CO-DEPENDENT SPECIES

Fishing and otters

Otters, *Lutra lutra*, are most abundant where their fish prey are numerous (Kruuk, 1995). Because of the large heat costs of aquatic hunting, fish have to be abundant if the cost to otters of foraging is not to become prohibitive. In several areas otters appear close to their energetic limits, implying that a reduction in fish density would make these areas unsuitable. Although otters consume many unharvested species, they regularly rely on commercial fish species, such as salmon, *Salmo salar*, and eels, *Anguilla anguilla*. There are circumstances, therefore, in which otters and people may compete; one-fifth of 102 otter naturalists worldwide identified declines in fish abundance, arising partly from fishing, as a threat to otters (Kruuk, 1995).

Fishing and seabirds

Long-lived seabirds have several adaptations to the often high annual variability in their marine invertebrate and fish food supplies. When food is scarce, they may refrain from breeding or change their breeding sites or diet (Klomp & Furness, 1992; Furness & Camphuysen, 1997). In the North Sea, the lesser sandeel *Ammodytes marinus* is a vital – in some cases, the only – prey for most seabirds during summer, as well as for many commercially important fish and marine mammals. It is also the target species of the largest single species fishery in the North Sea (Monaghan, 1992). When sandeel abundance around Shetland collapsed, breeding success in kittiwakes *Rissa tridactyla* and Arctic terns *Sterna paradisaea* fell dramatically (Monaghan, 1992). Great skuas *Catharacta skua* and Arctic skuas *Stercorarius parasiticus*, which kleptoparasitize terns, increased their foraging effort, but not entirely successfully (Figure 5.1). There were long-term effects leading to reduced body masses or a failure to breed in subsequent years (Hamer *et al.*, 1991; Phillips *et al.*, 1996). Even in the common guillemot *Uria aalge*, whose breeding success did not collapse, the parents' increased foraging effort did not fully offset the reduced food density (Monaghan *et al.*, 1994, 1996; Uttley *et al.*, 1994).

Marked declines in the populations of Arctic terns, kittiwakes and, at some colonies, of guillemots, great skuas and Arctic skuas, in Sheltand have been recorded since the late 1970s (Monaghan, 1992). It appears that reduced fish stocks, through their effect on reproductive success, breeding frequency and recruitment, affected the populations of several species of seabirds in Shetland (Wright *et al.*, 1996). Although the extent of the effects differed between species, depending on their ability to buffer themselves against reduced food abundance, this is a general finding. Breeding failures and population declines in 12 seabird species worldwide have been associated with stock collapses in their main fish prey species (Hunt *et al.*, 1996).

The most likely direct detrimental influence of fishing on seabirds is when a fishery takes the small fish they eat. However, few studies have demonstrated that fish declines associated with seabird declines were mainly a result of harvesting rather than to other causes (Wright *et al.*, 1996). Most North Sea sandeels are fished offshore whereas seabirds forage inshore, and the degree to which sandeels from the two areas intermix is unknown. Because many seabirds eat sandeels before they are large enough to be harvested, any effect of fishing on seabirds would presumably

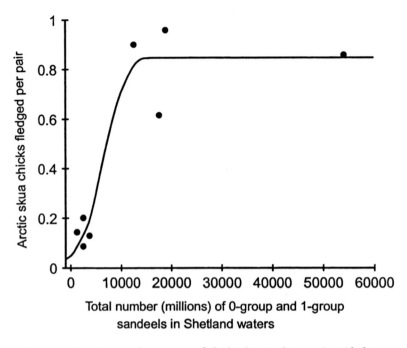

Figure 5.1 How the breeding success of Shetland arctic skuas varies with the abundance of their main sandeel prey [from *Ibis*; Phillips *et al.* (1996a)].

be through reducing spawning stock, and hence egg production. The main unanswered question is whether fishing reduced sandeel stocks to a level at which recruitment was affected, and this is difficult to answer when local recruitment may have been boosted by immigration from unknown stocks elsewhere (Wright *et al.*, 1996). If the effect of sandeel harvesting on seabirds is to be fully understood, greater integration is needed between fishery and seabird population studies and in the spatial scale at which they are conducted (Wright *et al.*, 1996).

Seabirds may also benefit from fishing because many fisheries discard unwanted material overboard (Hudson & Furness, 1988). In the North Sea, 790 000 t of offal, fish, echinoderms and benthic invertebrates were discarded in 1 year – equivalent to 22% of the total landings – and, of this, 310 000 t, or nearly 40%, were eaten by seabirds (Garthe *et al.*, 1996). Although the effect of this massive discard consumption on seabird demographic rates and population size cannot be calculated, it is sufficient food to support 5.9 million individual seabirds a year, implying that discards may have a substantial beneficial effect on numbers.

Fisheries may have indirect impacts on seabirds by altering marine community structure. Thus, bottom-fishing with mobile gear affects the benthic community structure and thus the food supply of fish (Collie *et al.*, 1997). Sometimes, these effects can be quite unexpected. Following a moratorium on groundfishing in Newfoundland in 1992, the absence of fishery discards in combination with a lowered abundance of the capelin *Mallotus villosus* not only reduced kittiwake breeding success directly through food shortage but also indirectly increased losses of eggs to great black-backed gulls *Larus marinus* (Regehr & Montevecchi, 1997). Fisheries usually harvest large-sized fish that may either eat or outcompete the smaller fish taken by seabirds. When large fish are depleted, smaller ones may increase to the benefit of seabirds. Improving fishing techniques during the 1950s and 1960s led to a collapse of the spawning stock of herrings *Clupea harengus* in the Barents Sea, which persisted right through the 1970s. Krasnov & Barrett (1995) suggest that the subsequent increases in kittiwakes and common guillemots were a result of the increase in the abundance of a major prey species, the capelin, when herring, the capelin's main competitor and a major predator of its larvae, had been depleted. Through its knock-on effect on the community, the herring fishery may thereby have increased the food supply of some seabirds.

Bait digging and shorebirds

Polychaete worms play a large role in marine communities and a small number are harvested as live-bait for angling, the lugworm *Arenicola marina* being among the most important. It lives to a depth of 1 metre in silty sediments, which it ingests to extract organic matter. Lugworms periodically back up to the surface to defaecate, which is when they are vulnerable to attack by vertebrate predators, such as the bar-tailed godwit *Limosa lapponica*.

Unrestricted lugworm harvesting over 8 weeks removed some 4 million lugworms from 0.4 km² of Northumberland flats, and few remained (Olive, 1993). However, numbers recovered within a few months through immigration. If harvesting is continuous, however, the average long-term abundance of worms must be reduced over large areas as fishers successively deplete patches and the lugworm source areas lose emigrants. Bait-digging may also affect non-target organisms that are food for vertebrates.

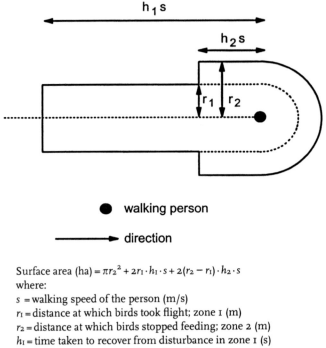

Surface area (ha) = $\pi r_2^2 + 2r_1 \cdot h_1 \cdot s + 2(r_2 - r_1) \cdot h_2 \cdot s$
where:
s = walking speed of the person (m/s)
r_1 = distance at which birds took flight; zone 1 (m)
r_2 = distance at which birds stopped feeding; zone 2 (m)
h_1 = time taken to recover from disturbance in zone 1 (s)
h_2 = time taken to recover from disturbance in zone 2 (s)

Figure 5.2 Surface area without birds after disturbance from a single person [from *Wader Study Group Bulletin*; Smit & Visser (1993)].

For example, lugworm-digging can substantially and extensively reduce cockle *Cerastoderma edule* stocks in the long-term (Jackson & James, 1979). However, by exposing buried prey, it can also provide a food bonanza for scavenging birds in the short-term.

The direct and indirect effects of lugworm and cockle reductions on the survival and reproductive rates of co-dependent vertebrates have not been estimated. Nor have the possible wider ecological effects of the release of toxic pollutants from the disturbed sediments been fully evaluated (Howell, 1985). However, studies have been made of the disturbance arising from bait-digging. By recording the distance at which various birds first stopped feeding and then flew away as a person approached, the effective area denuded of birds can be calculated (Figure 5.2); this may be a maximum estimate as birds may approach stationary bait-diggers more closely. In Northumberland, three bird species, two of which did not consume lug-

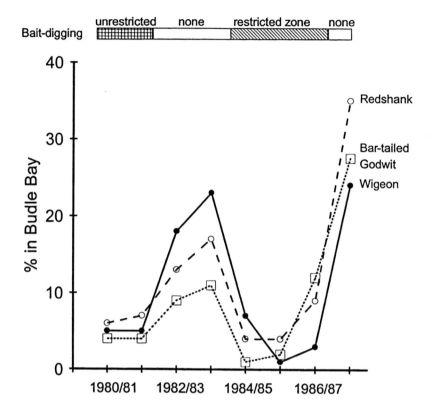

Figure 5.3 The percentage of the local populations of three bird species (redshank *Tringa totanus*, bar-tailed godwit and wigeon *Anas penelope*) that were present in Budle Bay, an area in which bait-digging was, successively, (1) allowed to proceed without restriction, (2) banned entirely, (3) allowed only in a restricted zone and, finally, (4) again banned altogether; these periods are shown in the horizontal bar at the top [from *Wader Study Group Bulletin*; Townshend & O'Connor (1993)].

worms themselves, decreased locally during periods when the amount of lugworm harvesting was high but not when it was low (Figure 5.3). Such studies show that harvesting can affect the local distribution of birds, but do not measure its impact on survival or reproductive rates, and thus population size.

Shellfishing and shorebirds

Despite huge and unpredictable fluctuations in abundance, cockles are fished widely in Europe. Traditional fishing by hand-raking at low tide, which both depletes the cockles and disturbs the birds, is now less common. Most fishing is done over high tide. Sometimes a vessel rotates around its anchor so that propellor turbulence 'blows' the cockles from the sand ready for collection by hand at low water (Franklin & Pickett, 1978). Dredging from a boat, which does not disturb the feeding birds, is more usual; a modern vessel can dredge 4 ha daily. This may sometimes temporarily benefit birds because some cockle shells are smashed, allowing easy access. However, the potential for the fishery to deplete the birds' food supply is considerable.

Some studies have measured depletion by shellfishing and its impact on co-dependent oystercatchers *Haematopus ostralegus*, but not yet on other shellfish-eating shorebirds, notably the knot *Calidris canutus*. Sometimes, there may be little direct effect; in the Baie de Somme, the minimum legal size of fished cockles exceeds the size taken by oystercatchers (Triplet & Etienne, 1991). However, generally, the minimum fishable size falls within the range taken by oystercatchers, so fishing immediately depletes the birds' food supply. Shellfishing may remove 45–70% of cockles in years of medium to high cockle abundance in the Burry Inlet (Horwood & Goss-Custard, 1977), up to 36% in years of low cockle abundance in the Dutch Wadden Sea and may regularly halve the oystercatcher food supply in the Oosterschelde, the other major Dutch cockle fishery (Lambeck *et al.*, 1996). There is good evidence that, by depleting the food supply, such high fishing rates may increase oystercatcher winter mortality rates, especially in cold winters (Clark, 1993; Camphuysen *et al.*, 1996; Lambeck *et al.*, 1996).

Mussel beds are also important feeding areas for oystercatchers and mussel-fishing may both hinder and help the birds. Subtidal mussels are sometimes dredged up and put on the foreshore for short-term storage, where birds can consume them over low tide. Elsewhere, transplantation is in the reverse direction; intertidal mussel beds are moved to subtidal lots where growth conditions are optimal, but oystercatchers cannot reach them. After 3 years of poor spatfall in the Dutch Wadden Sea, virtually all mussels were fished. Although oystercatcher numbers stayed high because birds turned to other bivalve prey and lugworms (Beukema, 1993), their mortality rate increased (Camperhysen *et al.*, 1996).

There may also be several indirect effects of cockle fishing. Blowing and dredging disturbs sediments and may kill 10–50% of juvenile cockles (Franklin & Pickett, 1978). This may reduce the cockle population in the longer-term, thus indirectly affecting oystercatchers and other predators of large cockles, such as the eider duck *Somateria mollissima*. Other bird and fish consumers of small cockles, such as knot, plaice *Pleuronectes platessa* and flounder *Platichythes flesus*, could also be affected, both directly and indirectly. Furthermore, the whole benthic community could be altered by sediment winnowing and the killing of vulnerable invertebrate species. Benthic community changes would affect the many species of fish and birds that forage on intertidal flats. However, dredge trails mostly infill within 8 months and there is no evidence that dredged sediments deter settling larval cockles (Franklin & Pickett, 1978). Measurable effects on most other invertebrate species disappear over 1–4 years (Lambeck et al., 1996), except possibly tube-dwelling polychaete worms, such as *Lanice conchilega*, which is eaten by several vertebrates. Present evidence suggests that, unless fishing is intense and sustained, co-dependent vertebrates are more likely to be affected by food supply reduction and disturbance than by longer-term changes in the benthic community.

In contrast, mussel culture and fishing may have important knock-on effects on the ecosystem, as the Oosterschelde estuary study showed (Nienhuis & Smaal, 1994). Mussels are powerful filter-feeders and can filter much of the water column in some estuaries over a single high tide. In doing so, they greatly deplete the phytoplankton and reduce its production. This then reduces the growth rates of the mussels themselves (van Stralen & Dijkema, 1994) and their value as food to their consumers. It also decreases the plankton available to other suspension-feeding invertebrates and so reduces the food supplies of vertebrates that feed on them. Ecosystem modelling in the Oosterschelde showed that doubling the mussel biomass would lead to a long-term increase of only 41% in the biomass of suspension-feeders as a whole, including mussels, because of intra- and inter-specific competition for phytoplankton (van Stralen & Dijkema, 1994). However, plankton biomass and primary production would decrease by only 20% and 14% respectively, much less than the increase in the biomass of suspension-feeders. The reason for this apparent discrepancy between the effect on consumers and producers is that mussel faeces are remineralized rapidly, accelerating the release of dissolved inorganic nutrients available to phytoplankton. Some 20–75% of the total nitrogen mineralization in Oosterschelde soft sediments occurs in mussel beds that cover only 10% of the estuary floor; hypothetically, benthic filter-feeders could enhance primary production in the water column (Prins & Smaal, 1990). Increasing mussel stocks through protection from predators [for example, by building fences proof against shore crabs *Carcinus maenas* (Davies et al., 1980)], or importing mussels from other estuaries, or decreasing their abundance by increased harvesting, may change ecosystem functioning with consequences, not only for the vertebrates that eat these shellfish, but also for those that consume other benthic organisms.

Fishing can also have far-reaching effects on the community structure of rocky shores (Hockey, 1994). In southern California, the large territorial limpet *Lottia gigantea* has been harvested for centuries and is a major prey species of the American black oystercatchers *Haematopus bachmani*. When abundant, its grazing limits macroalgae while its domination of space limits smaller *Lottia* species (Lindberg *et al.*, 1999). *L. gigantea* has been so depleted by fishing in some areas that the smaller species have increased in abundance to levels where they limit macroalgae instead. Oystercatchers have essentially the same effect. Interestingly, Lindberg *et al.* (1999) found areas where human disturbance reduced oystercatcher numbers considerably but no limpet harvesting occurred. Here, the large *L. gigantea* dominated and macroalgae were heavily grazed. Disturbance alone, then, with no fishing can affect community structure on rocky shores where, in many parts of the world, oystercatchers are very important consumers (Hockey, 1994).

Summary of the effects of fishing

This review illustrates the direct and indirect impacts that fishing, and other forms of harvesting, may have on co-dependent vertebrates. It may directly deplete the main food supply, although demonstrating this is often difficult because it has not often been possible to make the necessary measurements and to build prey population models. The co-dependent vertebrates may respond behaviourally to food depletion in various ways, including diet shifts, re-distribution and changes in priorities, but only sometimes has any possible effect on population demographic rates, although not population size, been monitored. Fishing may sometimes disturb the vertebrates but, again for practical reasons, only the behavioural, not demographical or populational, consequences have been measured. Some fishing practices may directly benefit vertebrates, at least in the short-term, but any demographical and populational benefits have not been estimated. Fishing may indirectly affect some vertebrates through knock-on ecosystem effects on both target and non-target prey species. These arise from habitat modification, changes in ecosystem productivity and changes in community composition. However, once more, practical difficulties have prevented their effects on the demographic rates and population size of co-dependent vertebrates from being pursued. We know a great deal about possible effects of fishing on co-dependent vertebrates but have difficulty in quantifying the consequences at the population level.

MODELLING THE EFFECT OF SHELLFISHING ON SHOREBIRDS

A behavioural model

In combination with population models, behavioural models can help to understand and predict the demographic and population consequences of fishing for co-dependent vertebrates. In brief, behavioural models describe how the vertebrates would be expected to respond to the direct and immediate changes in their feeding conditions arising from depletion and disturbance by the fishery. Individuals-based models can additionally predict the effect on the food consumption rates of individuals and therefore the proportion that will starve. This demographic rate can then be inserted into population models to estimate the immediate and direct effects of fishing on population size. This is illustrated using an empirical model of the interaction between mussels, shellfishermen and oystercatchers on the ten mussel beds of the Exe estuary (R. A. Stillman, unpublished data).

The immediate direct effects on mussel-eating oystercatchers of mussel fishing depend on the technique employed. When fishing over high tide by dredging from a boat or at low tide by raking, strips of mussel bed are removed and mussels of all size-classes are taken, the large ones being selected later for sale. By removing whole sections of mussel bed, 'high-tide stripping' and 'low-tide stripping' reduce bed size. Mussels are also collected at low tide by picking saleable large ones and leaving the small ones on the bed. With this 'low-tide thinning', the density of the large mussels, but not bed size, is reduced. Whether stripping or thinning, fishermen disturb oystercatchers only at low tide; at high tide when the mussel beds are submerged, birds are either at roost or continue feeding in fields. Stripping and thinning reduce the abundance of the adult mussels under which young mussels find protection from crabs (McGrorty *et al.*, 1990), thus reducing recruitment to the mussel population and its long-term viability (Goss-Custard *et al.*, 1996b). The three mussel-fishing techniques have some similar and some dissimilar direct effects on oystercatchers.

An individuals-based, physiologically structured, game-theoretic ideal free model of oystercatcher distribution predicts the response of oyster-catchers to shellfishing and tracks the effects of their responses on their energy consumption rate, body condition and survival (R. A. Stillman, un-published data). Individual model birds decide each day on which mussel bed they would maximize their intake rate over low water, and move there. To choose a bed, each bird must know its potential intake on each bed under the prevailing conditions. The potential intake rates of each individual on each bed are calculated in two steps. First, the interference-free intake rates of individuals of differing efficiencies are calculated from the densities and energy contents of the mussels of different ages, and therefore sizes, present on each bed using a modified Charnov (1976) rate-maximizing foraging model. Second, the reduction in intake rate arising from oystercatcher interference is calculated for each bed from the individual's local dominance, given the density and fighting abilities of the other individuals present on each bed. The potential intake rate is calculated by subtraction and the bird spends the low water period on the bed where its intake rate is highest.

In every daily iteration from September to mid-March, each individual decides where it should feed. Many birds continually change mussel bed as the relative quality of beds alters through fishing and oystercatcher depletion. Individual oystercatcher daily energy requirements are calculated from the air temperature. Energy surplus to an individual's daily requirements is stored as fat, until a target mass is reached. Birds failing to obtain enough energy on mussel beds over low tide feed upshore on other prey, such as cockles, as the tide advances and recedes and also, over high water, on earthworms Lumbricidae. An individual metabolizes fat reserves when it fails through foraging to meet its current daily requirements and dies if its reserves fall to zero.

Mathematical formulation of the earlier version 2 of the model is given elsewhere (Clarke & Goss-Custard, 1996). Apart from revised estimates of several function shapes and parameter values, the present version 3 includes two important changes for predicting the effect of fishing. First, the foraging submodel calculates intake rate from the densities and flesh-content of each of ten size-classes of mussels rather than from overall mussel biomass density. The effect on oystercatcher intake rate of the fishery selecting large mussels can now be predicted. Second, the intake rates on supplementary foods used by some oystercatchers when mussel beds are covered at high tide were not available for version 2, and mortality predictions had to be re-scaled accordingly. Predicted mortality rates are no longer

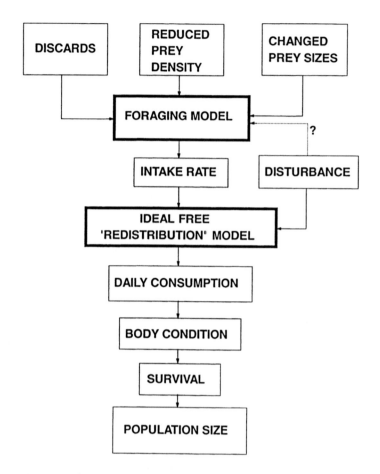

Figure 5.4 Flow-chart of the behaviour-based model for predicting the effect of shellfishing on the body condition of individual oystercatchers and on the survival rates and size of the population. The two behaviour submodels are outlined with heavy lines.

re-scaled in version 3 because field estimates of supplementary intake rates are now available.

The model includes three features that are particularly relevant for predicting the immediate and direct impact of shellfishing on oystercatchers (Figure 5.4). The rate-maximizing prey choice submodel predicts how the daily consumption rate of each individual would be affected by changes in the density and age-structure, and thus size class structure, of the mussel population following the fishing of large mussels. The ideal free distribution submodel predicts the reduction in each bird's consumption rate that arises when birds redistribute over the beds because of disturbance from fishermen or because of bed size reduction by fishing. Finally, by combining these two outputs, the model calculates the energetic consequences for each bird of the change in its consumption rate arising from its responses to fishing and so predicts how overwinter survival would be affected. The behavioural models provide a reliable basis for prediction because the choices made by model individuals are based on decision principles, such as rate maximization, that are unlikely to change with fishing. The consumption of discarded mussels often left by fisheries along the high tide mark can also be included in the model, although this has not yet been done.

The model (version 3) satisfactorily predicts current oystercatcher intake rates, distribution and survival rates (R. A. Stillman, unpublished data) so can be used with some confidence to predict survival rates under various shellfishing regimes. Fishing depletion is included by subtracting daily harvest from the current density of mussels. Fishing and disturbance only occur on spring tides and during daylight hours whereas the birds feed at night and on neaps. Fishermen in the model move during the winter to the mussel beds on which the yield is currently greatest. The effect of disturbance is included as illustrated in Figure 5.2, although changes over the winter in oystercatcher 'tolerance' are also incorporated.

Simulations show that the mere presence at low tide on the largest and densest mussel bed of up to 200 people who do not collect mussels has little effect on survival (Figure 5.5). The birds simply re-distribute to other beds where any shortfall in energy consumption is made up by feeding at night, or upshore or in fields when the mussel beds are covered by the tide. When people collect mussels, oystercatcher winter survival is not much affected by low tide thinning over the whole range of fishing effort explored. This is because the birds (1) maintain a high intake rate on fished beds by taking more small-sized mussels, (2) avoid fishermen by moving to un-fished areas, and (3) compensate for any resulting reduction in intake rate by feeding more at night, over high tide, upshore and at low tide on neap tides (R. A. Stillman, unpublished data). In contrast, with both high tide and low tide stripping, survival is greatly reduced at the higher levels of fishing effort. Stripping reduces mussel-bed size, forcing the birds to feed at higher densities and to suffer increased interference on both neap and spring tides. Low-tide stripping has the greater effect because it also disturbs the birds over low water. Although birds in the model compensate by feeding more at the times when fishermen are absent, this is not enough to

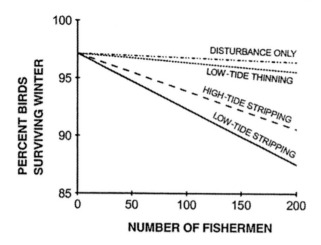

Figure 5.5 The predicted effect of increasing the numbers of shellfishers using different mussel-fishing techniques on oystercatcher overwinter survival on the Exe estuary. In low-tide thinning, large mussels are picked by hand over low water. In high-tide stripping, mussels of all sizes are trawled over high tide, the large mussels being selected later for sale. In low-tide stripping, mussels of all sizes are raked up over low tide. For comparison, the effect is also shown of the presence of up to 200 people, who do not collect mussels, on one of the largest mussel beds where large mussels are most abundant.

prevent increased numbers of birds from dying as fishing effort increases. Although not shown here, the inexperienced young birds suffer the greatest reduction in survival.

Modelling longer-term effects

Although the impact of fishing may appear small within one winter, the cumulative effects over several years may be much more pronounced. For example, low tide stripping could substantially reduce mussel bed size, and thus the protective space available to settling larvae, unless mussels were replaced, either by natural increases in area or by restocking. To illustrate how large might be the cumulative effects of shellfishing, multiple year simulations were run with intermediate levels of fishing effort using low-tide stripping and thinning, the two extreme scenarios.

As the model is age-structured, the year-by-year effect on bird numbers is included by: (1) reducing adult numbers returning each successive autumn by the numbers that died during the preceding winter, (2) adding the immatures that survived the winter and became mature, and (3) adding the juveniles recruited each summer per returning adult, assuming a density-dependent per capita reproductive rate arising through competition for breeding territories (Goss-Custard et al., 1996a). For tracking the year-by-year changes in the mussel population, a traditional demographic population model was added, based on a study of density-dependent and density-independent rates of recruitment and mortality on each mussel bed (McGrorty et al., 1990; McGrorty & Goss-Custard, 1991).

Low-tide thinning over 10 years has little effect on population size (Figure 5.6a). Although thinning depletes the mussel stocks within each winter, young mussels can be recruited during the summer because there are many larger mussels amongst which they can find protective space. The recruitment, and the growth of individual mussels, are sufficient to maintain the mussel population year on year at the levels of fishing effort explored. In contrast, low-tide stripping rapidly reduces the mussel population because fewer larvae can be recruited each year as mussels of all sizes, and the protective space they provide, are fished away; in effect, the beds shrink (Figure 5.6b). Even in the rather mild winters used in these simulations, oystercatcher numbers fall rapidly as the mussels disappear because birds cannot compensate fully by feeding in fields and upshore areas. The results show that the consequences of shellfishing (1) depend on the fishing method used, (2) may not be apparent for some time and (3) can, eventually, be much more severe than might be anticipated from the effects predicted for one winter alone.

These simulations show what happens over a few years but do not summarize the long-term effect on oystercatcher equilibrium population size. This work is in progress. A traditional demographic population model is being used (Goss-Custard et al., 1995e), adapted to simulate a discrete population wintering on the Exe and breeding elsewhere. The influence of shellfishing is being incorporated by including the density-dependent over-winter mortality functions that are predicted by the individuals-based model, as has been done with an earlier version of the model to predict the effects of habitat loss (Goss-Custard et al., 1996a). As the cumulative effect on long-lived animals of small decreases in survival, especially when juvenile survival is affected (Goss-Custard & Durell, 1984), has a disproportionate effect on equilibrium population size (Goss-Custard et al., 1996a), it is to be expected that the equilibrium population size will be much affected at higher levels of fishing effort, despite the sometimes small increase in mortality predicted for each winter.

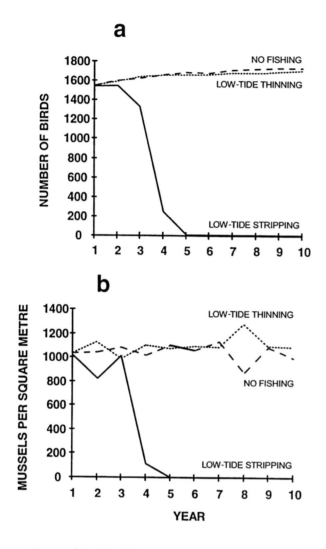

Figure 5.6 Predicted changes over ten successive years in the autumn sizes of the Exe estuary (a) oystercatcher population and (b) adult mussel population caused by intermediate levels (100 fishers) of low-tide thinning and low-tide stripping. For comparison, the effect of a disturbance is shown, as in Figure 5.5.

To summarize, the behaviour- and individuals-based model enables the immediate effect on oystercatcher survival to be predicted when shellfishing disturbs feeding birds and depletes their food supply. The long-term population consequences can be worked out by inserting the predicted density-dependent survival rates into a demographic oystercatcher population model. The long-term consequences on the mussels can also be followed with a demographic population model. However, the wider ecosystem effects still need to be taken into account.

The more immediate ecosystem effects could be included by modelling the particular food chain to which the co-dependent vertebrate belongs. Goss-Custard & Willows (1996) suggest that coupling the individuals-based, physiologically structured oystercatcher–mussel model with a physiologically structured mussel growth, survival and reproduction model would enable the effect of changes in phytoplankton abundance to be worked out at each level of that important food chain. This approach has the advantage of the predicted responses of the component species to environmental change being derived from basic properties of the organisms that are unlikely to change in the new environments (Goss-Custard & Willows, 1996). However, predictions would next be needed for the phytoplankton abundance, and not just for the mussels; phytoplankton abundance, and the influence of mussels upon it, affect the whole system. At present, these wider consequences can best be traced with ecosystem models of the kind developed for the Oosterschelde (Klepper *et al.*, 1994; Scholten & van der Tol, 1994). Although these are large and expensive projects, and their success can sometimes be limited by an inability to parameterize with sufficient precision some critical, and often non-linear, functions, better predictive tools have yet to be devised.

ACKNOWLEDGEMENTS

We are very grateful to the following for helping us with the review: Rob Willows, Bob Furness, Mick Marquis, Mike Kendal and Peter Olive. We are also grateful to the editors of the *Ibis* and the *Wader Study Group Bulletin* for permission to reproduce Figures 5.1, 5.2 and 5.3.

PART III

Habitat loss and fragmentation

Butterfly movement and conservation in patchy landscapes

CHRIS D. THOMAS, MICHEL BAGUETTE & OWEN T. LEWIS

INTRODUCTION

Knowing how far individual animals move during their lifetimes is essential to the development of practical conservation programmes. The more mobile an animal, the larger the area over which conservation management is likely to be required. Patterns of movement are determined by complex behaviours such as foraging, mating, and responses to habitat margins, and these behaviours must be understood if we are to manipulate mobility. Alas, the behaviour of movement has not yet been studied in sufficient detail in rare species that it has been possible to translate the results into practical conservation measures.

In this chapter, we hope to provide a stimulus towards this end. We describe patterns of butterfly dispersal and explore why an understanding of dispersal is important in conservation. We then identify some of the principal behavioural factors affecting dispersal rates, and finally consider how behaviour might in the future be manipulated to achieve conservation goals. We hope that, 10 years from now, it will be possible to write a completely different article, describing how the behaviour of movement has been manipulated successfully to achieve practical conservation goals.

SIMPLE PATTERNS OF MOVEMENT AND POPULATION STRUCTURE

Although we are a long way away from understanding the movements of rare insects in terms of detailed behavioural responses to a multitude of potential stimuli, we can nonetheless describe where animals move, and this may prove sufficient to develop preliminary management plans. We concentrate on butterflies because they represent the only insect group for which movement and population structure have been studied in a number of rare species. For most butterflies, it is possible to recognize distinct areas of habitat where local breeding is concentrated. These may be woodlands, meadows, or simply clumps of a host plant. Most butterfly species inhabit networks of such habitat patches, with breeding scattered across the landscape. In almost all butterfly species, caterpillars move short distances compared to adults, so immature stages are normally restricted to one such habitat patch, whereas adults can *potentially* range more widely. This conceptual separation of the landscape into breeding (habitat patches) and non-breeding areas is usually useful. We can then ask (1) what proportion of adult butterflies (and the eggs they lay) emerging in a particular habitat patch leave that patch, the 'emigration fraction' and (2) how far do emigrant individuals then move, the 'distribution of dispersal distances'?

The emigration fraction

Emigration fractions are usually estimated by marking individual insects with numbers or colour codes, releasing them where they were first caught, and observing where marked individuals are subsequently recaptured. The fraction of individuals and eggs emigrating from the natal habitat patch varies from 1.0 in highly mobile species, to about 0.01 in some relatively sedentary species (e.g. Arnold, 1983; Lewis et al., 1997). Any intermediate fraction is possible (Thomas & Hanski, 1997).

The overall fraction of emigrants is an informative descriptive statistic, but it is not very useful if we want to make predictions, mainly because emigration fractions vary from patch to patch. In almost all studies published so far, relatively high proportions of individuals emigrate from small patches, and lower proportions from larger areas of habitat (Arnold, 1983; Kuussaari et al., 1996; Hill et al., 1996; Sutcliffe et al., 1997; Baguette et al., 1996, 1998). This is exactly what would be expected if butterflies were simply behaving as randomly moving propagules (not that they are). The *number of individuals* emigrating is likely to increase with the patch perimeter (proportional to patch radius, r, if it is a round patch), whereas the number of individuals emerging in a patch can be expected to scale with patch area (increasing with patch r^2). Thus, the *fraction* emigrating should be a function of r/r^2, and should decline with increasing patch area.

Quantifying emigration fractions exactly is extremely difficult because it is usually impossible to distinguish between insects that have left a patch

Figure 6.1 A silver-spotted butterfly, *Hesperia comma*, bearing a patch-specific mark to identify its place of origin in a mark–release–recapture programme. Such studies show how dispersal determines the minimum area that can sustain a population. Populations in patches smaller than 0.67 ha experience too great a loss through emigration to persist (photograph by Owen Lewis).

and those that have died *in situ*. Most researchers simply record butterflies that successfully immigrate into another patch of habitat because they can not search the entire landscape thoroughly. The result is that published emigration fractions are underestimates – insects that emigrate and perish without finding another habitat patch are usually ignored. To try to correct for this, we first measured how far adult butterflies move naturally (by marking and recapturing) (Figure 6.1). Using the resulting empirically measured distribution of dispersal distances, we then imagined that the insects moved in random directions, having emerged at random locations within a habitat patch (Thomas & Hanski, 1997; Thomas *et al.*, 1999). The result of this simulation was an estimate of how many insects were still in their natal habitat patch, and how many had left, and thus the emigration fraction. Emigration fractions were relatively high from small patches, which have high perimeter to area ratios (Figure 6.2). Perimeter to area ratio is a better predictor of emigration fraction than area alone. However, for realistic patch shapes, area is the main determinant of perimeter to area

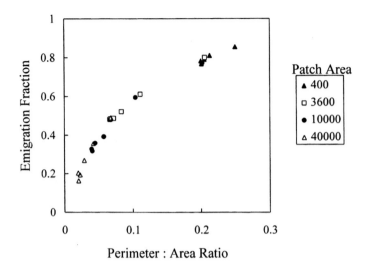

Figure 6.2 Plot of simulated emigration fractions against patch perimeter: area ratio, for patches of four areas, using the negative exponential distribution, $I = e^{-kD}$ where I is the (arcsine square root transformed) proportion of individuals of either sex travelling $\geq$ distance D, in m. The value $k = 0.03$ was used, as a value appropriate for *Plebejus argus*. For each of the four patch areas (m^2), the perimeter: area ratios were varied by altering patch shape.

ratio. The estimates of emigration that we obtained were a little higher than those measured empirically, but this is what we would expect, given that some emigrants are never seen again.

Butterflies are not really moving at random, but this extreme assumption may give approximately the right answer because the empirically measured distribution of dispersal distances is itself the product of large numbers of individual behavioural decisions. For example, the distributions of distances moved that were used in the simulations include data from insects that have encountered patch boundaries and then turned back. Some kind of estimate of dispersal is critical in conservation (below), and we have to try and obtain a reasonable estimate for a reasonable amount of effort (in this case, *c.* 3 person months in the field, and 2 person months of analysis). Despite some quantitative uncertainty, the results should be qualitatively robust, even if there are strong behavioural responses to the habitat edge (see below). Small patches with high perimeter to area ratios have high per capita emigration rates.

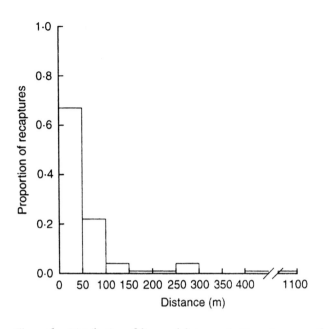

Figure 6.3 Distribution of dispersal distances in *Hesperia comma* (from Hill *et al.*, 1996). *n* = 133 between-patch movements.

The distribution of dispersal distances

The range of mobilities in butterflies is vast; from monarchs (*Danaus plexippus*) and painted ladies (*Cynthia cardui*) that migrate across entire continents, down to species like silver-studded blues (*Plebejus argus*) that rarely move further than a few tens of metres from where they emerged (e.g. Scott, 1975; Lewis *et al.*, 1997). The distance moved by each individual will be determined by its morphology and physiological status, such as the number of eggs in the abdomen, and by a complex set of behavioural interactions with other individuals of the same species, with other species, and with the physical environment. Nonetheless, when all of the movements from all individuals are added together, they usually conform to some general theoretical pattern, such as a negative exponential or negative power function (Hill *et al.*, 1996). For both of these theoretical distributions, many individuals move rather short distances, and some move considerably further (the negative power function has a longer 'tail' than the negative exponential, for a given mean distance moved) (Figure 6.3). Thus, immigration rates decline with increasing patch isolation.

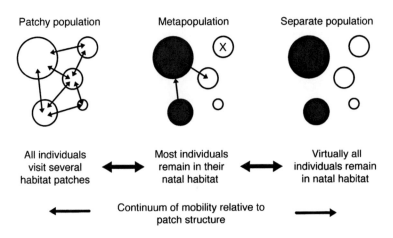

Figure 6.4 Schema of population structure (modified from Harrison, 1991). In patchy populations, individuals move in and out of patches of habitat (open circles) freely. In metapopulations, most individuals remain within local populations (solid), but a few move between local populations, or colonize empty habitat (open); some local populations become extinct (X). In separate populations, emigrants that do leave existing populations (solid) fail to arrive in other populations, or in empty habitat (open).

Population structure and its consequences

In the past, the overall exchange rate of individuals between local populations has been used to describe the structure of populations – *patchy populations* when there is a high percentage exchange, *metapopulations* with a low percentage exchange, and *separate populations* when there is virtually no exchange (Harrison, 1991, 1994). Because every kind of intermediate is possible (Figure 6.4), there is no point arguing about exact definitions; the more precise the definition, the more exceptions.

Headline values for percentage exchange hide as much as they reveal because immigration and emigration rates vary from patch to patch within each population system. The combined information on (1) patch-specific variation in emigration rates, (2) distributions of dispersal distances for emigrants from each patch, and (3) the locations of each habitat patch, provide a much richer picture of the overall spatial structure. Small isolated patches have high emigration fractions (if they are populated at all) and receive few immigrants; large isolated patches have lower emigration fractions and few immigrants; small connected patches have high emigration and immigration fractions; large connected patches have lower emigration

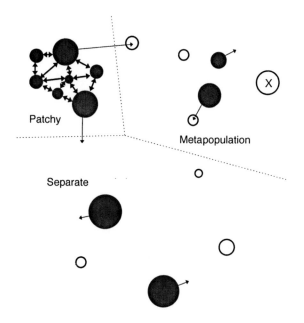

Figure 6.5 Schematic diagram to show how one species in one hypothetical landscape may exhibit different theoretical types of population structure in different areas, depending on the areas and spacing of breeding habitats, relative to the distribution of dispersal distance. Mixed population structure may be common in nature. Solid circles indicate populated habitat; open circles are empty habitats. Arrow width varies with number of individuals moving between patches. X shows an extinction.

fractions, and receive many immigrants. This sort of pattern has been observed in both the ringlet butterfly, *Aphantopus hyperantus* (Sutcliffe *et al.*, 1997), and the silver-spotted skipper, *Hesperia comma* (Hill *et al.*, 1996). Some of the habitat patches in these systems are characterized by a rapid flow of individuals in and out (characteristic of patchy populations), whilst others contain distinct local breeding populations (characteristic of meta-populations) (Figure 6.5). A mixed population structure may be quite normal for many species of butterflies in fragmented landscapes, simply because there is so much variation in patch area and isolation.

Given a good, quantitative description of movement patterns, it may be possible to make predictions. For example, suppose that conservation managers are planning to create a new patch of habitat for a particular species, but the exact location has not yet been decided. If they can estimate approximate numbers of insects leaving each existing population, and the distances moved by these, possible rates of immigration could be estimated for any prospective location before expensive restoration management is carried out. Similarly, they could predict whether the 'new' population would allow further spread of the species to patches that were previously too isolated.

DISPERSAL RATES

High dispersal rates may be bad

We concluded above that populations in isolated habitat patches suffer from a net loss of individuals. Emigrants are lost in the surrounding countryside and they are not replaced by immigrants. In the 1980s, many researchers emphasized the dangers posed by high emigration rates from isolated populations (e.g. Game, 1980; Buechner, 1987; Stamps et al., 1987). Can local breeding within isolated populations repeatedly replace the fraction of the population that is lost to emigration, or will the constant drain of emigrants cause a population to dwindle to extinction? This situation is analogous to harvesting a natural population – what fraction can be removed each year or generation before the population collapses and becomes extinct?

We also concluded that emigration fractions are relatively high from small patches. Therefore, is there a minimum threshold area for isolated habitats, below which a population is so leaky that it is bound to become extinct? We tried to predict this area for two butterfly species: the silver-spotted skipper and silver-studded blue. We measured dispersal distances by marking and recapturing adult butterflies (Hill et al., 1996; Lewis et al., 1997), and used the simulation approach described above to estimate the emigration fraction for habitat patches of different areas (Figure 6.2; Thomas & Hanski, 1997; Thomas et al., 1999). We needed two further pieces of information for both species; first, natural densities in very large areas of habitat, where emigration would have little influence on density (available for both species) and, second, the rate of population increase during periods immediately following the colonization of empty habitat (available for the skipper; we used the value from a related butterfly for the silver-studded blue). The latter defines the maximum potential population productivity when it is reduced to low densities by the loss of emigrants. In both cases, we predicted that isolated populations would collapse if more than about two-thirds of the population emigrated every generation. (The emigration fraction that causes a population to collapse depends on the intrinsic rate of population increase, and vertebrate populations may not be able to survive nearly such high emigration rates.) Even for these butterflies, managers should actually aim to retain much more than one-third of the individuals within isolated reserves, because population growth rates and emigration fractions will vary from year to year. The silver-studded blue is a very weak disperser, so the predicted threshold area was about 0.05 ha, whereas the more dispersive skipper was predicted to require a threshold of about two-thirds of a hectare. Empirical data on the distributions of both species in semi-isolated patches of habitat corresponded closely to the predicted values (Thomas *et al.*, 1999). Populations are not viable in smaller patches of breeding habitat because too many butterflies emigrate and perish in the surrounding countryside – those that are left behind cannot produce enough offspring to replace the emigrants. The more mobile the butterfly, the larger the area it requires.

If isolated populations can be endangered by high emigration rates, can losses during dispersal also threaten more complex population systems in patchy landscapes? Lande's (1988b) models showed that spotted owl populations may fail to survive if mortality of dispersing juvenile birds is too high in fragmented forests. Similarly, Hanski & Zhang (1993) showed that metapopulations may collapse if there is too much migration, just as they may fail to persist if migration rates are too low. The reason is exactly the same as before. Local breeding within remaining habitat patches cannot replace the high fraction of the entire metapopulation that dies during dispersal between patches. There is no unequivocal empirical example of this phenomenon, but this is hardly surprising, given the difficulty in documenting mortality during dispersal. The phenomenon is likely to be most important for relatively mobile species, because they are likely to have high emigration rates from every breeding patch. However, high emigration does not necessarily lead to high mortality provided that there are enough patches of habitat in the landscape – mobile species may be more likely to find these successfully.

Low dispersal rates may be bad

In the last 10 years, population biologists and landscape ecologists have tended to emphasize the potential benefits of dispersal in patchy landscapes (e.g. Hanski & Gilpin, 1997). Local breeding populations of butterflies become extinct for all sorts of reasons, including changes in habitat quality and runs of bad weather (Thomas & Hanski 1997). Dispersal promotes metapopulation persistence in networks of habitat patches, (1) by recolonizing patches of habitat where a population has just become extinct, (2) by colonizing fresh patches of habitat that become available (e.g. when new habitat is created through successional processes), and (3) by reducing the rate of local extinction, through the rescue effect (a few immigrants arrive and 'prop-up' a small breeding population that is teetering on the verge of extinction) (Gilpin & Hanski, 1991; Hanski & Gilpin, 1997). Rapid dispersal increases the colonization rate in (1) and (2), and decreases the extinction rate in (3). Since the balance of extinction and colonization determines whether or not a metapopulation system will survive, these processes all increase the chances of persistence.

The consequences for conservation have been described elsewhere, as have estimates of how much habitat is required for metapopulations and populations to persist for some given length of time (e.g. Thomas, 1994, 1995; Hanski & Gilpin, 1997; Thomas & Hanski, 1997; Thomas et al., 1999). Much has already been written on this topic, so we will give just one example of the practical implications of understanding butterfly metapopulation dynamics.

As conservation biologists, we are often charged with attempting to devise recovery programmes for species that have become highly localized. One element of these programmes is usually the establishment of new populations. New populations are established either through natural colonization, or by artificially (re)introducing a species to suitable habitat that is unlikely to be colonized naturally, because it is deemed to be too far away from existing populations. This issue has arisen in the conservation of the silver-spotted skipper, *H. comma*.

The silver-spotted skipper reached its minimum distribution in England in the mid to late 1970s. Most of the country's chalk grassland had already been ploughed up, and most remnants of chalk grassland had become too overgrown, mainly because myxomatosis killed off the rabbit populations that previously maintained the short, sparse turf required by this skipper (Thomas et al., 1986). Recovery of rabbit populations and conservation management (restoring traditional livestock grazing) has increased habitat availability again, and skipper populations have been re-expanding slightly since the early 1980s (Thomas & Jones, 1993). We wanted to know whether an understanding of dispersal and colonization could be used to predict where this butterfly would recover, where it would not, and where it might be necessary to reintroduce skippers. Simulation modelling led to the conclusion that there was one area in south-east England where recovery was almost complete (A, Surrey), another where recovery was still taking place and is expected to continue for several decades (B, East Sussex), and a third area where there was a substantial network of habitat patches that appear very suitable, but which would be very unlikely to be recolonized naturally within 100 years (C, North West Kent: Thomas & Jones, 1993; Hanski & Thomas, 1994; Hanski, 1994; Hill *et al.*, 1996). The main priority in area A can be identified as habitat management because the emphasis is on maintaining populations, rather than on further colonization. The main priority in area B is to ensure that habitat quality and quantity is maintained around the edges of the current distribution, to encourage further spread. The main priority in area C is re-introduction. A good understanding of colonization rates was required to conclude that re-introduction is required in area C, but unnecessary in area B.

BEHAVIOUR

The previous sections have illustrated that a general understanding of insect movement patterns may be important in conservation programmes, and there may be circumstances in which a conservation manager may wish either to increase or to decrease rates of movement. However, to manipulate the amount of movement successfully is likely to require some knowledge of the behavioural factors that determine whether an individual will emigrate from its natal habitat patch, which direction it will then take, and how far it will go.

Most movement by butterflies takes place in the adult phase. Adult males need to obtain adult resources (e.g. flower nectar, minerals from damp soil) and matings. Adult females need to find locations to lay their eggs, as well as adult resources and matings. Finding eggs sites may involve some kind of assessment of host plant quantity and quality, and perhaps an assessment of population density (e.g. avoiding plants which already have eggs). Therefore, we might expect dispersal to be related to resources for adults and immature stages, and to population density. Because many of these variables are correlated in nature, few published examples provide a rigorous test of the effects of each factor. Therefore, we have chosen to describe a few examples in a little detail, rather than attempting a wide-ranging review.

LARVAL RESOURCES

Adult females require high quality locations for laying eggs. Adult males should also be attracted to these areas because virgin females may emerge there, and mated females may be intercepted whilst ovipositing (in species where females mate more than once). All else being equal, adults of both sexes might be expected to remain in areas that contain high-quality resources for larvae, and leave areas where larval resources are poor, rare, or absent. There are few published examples that clearly demonstrate direct effects of larval resources on adult movement because areas with plentiful larval resources usually also contain high densities of adult butterflies, and it is difficult to tell whether the butterflies are responding to habitat quality or adult density.

An unusual example was provided by a study of *Euphydryas editha* butterflies, which took advantage of the fact that different females in the same population preferred to lay their eggs on different plant species (Thomas & Singer, 1987). Egg-laying preferences are not learned in this species, and appear to have some degree of genetic basis (Thomas & Singer, 1987; Singer et al., 1988; Singer & Thomas, 1996). This meant that different females that flew together at the same time in the same habitat, with the same experiences of adult density and adult resources, differed in their perception of the larval resources available in different areas. Some females preferred to lay their eggs on *Pedicularis semibarbata*, a plant that forms rosettes about 30 cm across, whereas other females preferred to lay on *Collinsia torreyi*, a spindly, annual plant (a third class of females did not distinguish between the plants – we ignore them here). Preferences were determined by a behavioural test, recording the responses of each female when placed on the two species of plant (Singer, 1982). The study site, in the Sierra Nevada in California, contained two areas of habitat separated by 50–100 m; one area was a granite outcrop and the other was a forest clearing. The outcrop contained *Pedicularis semibarbata*, and the clearing contained high-quality *Collinsia torreyi* (quality was judged by the survival of larvae). Female butterflies were captured in each area, given individual marks with a felt-tip pen, and released where they were caught. They were then recaptured 1–14 days later, taken into captivity, and the preference of each female was determined. It was found that females were likely to have left areas without preferred hosts and to have stayed in areas where preferred hosts were present ($P < 0.001$), whilst experience of other variables (e.g. male density, nectar resources, microclimate and shelter) was held constant (Thomas & Singer, 1987).

Clear experimental examples of habitat-dependent emigration are rare, although there are plenty of observations that are consistent with this interpretation. For example, Murphy & White (1984) reported mass emigration of *Euphydryas editha* from populations where there were very few remaining host plants, and Gilbert (1986) suggested that dry-lands butterflies may initiate migrations in response to deteriorating habitat quality. Even though there are few clear examples in the literature, we believe that variation in larval resource quality and quantity almost always influences female movements, and the same phenomenon might also be widespread in males. The lack of evidence is very surprising, considering that larval resources are usually used to define habitat, and that responses to habitat may be crucial during colonization. For example, mated adult females (potential colonists) that intentionally or accidentally emigrate from their natal habitat find themselves in a landscape without suitable larval resources. They are then likely to move considerably longer distances than would normally be observed within habitats, flying on until they either find larval habitat elsewhere or perish (Harrison, 1989).

Adult resources

Adult resources are important to power flight, enhance longevity, and increase reproductive output by both females (egg size, number, quality) and males (spermatophore size and quality – nutrients are passed from males to females during mating). Therefore, one would expect both males and females to remain in breeding patches with high densities of adult resources, all else being equal. Again, critical experiments are rare. The best study involved the release of 882 freshly emerged Glanville fritillary butterflies, *Melitaea cinxia*, into 16 out of 64 empty habitat patches on an island in the Baltic, where the fritillaries did not occur naturally (Kuussaari *et al.*, 1996). This experiment revealed that several factors were important determinants of emigration: patch area (higher rates from small patches, above), population density and marginal vegetation (below), and the abundance of flowers. Emigration rates were significantly reduced from patches with high flower densities for males ($P < 0.001$), but only marginally for females ($P = 0.084$).

Flower abundance may also influence immigration rates, either because dispersing butterflies are attracted to flowers from some distance, or because immigrants may leave again quickly (before a researcher spots them) from patches with low flower abundance. Whichever the mechanism, Kuussaari *et al.* (1996) found that male Glanville fritillaries were significantly more likely to immigrate into potential breeding habitat that contained high flower abundances.

In a slightly different context, Murphy *et al.* (1984) found that the distribution of checkerspot, *Euphydryas chalcedona*, eggs on larval host plants tended to be associated with areas with high nectar abundance, indicating that female distributions and oviposition may be influenced by adult resource density, and the same phenomenon has been observed in a sphingid moth (Karban, 1997). Overall, the above results suggest that adult resources do influence adult movements and local distributions.

Edges

How any animal responds to the boundary between breeding habitat and the remainder of the landscape is a critical determinant of emigration. Thomas (1983) followed gatekeeper butterflies, *Pyronia tithonus*, in imaginary 5 m square boxes, one side of which was the boundary between species-rich grassland (where they bred) and species-poor pasture (where they did not breed). The boundary was a relatively inconspicuous, two-strand, barbed-wire fence. The species-poor turf was slightly greener and more uniform, but the overall vegetation height was similar. Adult butterflies left the species-rich grassland less frequently than expected: all of the butterflies that did leave visited thistle flowers in the species-poor pasture, before returning to the breeding habitat.

Sutcliffe and Thomas (1996) followed ringlet butterflies in a woodland clearing, bounded by dense woodland (77.4% of boundary), by areas of lesser restriction (14.1% small gaps or tall, mesh deer fence), and by an open track (8.5%). These boundaries accounted for 2%, 53% and 45% of emigrants respectively, indicating that an adult ringlet was about 185 times more likely to leave into the open track than into the dense woodland, per unit length of boundary. When Kuussaari *et al.* (1996) released Glanville fritillaries on rocky outcrops and in meadows, these butterflies were also reluctant to leave across woodland boundaries: emigration rates were higher from habitat patches that were surrounded by open vegetation than from patches surrounded by woodland. However, Sutcliffe & Thomas (1996) noted that some other butterfly species did not respond in the same way as ringlets, readily flying up and over the woodland, and grassland silver-spotted skippers may cross scrub barriers quite readily (Thomas *et al.*, 1986; Hill *et al.*, 1996). Some generalizations may be made, but boundary effects are likely to be species-specific.

Corridors

Effects of landscape patterns on the exchange of individuals between breeding areas are also poorly understood, partly because the diversity of landscape structures between patches is so great. Sutcliffe & Thomas (1996) found that exchange rates of ringlet butterflies between woodland clearings was predicted better by distance-via-rides (open grassy tracks, bordered by woodland) than by straight-line distance. Presumably, most exchanges occurred when adults flew out of clearings into open rides (above), and then flew along the rides until they arrived in another clearing.

However, it is unclear whether favourable, intervening habitats always increase exchange rates between breeding areas. If butterflies typically fly much further when they find themselves in 'non-habitat', and we suggest they do, then butterfly dispersal may actually be slowed down by complex landscapes where they are distracted by hedgerows, tracks, ditches and a rich tapestry of other non-habitats. We are not arguing against the value of complex landscapes, but much more work is required (e.g. Dover, 1994; Fry & Robson, 1994) before we draw firm conclusions about butterfly movements in relation to landscape structure.

Sex and density

We deal with these two factors together because males and females are expected to show different responses to density. We start with females. Females must mate, usually soon after emergence, so one might imagine that females would be drawn to areas with high densities of males if they do not immediately find a mate where they emerge. Thereafter, mating will occupy only a small proportion of an adult butterfly's time budget (females of some species only mate once), and the unwanted courtship attentions of ardent males may cause mated females to leave areas of high male density, everything else being equal (Odendaal et al., 1989; Baguette et al., 1998). Because the best areas for egg-laying are often populated by high densities of both males and females, females may choose to endure frequent courtship so as to gain access to egg-laying sites. Indeed, it has been suggested that some butterflies may actually use high adult densities to identify the location of high quality habitat for immature stages (Gilbert & Singer, 1973; Kuussaari et al., 1996).

Wickman (1988) provided evidence that virgin females may move to areas of high male density in the wall butterfly, *Lasiommata megera*. He showed that released virgin females tend to fly uphill (males aggregate on hilltops), whereas mated females do not. Male mating aggregations occur in many butterflies, so this phenomenon is probably quite widespread, especially in low density populations and species (Shields, 1967; Lederhouse, 1982; Alcock, 1987; Wickman, 1988).

General mark–release–recapture work can usually be interpreted as showing the movements of mated females because most female butterflies are already mated when they are first caught. Mark–release–recapture data provide conflicting results. Dispersal distances or emigration may increase with density (e.g., Dethier & MacArthur, 1964; Shapiro, 1970), but the reverse is true in other studies (e.g. Gilbert & Singer, 1973; Brown & Ehrlich, 1980; Arnold, 1983). As stated above, interpretation is hampered by probable positive correlations between habitat quality and adult density.

The potential effect of male density on female emigration has been examined and modelled for the bog fritillary, *Proclossiana eunomia* (Baguette *et al.*, 1996, 1998). Limited direct observations showed that males tended to leave patches when females were not available, and that mated females sometimes emigrated directly after being harrassed by a male. Simulation models (males continued searching in a patch if they had encountered females; mated females emigrated after a variable number of encounters with males) predicted asymmetrical effects of population density on emigration rates in males and females. At low density, males tend to emigrate in search of females, but mated females are predicted to stay. As overall density increases, male emigration decreases as more females are encountered, whereas emigration by females increases, as they are pestered by more males. Empirical mark–release–recapture data, obtained in two years, conformed to this expectation. The two years differed in overall population density by a factor of two. Comparing the two years, a significantly higher proportion of females moved between patches (> 200 m apart) in the *higher* density year (15% versus 5%), and females moved significantly longer distances in the same year. Males showed the reverse pattern, with significantly more moving between patches in the *lower* density year (8% versus 4%), with longer distances achieved in the same year. Thus, dispersal between patches was female-biased in the high-density year (female 15%, male 4%) and male-biased in the lower-density year (female 5%, male 8%).

Not all studies show density-dependent emigration by females. When Glanville fritillaries were released into empty habitat at two different densities, there was no significant effect of release density on the proportion of females that stayed, and females actually tended to remain longer in patches with higher release densities of both males and females (Kuussaari *et al.*, 1996). This tendency of females to aggregate may not be typical for butterflies as a whole, given that Glanville fritillaries are likely to be somewhat distasteful to predators (e.g. they have conspicuous, gregarious larvae, which predators may learn to avoid). Thus, fitness may be increased by aggregation when overall densities are low or moderate (the same explanation may apply to aggregation in *Euphydryas*; Gilbert & Singer, 1973).

Males need to maximize potential encounters with receptive females, which usually involves moving to, or remaining in, areas of high population density. Male *Euphydryas editha* seek out areas with high densities of emerging females (Singer & Thomas, 1992), whilst other butterflies may move to rendezvous locations, such as hilltops, when densities of receptive females are too low for males to search for them directly (e.g. Shields, 1967; Lederhouse, 1982). Kuussaari *et al.* (1996) found that male Glanville fritillaries remained in patches where more butterflies had been released, and tended to stay in high-density patches for longer.

We expect that most detailed studies will find some effect of population density on movements of either male or female butterflies. However, we also expect that many different patterns will be revealed, depending on age, mating status, average population density, average mobility, mate-location strategy, reproductive strategy, larval aggregation, larval distastefulness, and adult distastefulness. These factors are undoubtedly very important, but most of them have only limited implications for practical conservation, because management goals usually aim to generate densities that are as high as possible.

DISPERSAL BEHAVIOUR AND CONSERVATION

Certain generalizations can be made about patterns of butterfly movements, including the relationship between emigration and patch area, and distributions of distances moved. These coarse generalizations allow us to make very approximate predictions about colonization and dispersal in landscapes that have not been studied in detail. These predictions are often useful in the development of conservation programmes. However, if we want to predict finer differences in emigration between patches of, say, 0.5 ha and 0.8 ha, we must first understand the behavioural determinants of movement. In the remainder of this chapter, we give a few examples of how knowledge of behavioural responses may allow us to manipulate butterfly dispersal, and so achieve conservation goals.

Isolated populations

As described above, isolated populations may collapse if the emigration fraction is too high – emigration that reduces local density may also make the population more susceptible to fluctuations caused by the weather. Assuming that there is no empty habitat nearby 'waiting to be colonized', the task is to retain as high a proportion of the population within the breeding habitat as possible. Given the behavioural responses described above, this may be achieved by enhancing the quality of egg-laying sites, providing high-quality adult resources within the breeding habitat, and surrounding the habitat with a barrier that minimizes emigration. The effects may be great. An open field populated by ringlet butterflies might have a hundred-fold higher emigration rate than would the same field if it was entirely surrounded by dense woodland (above).

Metapopulations with low connectivity

Where local breeding areas have low exchange rates, it may be worth attempting to manipulate dispersal behaviour. Paradoxically for a system with low connectivity, one may actually want to reduce emigration rates, so as to minimize the proportion of the population that dies outside breeding patches. For small patches that are below or not far above the individual patch-area threshold, the best strategy may be to erect barriers to ensure that the emigration fraction does not exceed the critical level that cannot be replaced by local breeding (just as for single isolated populations). For patches that are *at least* one order of magnitude larger than the minimum area threshold, one may actually want to facilitate emigration (and thus colonization within the entire system) by leaving the patch edge open.

Of those individuals that do emigrate, one wants to maximize the proportion arriving successfully in breeding habitat elsewhere, whether or not it is already occupied. The best general strategy is to increase the numbers and areas of target habitat patches in the landscape (large patches make larger targets for dispersing individuals). However, this option may not be available, and one may need to consider whether ecological corridors or other benign landscape features would both increase exchange rates *and* minimize the overall mortality rate of dispersing individuals. At the present state of knowledge, it is not obvious how to achieve this. High-quality breeding patches separated by wheat fields (or other habitats that provide no resources) may be reasonably well connected if butterflies show rapid and perhaps directed flight in hostile habitats. Whilst hedgerow networks are generally beneficial to wildlife, they may actually distract dispersing individuals and reduce exchange rates over relatively long distances (unless all hedgerows going in the 'wrong' direction are removed – not an activity we would advocate). There is considerable potential to manipulate butterfly exchange rates between areas of breeding habitats by exploiting and manipulating behavioural responses to intervening landscape elements. However, we do not know how to do it yet, and much more research is required.

Release programmes

Introduction of rare butterflies to new or former habitats has become an important element in conservation strategies. The ultimate success of such programmes is likely to depend on the quantity and quality of potential breeding habitat available in the landscape around release locations (Thomas, 1992). However, many attempts to establish butterflies in empty habitat fail in the first few generations. Often this is caused by poor habitat, but sometimes by bad luck, such as poor weather soon after the release. Given that insect populations show wide fluctuations in numbers, often caused by the weather, a key element of releases should be to attain large numbers of individuals rapidly. So, it is important both to release large numbers of individuals (if available) and to maximize the initial population growth rate. The latter may be achieved by releasing insects into fairly large, high-quality sites surrounded by barriers to emigration. For species which normally mate only once as adults, mated females should be released but no males, to avoid male harrassment inducing female emigration. For multiply mating females, some males should also be released.

Many temperate-zone butterfly species are protandrous: on average, males emerge a few days or weeks before females, so male population density peaks at the time of maximum female emergence, when most matings are available. Thus, the year after a release, when numbers may be very small, males may emerge and fail to find *any* conspecific females (and only one or two males) for several days, and may thus emigrate. By the time any females emerge, all of the males may have left, and the females may then follow suit, having failed to find a mate. Again, the main solution is to release large numbers (hundreds or thousands, rather than tens). If this is impossible, it may be more successful to release moderate numbers into a medium-sized area of breeding habitat, surrounded by a barrier to emigration, where adults will be able to find each other, than to release the same number into a very large site where individuals may find it difficult to locate one another in the next generation. The latter might have lower initial rates of population growth, despite having higher final population sizes.

CONCLUSIONS

Survival of populations and metapopulations in fragmented landscapes depends critically on rates of emigration and immigration for individual habitat patches. Emigration and immigration rates depend on individual responses to habitat quality and patch boundaries, to conspecific density, to landscapes that occur in between breeding populations, and to patch geometry and location. The interaction of adult behaviour and landscape patterns generates a great variety of dispersal rates.

The examples described in this last section are speculative, but indicate that manipulation of individual behaviour has considerable potential in conservation programmes for rare butterflies, and presumably for other animals. By considering behavioural responses to different landscape elements, and then manipulating the landscape, we believe that it may sometimes be possible to generate at least 10-fold differences in emigration and exchange rates. This may make all the difference in determining whether remnant populations and metapopulations will survive or perish in fragmented landscapes. However, the task of understanding and then manipulating these behaviours has hardly begun.

ACKNOWLEDGEMENTS

We thank Jane Hill and Odette Sutcliffe. The work was supported by NERC research grant GR3/9107 to C.D.T., and by European TMR network FRAG-LAND to C.D.T. and M.B.

Life history characteristics and the conservation of migratory shorebirds

THEUNIS PIERSMA & ALLAN J. BAKER

INTRODUCTION

The world recently lost, or is in the process of losing, two of the 112 extant species of migratory shorebirds (del Hoyo *et al.*, 1996). Over a span of only 30 years at the turn of the last century, the Eskimo curlew *Numenius borealis*, then the most abundant but also the most gregarious of the four curlew species in America, virtually disappeared owing to excessive hunting and destruction of their tallgrass prairie spring staging habitat (Gollop *et al.*, 1986; Gill *et al.*, 1998). In spite of some odd sightings over the last 50 years, it can probably be regarded as extinct.

At the close of the twentieth century, the slender-billed curlew *Numenius tenuirostris* of Eurasia and North Africa is about to follow its smaller congener's fate. Overhunting, and perhaps the modification of breeding, staging and/or wintering habitats, have led to its demise (Gretton, 1991). Other extinctions, of specific populations rather than entire species, are harder to notice. For example, since the 1950s an aberrant population of Pacific golden plovers *Pluvialis fulva* wintering in temperate Europe (rather than the tropical Indo-Pacific region), seems to have disappeared (Jukema, 1987).

Although it is no longer believed that the extinction of migrating shorebirds would induce plagues of mosquitoes in the world's tundra areas, the conservation of these birds and their long-distance migratory systems should remain a priority for several reasons. First, migrations are unique phenomena that deserve conservation for aesthetic and ethical considerations (Brower & Malcolm, 1991; Dingle, 1996). Second, migrating shorebirds sample habitats around the globe in the course of their seasonal travels, and thus serve as indicators of potentially serious ecological problems far away from regions with intense biological monitoring and understanding. Their very existence signals that some of the world's ecological integrity remains. In addition, long-distance migrants achieve feats of endurance and navigation barely mirrored by mankind's latest technological devices, and are hardly understood scientifically. An elucidation of the physiological, psychological and sensory mechanisms that enable bar-tailed godwits to make 11 000 km long trans-Pacific flights from Alaska to New Zealand and back (Piersma & Gill, 1998; R. E. Gill, unpublished data), will probably give us more basic biological insights than we can imagine.

That there is little reason for complacency, even in parts of the world where shorebird habitats have not only received the attention of conservationists but also legal protection, is illustrated by recent events in the Wadden Sea, The Netherlands, and in Delaware Bay, U.S.A. In spite of a plethora of conservation rules and regulations, industrial overfishing of mussel and cockle stocks in the Dutch Wadden Sea (removing biomass and modifying sedimentary habitats) is implicated in the dramatic declines in shellfish resources in the early 1990s (Piersma & Koolhaas, 1997). Dramatic consequences have been mass winter starvation and serious reductions of the Dutch breeding population of at least one migratory mollusc-specialist, the European oystercatcher *Haematopus ostralegus* (Camphuysen *et al.*, 1996; J. B. Hulscher & B. J. Ens, personal communication) (Figure 7.1). On the other side of the Atlantic, an economically very profitable harvest of adult horseshoe crabs *Limulus polyphemus* (especially gravid females), coming ashore to lay their trillions of eggs in the beaches of Delaware Bay, in 1997 had led to significant declines in horsheshoe crab populations and in the numbers of staging shorebirds. The shorebirds congregate there to refuel on the abundance of horseshoe crab eggs during a critical phase of their northward migration (Botton *et al.*, 1994), a phenomenon well advertized and attracting thousands of human spectators (Myers, 1986; Harrington, 1996) and ecotourism dollars.

The predicament of long-distance migrant shorebirds, and indeed most specialized organisms living in fragile environments, and the importance of understanding key behavioural traits to develop effective conservation arguments, were aptly summarized by Ulfstrand (1996):

> 'Behaviour, more than anything else, is what makes some organisms more and others less susceptible to the environmental calamities wrought by the exploding human population... The Red Queen may have accelerated a thousand times over the past couple of centuries. Unless a species has what it takes to keep her pace, its name will soon be on the species death roll.'

Figure 7.1 Oystercatchers, *Haematopos ostralegus*, and Eurasian curlews, *Numenius arquata*, foraging at high densities on an intertidal mussel bed in the Dutch Wadden Sea. This habitat was destroyed in The Netherlands by overfishing in the late 1980s and early 1990s and has still not recovered (photograph by Jan van de Kam).

Let's examine the life history characteristics of shorebirds, and discuss what this might mean in an ever faster changing world.

THE LIFESTYLE OF MIGRATORY SHOREBIRDS

To illustrate the lifestyle of a migratory shorebird, we have chosen the red knot *Calidris canutus*. While this bird occurs throughout the world, it has a circumpolar breeding distribution (Figure 7.2). It breeds on high Arctic and mountain tundra, the pairbond is monogamous, the males are breeding-site faithful (Tomkovich & Soloviev, 1994), and although the sexes share incubation, only the males take care of the precocial chicks during the 3 weeks from hatching to fledging (Whitfield & Brade, 1991). Birds from different parts of the breeding range migrate southward to different coastal wintering grounds (Figure 7.2; note that birds crossing the equator spend the 'winter' in the southern 'summer'), adult females making the trip before the males, and the males migrating before the juveniles. The southward and northward migrations are usually made in a few long flights

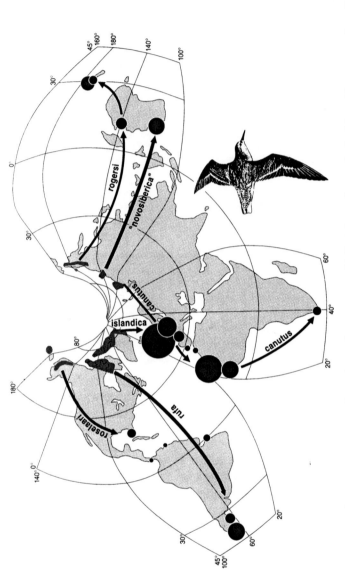

Figure 7.2 Intercontinental migration routes (with north–south axes arranged left–right) and annual schedules with respect to long-distance flights (triangles) and plumages (various shadings) in the five best-known migratory morphs of red knots, four of which currently are recognized as subspecies. This figure was compiled from Piersma & Davidson (1992), unpublished data by many colleagues and population estimates of, for example, Zwarts *et al.* (1998). Note that assigning the New Siberian Island-breeding and northwest Australia-wintering population to a separate group '*novosiberica*', and keeping the name *rogersi* for the red knots wintering in southeast Australia and New Zealand, is preliminary.

of 2000–8000 km, the birds often almost doubling body mass from 120 g to over 200 g during fuel storage before take-off (Piersma & Davidson, 1992).

At the time of writing, six different populations are recognized; five formally as subspecies on the basis of small differences in body dimensions and breeding plumage characteristics (Tomkovich, 1992; Piersma & Davidson,1992). The separate populations show an interesting variation in winter climates and migration distances (Piersma,1994), but all red knots winter in coastal areas with extensive intertidal flats where they can find shallowly buried bivalves and gastropods. Red knots appear to have a sensory (bill-tip) system specialized to locate buried bivalves (Piersma et al., 1998). Their digestive tract, with a large muscular stomach for crushing mollusc shells and an intestine able to withstand the stresses of rapidly passing shell fragments, is adapted to deal with mollusc prey (Piersma et al., 1993). The great intraspecific variation in migratory trajectories belies the finding that the different populations of red knot may share a quite recent common ancestry. The entire flyway system illustrated in Figure 7.2 may be of post-Pleistocene origin, as studies on allozyme variation and the sequence variation in the rapidly evolving parts of mitochondrial DNA suggest that the world population of red knots has been through a bottle-neck sometime over the last 10 000–30 000 years (Baker et al., 1994; Baker & Marshall, 1997). The species barely survived a near-extinction when massive icecaps melted, steppes and forests replaced Arctic tundra, and sea levels rose steeply (Piersma, 1994).

Obviously, red knots have many peculiarities not shared with other migratory shorebirds (see especially Robinson & Warnock, 1997), but on a more general level many characteristics are similar, especially amongst the plovers and the sandpipers that make up the bulk of migratory shorebirds (Piersma & Wiersma, 1996; Piersma et al., 1996). Which aspects of migratory shorebird life histories may be critical in a conservation context? Given the huge size of the available literature (see del Hoyo et al., 1996), during the following discussions only a selection of entries will be listed.

CRITICAL LIFE HISTORY CHARACTERISTICS

Low productivity

Clutch sizes of migratory shorebirds are never larger than four eggs (Walters, 1984). Only at southernmost breeding latitudes in the Northern

Figure 7.3 Different kinds of shorebird mortality. (A) shows the remains of a bar-tailed godwit and a grey plover that fell victim to large falcons (peregrine or lanner falcon) on the Banc d'Arguin, Mauritania, in February 1980. Note that only the flight muscles were eaten. (B) shows a series of sick, dying and dead red knots lined up near Lagoa do Peixe, southern Brasil, in April 1997. These birds died of 'self'-poisoning after heavy infections with acantocephalan helminths (*Parafilicollis* sp.) that punctured the intestinal lining (photographs by T. Piersma).

Hemisphere the season is long enough for clutches to be replaced after loss (Pienkowski, 1984), and in some of the smaller sandpiper species double-clutching occurs, with each parent taking care of one. Given that less than two of the four eggs eventually yields a fledgling chick, migratory shorebirds can safely be characterized as species with low fecundity (Evans & Pienkowski, 1984; Saether *et al.*, 1996).

Long lifespan

There are many causes of mortality in migratory shorebirds: from exhaustion during long migratory flights over open oceans or deserts, especially when hit by storms or other climatic hazards (see Bodsworth, 1956), predation by raptors in unfamiliar habitats (Whitfield, 1985; Cresswell, 1994; Figure 7.3A), hunting by humans (Jukema & Hulscher, 1988; Peach *et al.*, 1994), starvation on the ice (Davidson & Evans, 1982; Camphuysen *et al.*, 1996), poisoning by pesticides (Flickinger *et al.*, 1980), or lethal infections by parasites and pathogens (Woodard *et al.*, 1977; Figure 7.3B). Nevertheless, adult migratory shorebirds enjoy relatively high survival rates (Boyd, 1962; Marks & Redmond, 1996). If they survive the first non-breeding season, life expectancies of migratory shorebirds are 10 years or more. Studies of colour-marked individuals in north temperate wintering areas indicate that about two-thirds of the annual adult mortality takes place in winter (Evans, 1991). A special case is the bristle-thighed curlew *Numenius tahitiensis*. It winters on oceanic islands in the tropical and subtropical Pacific Ocean, where it enjoys the usual high survival. In the absence of terrestrial predators this curlew has evolved a very rapid wing-moult during which many individuals become flightless. Overwinter survival is thus easily reduced as a consequence of predation by the humans, dogs and cats that now live on these islands too (Marks & Redmond, 1994).

Trophic specialization

Most plovers are visual hunters of surface prey. Sandpipers have many feeding strategies. Some also use visual foraging techniques (buff-breasted sandpipers *Tryngites subruficollis* and ruffs *Philomachus pugnax*, for example), while many others mainly probe for prey (Piersma *et al.*, 1996). Probers are also specialized in various ways. Whereas snipes and many small sandpipers are equipped to detect the vibrations of subsurface crawling prey (Gerritsen &

Meijboom, 1986), red knots are probably unable to feel such vibrations. Instead, they are able to detect the presence of static objects such as molluscs buried in soft sediments by the Herbst corpuscles in their bill-tips, which may perceive actively formed pressure gradients (Piersma *et al.*, 1998). All these sensory specializations seriously constrain the options open to migratory shorebirds in regard to choice of foods and habitats.

Gregariousness

Food for migratory shorebirds usually occurs in patches, and such patches are often cryptic. One of the reasons that migratory shorebirds tend to occur in dense flocks when not breeding, other than to decrease individual predation risk (Cresswell, 1994), may relate to locating profitable feeding patches. Flying in flocks during long-distance migration may help to reduce the costs of flight and to reduce errors in navigation (Piersma *et al.*, 1990).

Immunospecialization

Recently, Piersma (1997) suggested a trade-off between investments in immunofunctioning and sustained exercise. This physiological trade-off might determine the year-round use of particular types of habitat by shorebirds. Some migratory shorebirds appear restricted to parasite-poor habitats (high Arctic tundra, exposed sea-shores) where small investments in immunomachinery may suffice. However, such habitats are few and far between, necessitate long and demanding migratory flights in the course of an annual cycle, and are often energetically costly to live in (Wiersma & Piersma, 1994). Species opting evolutionarily for parasite-poor habitats may be rather susceptible to parasites and pathogens as a result of nutritional investments in sustained exercise (including thermoregulation) rather than immunocompetence. Lack of immunocompetence may also be genetic; species like red knots may have failed to build up adaptive variation in the major histocompatability gene complex after population bottlenecks (Hamilton & Howard, 1997).

Sometimes strongly sexually-selected

For landbirds introduced in Hawaii it has been shown that sexually dimorphic species are less likely to become established than sexually monomorphic and presumably less intensely sexually selected species (McLain *et al.*, 1995). Migratory shorebirds show great variation in the extent of sexual and seasonal plumage dimorphism. Ongoing studies on bar-tailed godwits *Limosa lapponica* and ruffs suggest that the timing of moult and the extent of nuptial plumage are indicative of individual quality with reference to timing of fuel storage and migration, physical condition and parasite loads (Piersma & Jukema, 1993; Piersma *et al.*, 1996a; Jukema *et al.*, 1995; T. Piersma, unpublished data). Colourful nuptial plumages may thus be slight nutritional handicaps revealing body condition or viability (Andersson, 1994; Møller, 1994) rather than costly self-reinforced traits that compromise viability in a migratory context (McLain, 1993). In a different context, Höglund (1996) suggested that lekking species such as great snipe *Gallinago media* and ruff may require larger patches of breeding habitat than non-lekking species such as black-tailed godwits *Limosa limosa*, and would thus be more prone to local extinction.

Long flights

The scarcity of suitable feeding habitats and seasonal variations in habitat availability forces migratory shorebirds to routinely overfly large stretches of unsuitable land and ocean (Piersma, 1987; Alerstam, 1990; Williams & Williams, 1990). Continuous flights of up to 10 000 km or longer appear to be made. There is evidence that successful execution of these prodigious flights often necessitates tailwind assistance (Piersma & Jukema, 1990; Piersma & van de Sant, 1992; Butler *et al.*, 1997).

Metabolic 'adaptations' allowing these feats of endurance

The long-distance flights, and the associated episodes of rapid (re-)fuelling that migratory shorebirds must undertake to prepare for the flights, probably entails a multitude of metabolic adjustments. These include the building and breakdown, at appropriate times, of various functional parts of the machinery, such as the digestive tract and connected organs, and the heart and flight muscles (Weber & Piersma, 1996; Piersma & Lindström, 1997; Piersma & Gill, 1998). It may also include adjustments to the characteristics of the blood (Piersma *et al.*, 1996a), hormonal balance (O'Reilly & Wingfield, 1995; Ramenofsky *et al.*, 1995), and adaptations at even 'deeper' biochemical levels (Jenni & Jenni-Eiermann, 1999). Optimizing flight performance will compromise digestive performance (Piersma, 1998). Due to such physiological trade-offs, migratory shorebirds may be especially vulnerable to ecological contingencies during particular phases of the migration cycle.

Precise clock mechanisms to organize annual cycle

The existence of endogenous 'clocks' that organize annual migration schedules is well-established for songbirds (Gwinner, 1986; Berthold, 1996). Even though it is likely that similar clock mechanisms may help migratory shorebirds to perform on time, experimental evidence for endogenously controlled rhythms has been published only for red knots (Cadée *et al.*, 1996).

Orientation mechanisms

As juvenile migratory shorebirds find their way south from the breeding grounds without the help of their parents, they must possess some kind of inbuilt compass, and a clock. Again, such mechanisms have been well described for migratory songbirds (Berthold, 1996), but the experimental work with shorebirds has only just begun (Sandberg & Gudmundsson, 1996). Based on his experiences with sanderlings *Calidris alba* and the pollution of the wetlands on which they rely, Myers (1988) suggested that organochlorine pollutants such as pesticides might possibly interfere with orientation mechanisms, as a result of which a greater proportion of long-distance migrants would lose their way. Such interactions remain to be shown.

Geographic bottlenecks: relying on small numbers of wintering and stopover sites

The ecological specialization of shorebirds and the scarcity of suitable habitats not only necessitate long-distance flights in the course of an annual cycle, but also means that in winter and/or during migration large parts of, or even entire, populations of migratory shorebirds occur together at single sites (Myers *et al.*, 1987). This is especially true for species relying on ecologically predictable habitats such as coastal intertidal areas. For example (Figure 7.2), more than 90% of the red knots breeding in central Siberia winter at only two West African wetlands, the Banc d'Arguin in Mauritania and the Bijagos Archipélago in Guinea-Bissau (Davidson & Piersma, 1992). This entire population migrates through the north German Wadden Sea in spring. Similarly, no important spring staging site other than Delaware Bay on the North American east coast is known for the population of red knots breeding in the Canadian Arctic and wintering in southernmost South America. Red knots breeding on the New Siberian Islands may all winter along a 200 km long stretch of coastline in northwest Australia. Such restricted occurrences makes populations like those of the red knot very vulnerable to local ecological perturbances, such as reclamation of intertidal lands, oil spills, and overfishing or destruction of critical resources.

Reduced genetic variability as a result of repeated population bottlenecks

Based on allozyme data, Baker & Strauch (1988) concluded that most migratory shorebird species are genetically depauperate. Subsequent research has shown that the genetic impoverishment is more extreme in some species (red knot, for example; Baker et al., 1994) than others (dunlin *Calidris alpina* and ruddy turnstone *Arenaria interpres*; Wenink et al., 1994). Whether or not genetic uniformity makes species more susceptible to environmental perturbations (and is therefore a cause for concern in a world where mankind induces many changes at ever faster rates), is a hotly debated question (Avise, 1994; Haig & Avise, 1996). For the time being, the low levels of heterozygosity in some species should make us more cautious when evaluating possible negative effects of habitat loss or habitat changes. For example, Dolman & Sutherland (1995a) pointed out that if choice of wintering site is (partly) genetically determined, the lack of genetic variation might make a species less capable of adjusting to the loss of particular wintering areas by moving to others.

THE DEMOGRAPHY–KEY SITE QUANDARY

It is intuitively easy to accept that for species with low and variable fecundity and long life spans, such as migratory shorebirds, variation in adult survival matters more to population size than variation in breeding success (and see Charnov, 1986). In a comparative context, Saether et al. (1996) would call migratory shorebirds 'bet-hedgers', species that live in quite favourable habitats throughout the year, but with a large annual variation in

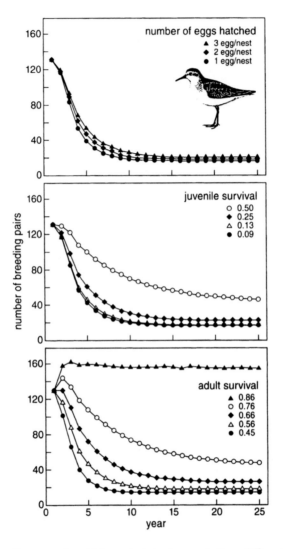

Figure 7.4 Sensitivity analysis to changes in three different parameters in a stochastic matrix population model of a breeding population size of semipalmated sandpipers in an area of tundra near Churchill, Manitoba, Canada. Explorations of the model to variation in hatching success (top), survival of juveniles throughout the first year of life (middle) and annual adult survival (bottom), are presented (median values of 1000 simulations; only for biologically realistic parameter values). Compiled from Hitchcock & Gratto-Trevor (1997).

the quality of the breeding habitats. They point out that such species will be dependent on access to high-quality survival (i.e. non-breeding) habitats (see Alerstam & Högstedt, 1982), and that the conservation of non-breeding areas should thus be a priority.

The fact that population size of migratory shorebirds is very sensitive to variations in adult survival rates is nicely illustrated by the 'stage-structured stochastic matrix population-model' built by Hitchcock & Gratto-Trevor (1997). This model successfully reconstructed a decline of a breeding population of semipalmated sandpipers *Calidris pusilla* observed during the 1980s in an area of tundra near Churchill, Manitoba, Canada. The model simulations showed that for biologically reasonable values of five variables (fecundity, juvenile and adult survival, delayed recruitment, and immigration), only adult survivorship, and to a lesser extent immigration rates, could account for the observed population decline (Figure 7.4). The model appeared highly sensitive to level of survival and the annual variability thereof, but it was not sensitive to (annual variations in) fecundity.

Adult survival is a critical variable in determining population size of migratory shorebirds, and this magnifies the importance of the quality of the limited non-breeding/survival habitats in the winter and migration seasons on which they rely. Indeed, the time seems ripe to extend the stalwart modelling efforts of Goss-Custard (1993) and co-workers (Goss-Custard *et al.*, 1995a,d,e). This team has built very sophisticated individual-based models of the population dynamics of wintering oystercatchers, based on the comprehensive empirical information collected at the Exe Estuary, southwest England. The family of models can be applied to investigate a large number of questions, ranging from the implications of different kinds and strengths of density dependence, to the effects of various kinds of habitat loss or deterioration. However, the models are specialized and not suitable to investigate the fate of migratory shorebird populations that have feeding ecological and distributional characteristics very different to oystercatchers, and that rely on chains of widely separated wintering and staging areas, rather than on single wintering and breeding sites.

An alternative approach is based on game theoretical ideas (Sutherland & Dolman, 1994; Sutherland, 1996a). It lacks the detailed empirical sophistication of the oystercatcher-model family, but incorporates the possibility that individuals must choose between alternative wintering sites, and that an individual's choice depends on relative migration costs and also on the choices that conspecifics make.

For the moment, it seems reasonably safe to state that the future of many species of migratory shorebirds depends on adequate conservation of relatively few wintering and staging sites, areas that are perhaps easier to protect than the immense patchwork of boreal and Arctic breeding grounds where the birds are more thinly spread. Although climatic changes may eventually modify the northern breeding grounds of migratory shorebirds on a large scale, at present the small wetland wintering and staging sites are most under pressure from ever expanding humanity.

EFFECTIVE POPULATION SIZE: POPULATION FLUCTUATIONS AND MATING SYSTEMS

Because the probability of contributing gametes to the next generation varies among individuals of a population, the effective population size N_e is usually much smaller than the total population size observed by counts, especially in birds (Avise, 1994; Frankham, 1995c). There are several aspects of shorebird biology that can reduce effective population size, most notably (1) fluctuations in population size from generation to generation, and (2) unequal numbers of males and females participating in breeding (e.g. Nunney, 1991). The long-term effective population size in a species is the geometric mean of population sizes through successive generations, and thus the effective population size will be much nearer to the smallest population size than the arithmetic mean population size. Drastic reductions in population size from hunting in the last 100 years almost certainly have contributed disproportionately to the small genetic variability observed in shorebirds (Baker and Strauch, 1988; Baker et al., 1994).

In addition, the 'natural' population fluctuations in shorebirds may reduce effective population size and erode genetic variability. For example, the well-known cycling of lemmings on the Siberian Arctic breeding grounds entrains a cycle of variable recruitment of young in which every third year tends to be much more productive than the other two (Roselaar, 1979; Underhill et al., 1989, 1993). This is because in the years of peak lemming abundance, predators take fewer shorebird young, and recruitment is correspondingly higher. Lemming cycles are longer and less prominent in the Canadian Arctic, so that variation in predation has less impact on shorebirds (H. Boyd personal communication).

In polygamous species of shorebirds, reduction in population size is expected if the number of breeding males differs from that of females (see

Pitelka *et al.*, 1974; Whitfield & Tomkovich, 1996; Reynolds & Szekely, 1997). This is because effective population size is given by:

$$N_e = 4 \ N_m \ N_f \ /(N_m + N_f),$$

where N_m is the number of males and N_f is the number of females (Falconer 1989). Unless the number of females equals the number of males, effective population size will always be less than census population size. Low levels of genetic variability observed in most species of shorebirds are consistent with the cumulative effects of these factors acting to reduce effective population size, and we are tempted to ask whether the trophic specializations, narrow habitat preferences, and even the long migrations to similar habitats in the south are consequences of restricted genetic variances for alternative lifestyles?

ECOLOGICAL NONLINEARITIES AND POPULATION DECLINES

Which of the supposedly critical life history features of migratory shorebirds helps us most to understand the recent (near-)extinction of two curlew species, the Eskimo curlew in the Americas and the slender-billed curlew in the Old World? In the case of the Eskimo curlew it is unlikely that changes on the Canadian breeding grounds (i.e. fecundity) or on the South American wintering grounds are responsible. Gill *et al.* (1998) argued that a highly increased adult mortality on staging sites caused the disastrous decline of the Eskimo curlew, a bird with a highly gregarious nature. Hunting took a particularly heavy toll, as Eskimo curlews congregated in a limited number of sites. They would have been attracted rather than be put off by the sounds and sights of conspecifics dying at the hand of hunters. Also, the rapid disappearance of a critical habitat (tallgrass prairie) and the extinction of an important prey (the Rocky Mountain grasshopper) may have increased non-hunting mortality during migration.

Whether the demise of the species was sped up by the so-called Allee effect (a decrease in population growth rates at low population size; see Saether *et al.*, 1996) is unclear, although one might argue that as the population dwindled, the birds would have found it increasingly difficult to locate profitable refuelling sites (that were unpredictably spaced) during migration. According to Bucher (1992), problems with food finding owing to lack of small food-finding (sentinel) flocks may have eventually made the passenger pigeon *Ectopistes migratorius* go extinct. The decline of the slender-billed curlew is generally attributed to high hunting pressure, perhaps in combination with a loss of wintering and staging habitats (Gretton, 1991). Again, it seems that the population dynamic predictions are borne out, in that high adult mortality caused the population to collapse.

The stories of the small curlew species give no role to several of the life history characteristics listed. Both were common inland species, and thus can be predicted to have been highly immunocompetent in view of the large number of parasites that they routinely encountered in freshwater and inland habitats (Piersma, 1997). In the context of the immunocompetence hypothesis, it thus seems unlikely that they were decimated by novel parasites or pathogens. Neither species shows much sexual or seasonal plumage dimorphism, nor do they display on leks, and thus they are not strongly sexually selected. The Eskimo curlew in particular probably showed many of the metabolic, rhythmic and sensory 'adaptations' necessary to make very long transoceanic flights (hunters sometimes called them 'dough birds', reflecting the incredible masses of white fat that enveloped their bodies before take-off). But this still does not exclude the possibility that the latter characteristics could have played a role in the fate of these migratory shorebird populations, especially as the interacting effects would be quite subtle. Mortality during migration and breeding success could all be strongly affected nonlinearly by small changes in fuel storage, and the resulting mortality would be hard to detect in the field.

Let us follow a shorebird population during its migration from the wintering to the breeding grounds (or vice versa), and follow the changes in fuel stores, as reflected in body mass, as the key variable (Figure 7.5). There would be variation in the rate of fuel storage between individuals (Figure 7.5B), but in populations that have had time to adjust to particular constellations of environments (Figure 7.5A), after migratory flights or on the breeding grounds few birds would end up with energy stores that are so small as to precipitate death by exhaustion (Figure 7.5C) or infection (Figure 7.5D) or prevent reproductive success (Figure 7.5E). Now consider a small man-made ecological change (for example, the depletion of shellfish stocks), negatively affecting the quality of the habitats where refuelling takes place (Figure 7.5B). Average refuelling rates would be slightly reduced (almost immeasurably so for the fieldworkers), but suddenly a much greater proportion of the individuals enters the range of energy stores at departure or after arrival, where the likelihood of death from exhaustion (Figure 7.5C) or death from disease (Figure 7.5D) steeply increases. Without much warning, adult mortality would have increased, with the population suddenly collapsing. One could additionally argue that the less immunocompetent and the more ecologically specialized and sexually selected the species is, the smaller the environmental perturbations would have to be for migratory populations to be affected.

Examples of such subtle mechanisms affecting shorebird populations may be all around us, but may go unnoticed because they are not looked for. The quite drastic population declines between 1980 and 1997 of several shorebird species wintering in West Africa (Zwarts et al., 1998) are possible examples. We are unaware of drastic changes in the benthic food supply for shorebirds on the wintering grounds in Mauritania and Guinea-Bissau, and there are no indications of large ecological changes on the Arctic breeding grounds either. Several populations sharing the same breeding grounds but wintering in Europe seem to do fine. However, on the main coastal (intertidal) stopover sites in Western Europe the conditions may be changing. In many areas the shellfishery industry has taken a much heavier toll in the years since 1980 than before (Piersma & Koolhaas, 1997). This may have negatively affected shellfish populations and positively affected the stocks of small polychaete worms.

Thus, it may not come as a surprise that the West African wintering populations of worm-eating species such as dunlin and curlew sandpiper Calidris ferruginea have increased in size (though, admittedly, rather few curlew sandpipers migrate through Western Europe), whereas a shellfish-specialist like the red knot shows a particularly strong decline (by about 40% of the original number; Zwarts et al., 1998). For the latter species it was shown how crucial the abundances of particular shellfish resources in the Dutch Wadden Sea were in determining feeding distribution and food intake rates, particularly for the population migrating to West Africa (Zwarts et al., 1992; Piersma et al., 1993a). During their short stopovers (when they have high food requirements to rapidly refuel), there is little time to find the best feeding sites. In view of the fact that those taking off with inadequate stores will face an immediate survival penalty (running out of fuel and not reaching the winter destination), the long flyers are likely to be affected first when food stocks crash.

PRESSING CONSERVATION ISSUES AND RESEARCH NEEDS

If we want to safeguard the world's migratory shorebird populations, we have to protect the interconnected chains of wetland sites from further

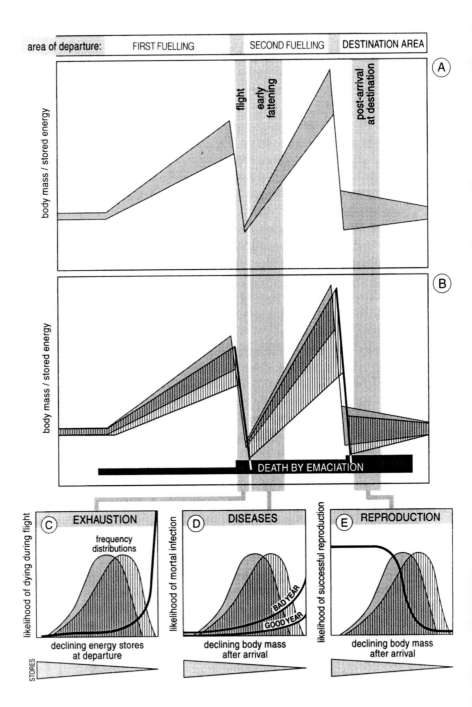

deterioration and disappearance, and the migratory populations from unnecessary additional mortality from hunting.

Given the issues of rapid climate change and sea level rise, it is particularly appropriate to implement and sustain early warning systems at key
sites, even when such sites are not under any immediate threat. Thus, the
tradition of waterbird counts and bird monitoring, e.g. the efforts organized
through Wetlands International, needs all the support it can get. At the
same time coverage should be expanded, not only to key areas outside regions with large populations of ornithologists (for example, the Banc
d'Arguin, Mauritania; and the large intertidal complexes in Patagonia and
Isla Grande, Argentina and Chile), but also for a much larger array of species (Piersma *et al.*, 1997). At present, only for about two-thirds of the 340
recognized populations of plovers and sandpipers (Charadriidae and
Scolopacidae) is an estimate available of the size of the total population
(Rose & Scott, 1997). For only a third of these 340 populations do data exist

Figure 7.5 Graphical model of the long-term and long-distance 'domino' effects
(on migrating shorebirds) of habitat deterioration affecting the rate of fuelling
(decreased food abundance → longer working days → reduced ratio
income/expenditure) at a departure area or a stopover site. Panel (A) shows the
changes in body mass (or relative energy stores) over time for a population of birds
that migrates from a 'departure area' to a stopover site for a second fuelling period,
to continue to an area of destination (see Piersma, 1987). The model could refer to
a northward migration (destination = breeding area) as well as to a southward
migration (destination = wintering area). There is a 'natural' amount of variation
among individuals in the rates of mass gains. Panel (B) illustrates the effects of
problems during the first and/or second fuelling episodes. The average rate of
mass gain decreases, the top speeds no longer occur, and the statistical
distributions of mass gain values get broader (see C, D and E for the frequency
distributions of body mass values at departure or after arrival for situations
without and with habitat deterioration) and lowest rates of mass gains in the
population get markedly lower. During the first and second migratory flight, the
accumulated energy stores no longer suffice and a proportion of birds dies from
exhaustion/emaciation before the second fuelling area or destination is reached.
Panel (C) illustrates how the nonlinear increase in the likelihood of dying during a
migratory flight with decreasing body mass at departure might disproportionally
affect mortality owing to migration. Panel (D) illustrates that birds arriving with
low body masses and trying to win back the mass and time delays might run a
disproportionally high risk of infection by dangerous parasites or pathogens, with
possible differences between years depending on environmental conditions.
Finally, panel (E) illustrates that if birds arrive in a poor energetic state on the
destination area, they would have a very small chance of reproducing successfully
if the destination area is the breeding area (or of surviving a moulting/wintering
period if the destination area refers to the wintering grounds).

on whether or not the population is increasing, stable or in decline. It is perhaps alarming that trends are negative in half of the cases listed by Rose & Scott (1997). Following slightly different tactics, the research, monitoring, education and conservation linkage between widely separated protected wetland sites developed and organized by the Western Hemisphere Shorebird Reserve Network (WHSRN) programme in the Americas, has also proven a valuable initiative (Harrington, 1996).

If we are to foresee increasing risks, we also need to find out much more about the factors that affect the survival of shorebird populations (see Ens *et al.*, 1994). Shorebirds differ from many other groups of avian migrants by their extremely long and demanding flights, gregariousness, restriction to limited numbers of sites, long lifespans and low recruitment. This chapter has tried to indicate that most life history characteristics may play a role in survival; different ones may be important in different species and at different times. We need studies on the linkages between life-history characteristics (including ecological and immunological specializations), population genetic parameters and potential susceptibility to habitat changes. We need to find out how pesticides and other pollutants influence migratory and breeding performance, disease resistance and orientation. We should admit that we are quite ignorant about the resilience of shorebird migration systems; it is currently impossible to predict when things go wrong because of the many subtle mechanisms at work (Figure 7.5). Finally, we need models to explore the boundaries of the measurable, and to determine the sensitivity of different populations and species to environmental and global changes.

ACKNOWLEDGEMENTS

We thank our friends and colleagues in the international Wader Study Group for many years of intense co-operation, innumerable insights and much fun. We gratefully acknowledge Dick Visser for preparing the figures, and Pablo Canevari, Bob Gill, Hugh Boyd, Petra de Goeij, Silke Nebel and Wim Wolff for constructive comments on a draft. T.P.'s shorebird research is supported by the PIONIER programme of the Netherlands Organization for Scientific Research (NWO). A.J.B.'s work on shorebirds is funded by the Natural Sciences Engineering Research Council of Canada and the Royal Ontario Museum Foundation. This is NIOZ publication 3249. We dedicate this paper to those that fight their lonely fights for the future of fragile biological phenomena such as the incredible shorebird migrations.

Ranging behaviour and vulnerability to extinction in carnivores

ROSIE WOODROFFE & JOSHUA R. GINSBERG

INTRODUCTION

Over the last 20 years, a great deal of research in conservation biology has been concerned with the dynamics of very small populations. A suite of theoretical studies have provided convincing evidence that stochastic effects that operate harmlessly in larger populations – effects such as genetic drift, local variation in climatic conditions, or individual variation in offspring sex ratios – may cause the extinction of populations that are reduced to their last few members (e.g. Soulé, 1987; Lande, 1988a). Mathematical models of such extinction processes are numerous (Boyce, 1992), but there have been very few attempts to quantify the role played by stochastic effects in the extinction of real populations. Thus there is little direct empirical support for specific model predictions.

Despite limitations in the supporting data, the theory of small population biology has been extremely valuable in directing *ex situ* conservation. For example, combatting genetic drift through the use of studbooks is a crucial part of the management of captive populations (Olney *et al.*, 1994). Because stochastic processes are also believed to threaten wild populations, it has been suggested that similarly intensive management should be applied to them; for example, maintaining genetic diversity by translocating animals between sites, or by inseminating wild females with semen collected from captive males (e.g. Seidensticker, 1986; van Dyk *et al.*, 1997).

In this chapter, we argue that attempts to combat stochastic processes will not avert the extinction of wild populations, because these processes are not the major threat to small populations. We show that species' behaviour is an important component of their vulnerability to extinction. Using data on large carnivores, we demonstrate that a deterministic process – high density-independent mortality caused by people – is the principle cause of local extinction. Even when populations are nominally protected inside reserves, species that range widely come into frequent contact with people, greatly reducing the protection afforded by reserves.

Large mammalian carnivores are an ideal group for investigating the vulnerability of small populations. Most species are top predators, and are therefore constrained to living at low population density (Colinvaux, 1980). As habitat becomes fragmented, contiguous populations are broken up into smaller units; this process leaves carnivore sub-populations that are smaller, in absolute terms, than those of other species at lower trophic levels. Stochastic effects may therefore have an especially marked impact upon carnivore populations. Furthermore, excellent data are available upon the past and current status of carnivore populations (e.g. Nowell & Jackson, 1996; Fanshawe et al., 1997).

WHAT ARE THE EFFECTS OF HABITAT FRAGMENTATION?

Humans modify the land they occupy, often converting natural landscapes into mosaics of cultivation, grazing and urbanization. As human populations rise, patches of natural habitat are left isolated within a matrix of human-altered habitat. Species' response to such habitat fragmentation depends in part upon their ability to survive in this matrix (Laurance, 1991). Some species thrive in human-altered habitat; others are able to cross it safely and move freely between patches of natural habitat. Species that can neither cross nor use such habitat, however, will become isolated in small populations occupying the remaining fragments of natural habitat, and will be vulnerable to local extinction.

Large carnivores are extremely sensitive to habitat fragmentation, because it is rarely possible for them to survive in human-altered habitat. Quite understandably, local people see predators as a threat to their livestock – and often to themselves – and poison, shoot or trap any that venture onto their land. Even where human population densities are extremely low, carnivores may be shot and trapped for their fur. A succession of studies have shown that large carnivore populations living outside protected areas are limited by human persecution (e.g. Fuller, 1989; Wielgus & Bunnell, 1994; Powell et al., 1996). More usually, large predators are completely extirpated where human density is high (e.g. Woodroffe et al., 1997). Human presence can therefore make large tracts of land inhospitable for large carnivores, even if there is no further modification of the habitat. Under such pressure, populations of large predators often survive only where they are protected from persecution inside national parks and reserves.

While it is no surprise that predators are often killed outside reserves, few people realize that protected populations are also vulnerable to persecution. In Table 8.1 we have summarized the results of 16 studies of large carnivores inhabiting protected areas, reporting the proportion of recorded adult mortality that is directly attributable to human. All but a handful of these studies used radio-telemetry to locate dead study animals, so it is unlikely that their estimates are strongly biased towards those killed by people. The impact of human-induced mortality is spectacular: of the 616 predator deaths recorded in Table 8.1, 455 (74%) were directly caused by people. Contact with human activity is the major cause of recorded mortality for seven of the eight species, even though all of the studies involved populations living in protected areas.

For most species, these deaths occur on reserve borders. Conflict with farmers on ranches bordering reserves caused nearly half the deaths recorded among adult lions *(Panthera leo)* originating in Etosha and Nairobi National Parks (Rudnai, 1979; Stander, 1991). Most of the Iberian lynx *(Felis pardina)* that were radio-collared in Doñana National Park were subsequently killed in traps and snares, and in road accidents, on neighbouring private land (Ferreras *et al.*, 1992). Grey wolves *(Canis lupus)* following elk *(Cervus elaphus)* migrating beyond the borders of Algonquin Provincial Park were legally killed by hunters and trappers outside the park (Forbes & Theberge, 1995), while spotted hyaenas *(Crocuta crocuta)* commuting to the western border of Serengeti National Park became caught in snares set illegally to catch wild ungulates (Hofer *et al.*, 1993).

Because carnivores range widely, these 'edge effects' can operate over large distances. Fifty-two per cent (16/31) of African wild dog *(Lycaon pictus)* deaths recorded from packs radio-collared in Hwange National Park were caused by traffic accidents on a road some 20 km outside the park. Meanwhile, poaching on the borders of Kruger National Park means that every wild dog pack studied within the park has had at least one member caught in a snare (M. G. L. Mills, personal communication) – truly remarkable in a reserve which, at 22 000 km², is roughly the same size as the state of Israel.

Table 8.1

The proportion of adult deaths caused by contact with human activity recorded by studies of large carnivores inhabiting protected areas

	Proportion of mortality caused by people	Species total
African wild dog[a]		
Hwange National Park	81% (31)	
Kruger National Park	47% (19)	
Moremi Game Reserve	7% (15)	
Selous Game Reserve	25% (4)	
Various Zambian Reserves	75% (36)	61%
Grey wolf		
Algonquin Provincial Park[b]	54% (26)	
Glacier & Banff National Parks[c]	95% (60)	83%
Lion		
Etosha National Park[d]	25% (4)	
Nairobi National Park[e]	54% (31)	
Serengeti National Park[f,g]	33% (27)	50%
Tiger		
Royal Chitwan National Park[h]	67% (3)	67%
Iberian lynx		
Doñana National Park[i]	75% (24)	75%
Spotted hyaena		
Masai Mara Game Reserve[j]	61% (18)	
Serengeti National Park[k,l]	42% (38)	49%
Black bear		
Great Smoky Mountains National Park[m]	50% (2)	
Pisgah Bear Sanctuary[n]	90% (21)	87%
Grizzly bear		
Selkirk Mountains[o]	71% (7)	
Yellowstone National Park[p]	89% (250)	89%

All data come from intensive studies, of which all but two used radio-telemetry to locate dead animals. The figures in brackets give the total number of mortalities recorded owing to all causes. Causes of death include legal hunting outside reserves, poaching, and persecution as well as accidental killing through road accidents and snaring. Data on Yellowstone grizzly bears also include legal hunting and control of problem animals inside the park, which together account for 130 of the 250 deaths (Peek *et al.*, 1987).

References: [a]Woodroffe *et al.* (1997); [b]Forbes & Theberge (1995); [c]Boyd *et al.* (1995); [d]Stander (1991); [e]Rudnai (1979) (animals not radio-collared); [f]Packer *et al.* (1988); [g]Schaller (1972); [h]Sunquist (1981) (animals not radio-collared); [i]Ferreras *et al.* (1992); [j]Frank *et al.* (1995); [k]Hofer *et al.*, (1993); [l]Kruuk (1972); [m]Garshelis & Pelton (1981); [n]Powell *et al.* (1996); [o]Wielgus *et al.* (1994); [p]Peek *et al.* (1987).

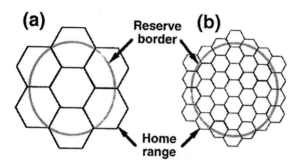

Figure 8.1 The effect of ranging behaviour upon the magnitude of edge effects. Shaded circles denote protected areas; hexagons denote home ranges. (a) When a species ranges widely, a large proportion of the population is exposed to reserve borders, and therefore to persecution; here six out of seven home ranges incorporate border areas (86% of the population). (b) When home ranges are smaller, the proportion of the population exposed to the border is also smaller; here it is 49% (18 out of 37 home ranges). Note that the size of the two populations could be identical if species (a) lived in groups, or occupied home ranges with a high degree of overlap.

CAN EDGE EFFECTS CAUSE EXTINCTION?

The impact of human activity outside reserves could well cause the extinction of carnivore populations inside. With such high levels of density-independent mortality occurring on the edges of reserves, border areas may become population sinks where predator populations can persist only if they are replenished by immigration from the safer core of the reserve. Very small reserves will contain no such core areas, and populations inhabiting them may therefore be driven to extinction.

Species' exposure to edge effects will be mediated by their ranging behaviour; we have illustrated this phenomenon in Figure 8.1. In the wide-ranging species depicted in Figure 8.1(a), six of the seven home ranges are in contact with the reserve border, so 86% of the population is at risk of persecution. In contrast, Figure 8.1(b) shows a species that occupies much smaller home ranges; here just 49% of the population (18/37 home ranges) is in contact with the border. Note that the two populations will not necessarily differ in size: if the wide-ranging species lives in groups, or has a high degree of home range overlap, the population density of the two species could be identical.

If edge effects really can cause extinction then, for a reserve of a given

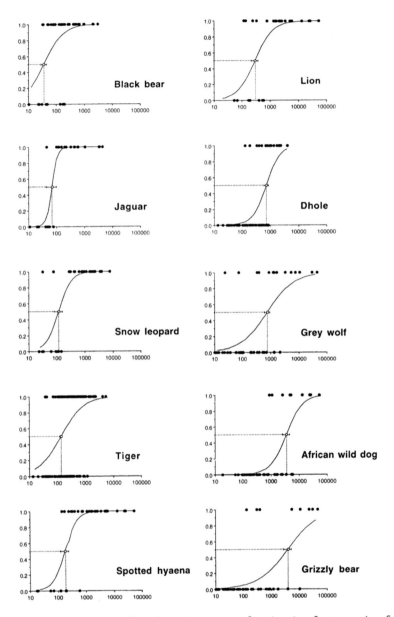

Figure 8.2 Persistence of populations in reserves of varying sizes for ten species of large carnivores; data sources are given in Table 8.2. Population persistence is related to reserve area for all species (Table 8.2). Solid curves show the probability of persistence predicted by logistic regression. Open circles show, for each species, the critical reserve size (± SE) at which the model predicts a 50% probability of population persistence.

size, wide-ranging carnivores should be more likely to disappear than those that restrict their movements to smaller areas. In contrast, if most extinctions of small populations are the result of stochastic effects, then the carnivores that live at lowest population density should be most at risk.

We investigated the relative importance of these two factors by compiling data upon population extinctions for ten species of large carnivore (Table 8.2). For each species, we chose a geographic region within the species' historic range in which suitable habitat has become fragmented. In all of these regions, people kill large carnivores ranging outside of protected areas (Table 8.1). Quantitative data on mortality were not available for dhole *(Cuon alpinus)*, snow leopards *(Panthera uncia)* or jaguars *(Panther onca)*, but persecution is recorded as a major threat to all three species (Ginsberg & Macdonald, 1990; Nowell & Jackson, 1996). For each region, we identified the protected areas in IUCN categories I–IV that fall within the former geographic range of the species, treating complexes of contiguous reserves as single protected areas (IUCN 1992a,b,c). We recorded the size of each protected area (reserve size), and the time elapsed between the date that it was gazetted and the date it was surveyed for carnivores (reserve age). We then determined the presence or absence of the species in each of these protected areas using a combination of published and unpublished data (Table 8.2). Because none of the species have highly specific habitat requirements, and all have experienced range contractions within the last century, absence from those protected areas containing suitable habitat can be taken as evidence of local extinction.

We investigated the relationship between reserve size, reserve age and carnivore extinction using logistic regression, a standard technique for the analysis of binary data (Cox, 1970). For each species, we constructed a full model containing both independent variables, then dropped them one by one to estimate the contribution of each to total deviance. The resulting changes in deviance are distributed as χ^2.

As shown in Table 8.2, all ten species were more likely to disappear from small reserves than from larger ones. Extinction was related to reserve age in only one species, the spotted hyaena. While the statistical effect of reserve size is very strong for all species, examination of Figure 8.2 reveals that species vary substantially in the size of the reserves in which they have persisted. We derived a measure of critical reserve size by using the logistic regression models to predict the area at which populations persisted with a probability of 50% (Figure 8.2). This measure is analogous to an LD50, the dose of a drug which, when administered to experimental subjects, kills

Table 8.2.

Results of logistic regressions on the presence and absence of large carnivores in protected areas that fall within their historic range

Species	Region	Number of reserves	Change in deviance owing to		Critical reserve size (km²)
			Reserve age	Reserve size	
African wild dog[a,b]	East Africa	46	3.15	26.59 ***	3606
Grey wolf[c,d]	Western Canada	44	0.0	19.82 ***	766
Dhole[e-g]	India	71	1.69	30.59 ***	723
Lion[a,unpubl.]	East Africa	32	1.39	17.61 ***	291
Tiger[e,h]	India	154	0.0	39.1 ***	135
Snow leopard[e,i]	India, Nepal & Pakistan	30	1.27	21.09 ***	116
Jaguar[c,j-l]	Central America	28	0.91	29.98 ***	69
Spotted hyaena[a,unpubl.]	East Africa	37	5.14 *	20.22 ***	179
Black bear[c,m]	California	45	1.48	13.05 ***	36
Grizzly bear[c,n,o]	Western Canada & North-West USA	54	0.78	18.48 ***	3981

The effects of reserve age and size (both log-transformed) were assessed by constructing a model containing both factors, then dropping them one by one to estimate the contribution of each to total deviance. The resulting changes in deviance are distributed as χ^2; asterisks denote the significance values associated with each. Critical reserve size is the area for which the fitted logistic model predicts a 50% probability of population persistence (Figure 8.2). References: [a]IUCN (1992b); [b]Woodroffe et al. (1997); [c]IUCN (1992c); [d]Hayes & Gunson (1995); [e]IUCN (1992a); [f]Johnsingh (1985); [g]Stewart (1990); [h]Dinerstein et al. (1997); [i]Green (1994); [j]IUCN (1982); [k]Melquist (1984); [l]Lopez Pizarro (1986); [m]LaRoe et al. (1995); [n]Peek et al. (1987); [o]Paquet & Hackman (1993); 'unpubl.' indicates that analyses were based upon unpublished data derived from correspondence with researchers and conservation organizations.

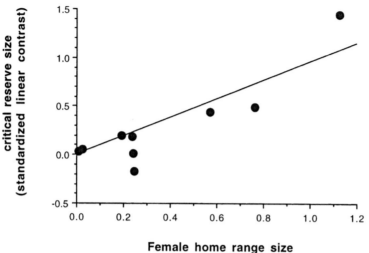

Figure 8.3 The relationship between phylogenetically independent contrasts in log (critical reserve size) and log (female home range size) calculated for ten species of large carnivore ($r^2 = 0.84$, $F_{1,8} = 42.1$, $P < 0.005$). The effect remains strong after controlling for the (non-significant) effect of population density ($t = 4.00$, $P = 0.005$).

exactly half of them. As shown in Table 8.2, critical reserve sizes ranged from 36 km² for the black bear *(Ursus americanus)*, through to 3981 km² for its congener, the grizzly bear *(Ursus arctos)*.

If a population's probability of extinction is principally determined by its size, then average population density should be a good predictor of critical reserve size. This is because the size of a population at the time it is isolated will depend upon population density and the area of the reserve. If, on the other hand, extinction is caused by edge effects, critical reserve size should be correlated with home range size, as long as reserve shape varies randomly with reserve area.

We tested these predictions using data on population density and ranging behaviour collected from the literature. For each species, we estimated the average adult population density and the average size of female home ranges within the regions for which we investigated population extinction (Table 8.3). We ensured that measures taken from closely related species were statistically independent by analysing phylogenetically independent contrasts (Felsenstein, 1985). We calculated contrasts from log-transformed data, using a composite phylogeny for the Carnivora and the

Table 8.3
Population densities and home range sizes of large predators

Species	Average density (adults/100 km^2)	Average female home range size
African wild dog[a]	2.7 (6)	823.1 km^2 (12)
Grey wolf[b–d]	1.1 (9)	684.6 km^2 (11)
Dhole[e]	10.6 (1)	68.8 km^2 (2)
Lion[f,g]	16.2 (12)	121.4 km^2 (59)
Tiger[h]	3.6 (3)	16.9 km^2 (3)
Snow leopard[i]	4.6 (6)	29.3 km^2 (2)
Jaguar[j,k]	6.8 (2)	18.8 km^2 (5)
Spotted hyaena[l–p]	62.5 (6)	34.9 km^2 (12)
Black bear[q,r]	68.0 (5)	16.6 km^2 (9)
Grizzly bear[s–w]	2.0 (5)	774.0 km^2 (22)

Wherever possible, data are taken from the region for which critical reserve sizes were determined. All data are averaged across studies; figures in brackets give the sample size of studies (for density) or individuals or social groups (for home range size). [a]Woodroffe et al. (1997); [b]Fuller & Keith (1980); [c]Bjorge & Gunson (1989); [d]Fuller (1989); [e]Venkataraman et al. (1995); [f]Creel & Creel (1997); [g]van Ordsol et al. (1985); [h]Sunquist (1981); [i]Jackson & Ahlborn (1989); [j]Rabinowitz & Nottingham (1986); [k]Tewes & Schmidly (1987); [l]Kruuk (1972); [m]Frank (1986); [n]Sillero-Zubiri & Gottelli (1992); [o]Creel & Creel (1996); [p]Hofer & East (1993); [q]Novak et al. (1987); [r]Gompper & Gittleman (1991); [s]Blanchard & Knight (1991); [t]Wielgus & Bunnel (1994); [u]Wielgus et al. (1994); [v]McLellan (1989); [w]Servheen (1983).

program CAIC (Purvis & Rambaut, 1994; Bininda-Emonds et al., 1999). All of our regressions of contrasts on contrasts were forced through the origin. Contrasts in population density and female home range size were not closely correlated ($r_8 = -0.69$), partly because some species are social, and partly because home range overlap is relatively high in species with very large home ranges (Blanchard & Knight, 1991; Woodroffe et al., 1997). When contrasts in both measures were entered into a multiple regression, average female home range size was a good predictor of critical reserve size, but population density had no effect (Figure 8.3; overall, $F_{2,7} = 20.6$, $P < 0.005$; effect of density, $t = 0.82$, $P > 0.4$; effect of home range size, $t = 4.00$, $P = 0.005$).

These results show that, in a reserve of given size, wide-ranging species are more likely to disappear than those with smaller home ranges, irrespective of population density. This indicates that edge effects do indeed cause local extinctions. These results do not imply that stochastic effects cannot cause the extinction of small populations of carnivores; they simply show that edge effects have a much greater impact.

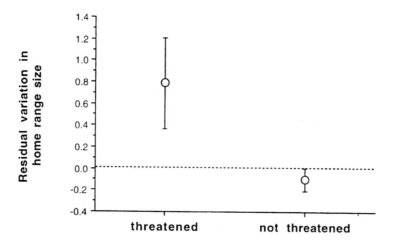

Figure 8.4 Using ranging behaviour to predict threatened status within the Carnivora. This graph shows residual variation in home range size after controlling for metabolic needs ($\pm$ SE), comparing carnivore species listed as 'threatened' in the *IUCN Red List of Threatened Animals* with those not so listed (IUCN, 1996). $t_{42} = 2.72$, $P < 0.01$. Residuals are calculated from data presented by Gittleman & Harvey (1982). This analysis excludes the giant panda *(Ailuropoda melanoleuca)* which, being folivorous, has a much smaller home range than other carnivore species. When the panda is included, $t_{43} = 1.98$, $P = 0.054$.

As well as causing deterministic extinction, edge effects may also make populations more vulnerable to stochastic processes. By increasing adult mortality, edge effects have the ability to limit both population size and population resilience. Edge effects could therefore compound the stochastic effects of small population size.

The potential for such interaction is illustrated by population viability analyses carried out for wild dog populations in southern Africa. Bruford *et al.* (1997) constructed a model to simulate the impact of rabies on specific wild dog populations. Rabies is a serious threat to wild dogs, which has caused the extinction of at least one population (Woodroffe *et al.*, 1997). Bruford *et al.* (1997) used real data to reconstruct the dynamics of two existing wild dog populations. One, in Kruger National Park, suffers high adult mortality (32% per annum, Bruford *et al.*, 1997), with 47% of deaths owing to snaring, shooting and road accidents (Woodroffe *et al.*, 1997). The other, in the Okavango delta, loses very few adults to human causes (Woodroffe *et al.*, 1997) and, probably as a result, has markedly lower adult mortality overall (26% p.a., Bruford *et al.*, 1997). Population simulations predicted that both populations would have extremely low probabilities of extinction in the absence of rabies. When rabies outbreaks were introduced to the model, however, simulations of the Kruger population were seven times more likely to become extinct than simulations of the Okavango population (Bruford et al., 1997). The higher adult mortality in Kruger limited the resilience of the simulated populations, greatly reducing their ability to recover from rabies outbreaks. In this way, edge effects may influence population viability even if they are not a direct cause of extinction.

DO EDGE EFFECTS CAUSE EXTINCTION IN OTHER ANIMALS?

Ecologists normally associate edge effects with plant and invertebrate communities rather than large generalist predators. Nevertheless, the role that edge effects play in promoting local extinctions of vertebrates may well have been under-estimated. Most studies that report relationships between population size and probability of extinction do not, in fact, measure population size directly; instead, they extrapolate the initial sizes of extinct populations from the areas they occupied before they disappeared, much as we have done (e.g. Brown, 1971; Bolger et al., 1991; Newmark, 1995). This is understandable, because direct measures of the sizes of now-extinct populations are rarely available (but see Berger, 1990). Such indirect studies cannot, however, exclude the possibility that some of the extinctions they record are owing, not to the impact of stochastic processes on small populations, but to edge effects operating in small habitat fragments.

Interactions between ranging behaviour and population persistence have been reported for the butterfly *Hesperia comma* (the silver spotted skipper, Thomas & Hanski, 1997). This species is restricted to small patches of habitat that meet its stringent requirements for larval food plants and microclimate. Each generation, a proportion of butterflies emigrate from the natal patch, and this proportion is related to the perimeter:area ratio of the patch. In very small patches, emigration is so high that populations rapidly disappear (Thomas & Hanski, 1997). Thus patch occupancy is related to patch size owing to behavioural, rather than demographic, effects. While carnivores have critical reserve sizes in the region 35–3500 km², however, critical patch size for *H. comma* is less than 0.01 km² (Thomas & Hanski, 1997).

Circumstantial evidence suggests that edge effects might also contribute to local extinction for other taxa. High levels of interspecific brood parasitism on forest margins have been linked to declines in migratory bird species nesting in fragmented habitats (Robinson *et al.*, 1995); the same process could trigger rapid local extinction if it also occurred among more sedentary birds.

CAN HOME RANGE SIZE BE USED AS A SURROGATE FOR THREAT?

With habitat loss occurring at such an alarming rate, there is an urgent need for measures that can be used to predict species' vulnerability to extinction. Various measures have been considered as potential surrogate measures of threat, with rather little success (Mace & Balmford, 1999). We tested whether home range size could be used to predict threat by modifying an earlier analysis that related species' home range sizes to their metabolic needs (calculated as group size multiplied by body size raised to the power 0.75, Gittleman & Harvey, 1982). Using Gittleman & Harvey's (1982) published data, we repeated their analysis and, for each species, calculated the residual variation in home range size not explained by their regression model. We then compared the value of these residuals for species classified as 'threatened' according to the IUCN criteria (i.e. Critical, Endangered or Vulnerable) with those for species not considered threatened by IUCN (1996). As shown in Figure 8.4, residuals differ significantly between threatened and non-threatened species. Threatened species have residuals with large positive values, indicating that they occupy home ranges larger than would be expected on the basis of their metabolic needs.

These results show that ranging behaviour can be used to predict threat in carnivores. In practice, home range size has rather limited use as a surrogate for threat because measuring it is both difficult and labour-intensive. Nevertheless, our results suggest that it might be valuable to investigate the vulnerability of small-bodied yet wide-ranging species in other taxa.

CONCLUSIONS – PRIORITIES FOR CONSERVATION

While apparently simple, our findings have extremely important implications for the setting of conservation priorities. By revealing the factors associated with real extinction events, they have the potential to predict which species are most likely to be at risk, and which management strategies are

Figure 8.5 Wild dog, *Lycaon pictus*, packs have enormous home ranges and are often persecuted if they move outside protected areas. This species thus requires large reserves, in excess of 10 000 km^2, to retain viable populations. Across carnivore species the minimum reserve size necessary to conserve populations is correlated with range area (photograph by Joshua R. Ginsberg).

most likely to avert extinction. In this section, we list three crucial messages for the conservation of large carnivores as well as other species.

Big predators need big areas

Our results indicate that wide-ranging predators are only likely to survive in very large protected areas. Species such as wild dogs and grizzly bears have already disappeared from most areas smaller than 3500 km²; to persist in the long term they require areas much larger than this. Edge effects reduce the resilience of wild dog populations inhabiting protected areas as large as 20 000 km² or more (see above), yet the majority of protected areas are very much smaller than this. Any measures which increase the area of contiguous land available to large carnivores will therefore benefit their conservation. At the most basic level, the integrity of existing large reserves must be maintained if populations of wide-ranging predators are to persist – this is not a trivial recommendation with plans underway to de-gazette parts of several National Parks in Africa. More positively, establishing cross-border parks, corridors and buffer areas will all contribute to carnivore conservation. Most important are local initiatives to reduce persecution of predators. For example, farmers who use traditional husbandry practices, in which livestock are accompanied by herdsmen and kept in corrals overnight, often lose fewer livestock to predators than do those adopting more modern techniques which are less labour-intensive (e.g. Boitani, 1992; Woodroffe et al., 1997). Under such circumstances, some of the substantial tourist revenues generated within reserves by large charismatic carnivores (Western & Henry, 1979) could be used to encourage traditional livestock management on reserve borders, simultaneously protecting predators and creating local employment.

Conservation 'flagships' must be chosen wisely

Large carnivores are often used as 'flagships' for conservation initiatives that are also expected to provide benefits for other species (e.g. Project Tiger, Operation LifeLion, Nowell & Jackson, 1996; Roelke-Parker et al., 1996). Top predators are appropriate flagships because the conditions that allow self-sustaining carnivore populations to persist are also likely to protect species at lower trophic levels. Our analysis suggests, however, that such protection cannot be expected to extend to other large predators. Despite assumptions that reserves established under Project Tiger would also protect the dhole (e.g. Ginsberg & Macdonald, 1990), critical reserve size for dhole in India is five times that for tigers (Table 8.2), and dhole have disappeared from many of the Project Tiger reserves. In the same way, an African reserve system designed to conserve lions would be unlikely to provide adequate protection for African wild dogs.

Treating the effects of small population size may not avert extinction

Finally, our results indicate that conservation measures that aim to combat stochastic processes such as genetic drift are unlikely to avert the extinction of wild carnivore populations. This is simply because extinctions are caused by factors that are independent of population size. While well-intentioned, such measures treat the effects, rather than the causes, of small population size (Caughley, 1994). In large carnivores the same factor that causes regional decline – persecution by people – also leads to local extinction. Controlling the numbers of predators that are shot, snared or poisoned by people is the only way to protect them adequately.

ACKNOWLEDGEMENTS

We would like to thank Sarah Durant, Laurence Frank, Georgina Mace, John Robinson and Pej Rohani for valuable comments and discussions, and Andrew Balmford, Michael Cant, Tim Clutton-Brock, James Deutsch, Laurence Frank, Nigel Leader-Williams, Phillip Muruthi and Derek Pomeroy for information on their sightings of lions and hyaenas in East Africa.

Habitat fragmentation and swarm raiding army ants

GRAEME P. BOSWELL, NIGEL R. FRANKS & NICHOLAS F. BRITTON

INTRODUCTION

Different species have different habitat requirements and so, viewed on the correct spatial scale, the distribution of all species is patchy (Andrewartha & Birch, 1954). This habitat patchiness can take quite dramatic forms such as fragmented forests or mountain tops, or can be as simple as leaves on plants. When these patches are totally disconnected, so that there is no flow of individuals among them, the population dynamics in one region will have no effect on the dynamics in other regions, so a species becoming extinct in one habitat patch will not affect extinction of that species in any other patch. However, when habitat patches have some degree of connectedness, the between-patch dynamics will play an important role (Hanski & Gilpin, 1991). This is essentially what metapopulation dynamics are all about.

Imagine small islands in a lake, all relatively near to one another. Suppose a particular plant is introduced to one of these islands. Eventually, once the plant is well established, it will release seeds that give rise to daughter plants. Assume these seeds are dispersed by wind. Some of the seeds remain on the island, some end up in the water but a few land on neighbouring uncolonized islands. Once these few lucky seeds have germinated and new plants are established, they too may contribute to the colonization of other islands. Of course, each plant has a chance of dying and it is the balance between the death rate and colonization rate that will cause the plant species either to persist or to go extinct relative to our group of islands. Intuition suggests that persistence is possible only if the colonization rate is greater than the death rate. Later we shall present a mathematical model to investigate this further (Box 9.1).

Metapopulation dynamics are not just confined to ecology, they can be used in many other fields of study, most noticeably the spread of infectious diseases (Grenfell & Harwood, 1997). Here towns and cities represent islands, each island being in one of two states, namely susceptible or infected. An infected town may pass on the disease to its susceptible neighbours. At the same time the disease in other towns could die out, leaving us with the classic model of islands undergoing extinction and colonization.

We might expect that the nearer the local extinction rate is to the colonization rate, the rarer the species (or disease) becomes. It is important to note at this stage that even if a species is rare, it does not mean it has little or no environmental impact. As an example, consider the prickly pear cactus *(Opuntia)* in Queensland, Australia (Andrewartha & Birch, 1954). Sometime around the turn of this century, somebody introduced *Opuntia* to Queensland. Within a few years, this cactus covered approximately a quarter of a million square kilometres of potentially productive land. Farmers wanted this plant to be brought under control and so the prickly pear's natural enemy, a certain species of moth *(Cactoblastis cactorum)*, was introduced in 1925. Within 10 years the cactus had all but vanished. All that now remains are very remote groups of plants, tens of kilometres apart from one another. What is remarkable here, however, is that only on a very small percentage of the plants can one find the moth. The moth is extremely rare, yet without it, hundreds of square kilometres of land would again quickly be covered by the cactus. The message one takes from this example is that even a very rare species can have a massive influence on an ecosystem.

Army ants are another example of a rare species acting as a metapopulation which can have a major impact on an ecosystem. We must be particularly concerned for them as their natural habitat, the tropical rain forest, is being rapidly destroyed.

Tropical rain forests are the richest ecosystems on Earth. For example, when techniques were developed in the 1970s and 1980s for collecting insects in new ways from the canopies of tropical rain forest, guesstimates of the total inventory of the world's biodiversity shot upwards from about 5 million species to 15 million or perhaps even 30 million species (Erwin, 1983, 1991; Stork, 1988, 1993, 1994; May, 1990). Most of this diversity is associated with the species richness of tropical rain-forest trees. For example, a single 50 ha plot of rain forest on Barro Colorado Island in Panama (see later) is home to 186 species of trees (Hubbell & Foster, 1983). Compare this with the estimated 35 native species of trees in the whole of the British Isles (Mitchell, 1974, p. 30).

The riches of tropical rain forest makes their destruction a greater crime. Approximately half of all of the tropical rain forest that graced this planet only a few centuries ago has been cut (Laurance & Bierregaard, 1997).

Almost everything about tropical rain forests is enigmatic. For example, they can be seen as immensely rich ecosystems living amidst potential deserts. Almost all the carbon and nutrients in these ecosystems are tied up in living material. If large swathes of the forest are removed the soil is so poor that long agricultural use is often impossible and the forest is also unable to recover fully (Gradwohl & Greenberg, 1988). Clearly, nutrient recycling is important within tropical rain forest and the focal animals of this chapter, the army ants, are perhaps best seen as top predators of the detritivore food chain (Franks, 1982a), which begins with leaves falling from the canopy of the forest and starting to decay in the leaf litter of the forest floor.

As we will see in the next section, even if just one species of army ant, *Eciton burchelli*, became extinct in the Neotropical rain forests of Central and South America, countless other species would ride with it on a one-way journey to everlasting oblivion. Local extinctions of such species are ever more likely.

Conservation biology in the tropical rain forest has largely become the study of 'Tropical Forest Remnants', to quote the title of a recent book edited by Laurance & Bierregaard (1997). Their use of the term remnants is appropriately redolent of the carnage we have wrought on these ecosystems. The remnant bits of tropical rain forest will be like so many tattered rags that remain after a pan-tropical jumble sale of the world's greatest biodiversity. We face the prospect of trying to rescue the few soiled and ragged fragments of what was once an immense natural richness. These remnants are likely to be all that remains to remind ourselves of our own relentless greed and short-sightedness. We could shrug our shoulders and try to avert our eyes from the problem or we could attempt some form of rescue. This chapter is devoted to the conservation of one magnificent predator in the rain forests of the neotropics. If we learn to preserve it and take the appropriate steps, countless other species are sure to benefit.

A DAY IN THE LIFE OF AN ARMY ANT

Contrary to myth and legend, army ants are not just specialists in carnage, they can be helpful to other animals and can be extremely obliging creatures for study (Franks, 1989; Gotwald, 1995). Their savage raids can maintain diversity (Franks & Bossert, 1983) and for a scientist interested in pred-ator–prey interactions, they are in some ways surprisingly easy to study (Partridge et al., 1996). For example, one can stand within a few cen-timetres of an army ant raid and they will continue their activities regard-less. This means that they can be studied in quantitative details as they go about their business, undisturbed, in their natural habitat. Moreover, one can sit next to the column of ants returning from the swarm raid to the nest (Figure 9.1(c)), or bivouac, and count all the 30 000, or more, prey items (mostly ants, cockroaches, spiders and scorpions) that they capture in a single day (Franks, 1982a). It is the equivalent of being able to sit, unno-ticed, in the throat of a large carnivore and count all of the food it swallows piece by piece. Moreover, because one can watch them wherever they go, it is also comparatively easy to record the spatial and temporal patterns of their raids (Franks & Fletcher, 1983).

The best-studied army ant population is that of *Eciton burchelli* in the lush lowland tropical forest of Barro Colorado Island, Panama (Figure 9.1(f), (j)). This population was studied by T. C. Schneirla from the late 1920s through to the 1950s (Schneirla, 1971), by Carl Rettenmeyer in the 1960s (Rettenmeyer, 1963), by Ed Willis in conjunction with his investiga-tions of the ant-following birds, also in the 1960s (Willis, 1967) and by Nigel R. Franks and his colleagues starting in the late 1970s (Franks, 1989). As a result of such work we know that there have been roughly 50 colonies inhabiting the 15 square kilometres of Barro Colorado Island throughout this time (Franks, 1982b). For an invertebrate, this is an excep-tionally low natural density. Even though colonies may contain 500 000 workers (Franks, 1985) they have only one queen. Hence, there are only about three reproductive individuals per square kilometre. *Eciton burchelli* colonies are rare. Even an experienced observer may take several days

Figure 9.1 (Facing page) (a) A bicoloured antbird (*Gymnopithys bicolor*). (b) How ant-following birds use swarms. Redrawn from Willis & Oniki (1978). (c) Some of the 30 000 prey items returning to the nest. (d) A typical day's swarm raid. Redrawn from Rettenmeyer (1963). (e) Raiding patterns over two statary and one nomadic phase. (f) Map of Barro Colorado Island in Gatun Lake, Panama, showing some raiding trails drawn to scale. (g) A cellular automata model for the army ants. (h) Equations for mathematical model. The first describes the number of colonies in a reserve of size K patches. Kx' denotes the number of colonies and n is the patch recovery time (in cycles). The second equation is the expected time to extinction. T is the expected time to extinction in years, Q expected queen lifetime and a, b and d are constants. (i) Percolation matrix with 40% of patches removed. (j) Barro Colorado Island. (Photographs (a) and (c) by N. R. Franks.)

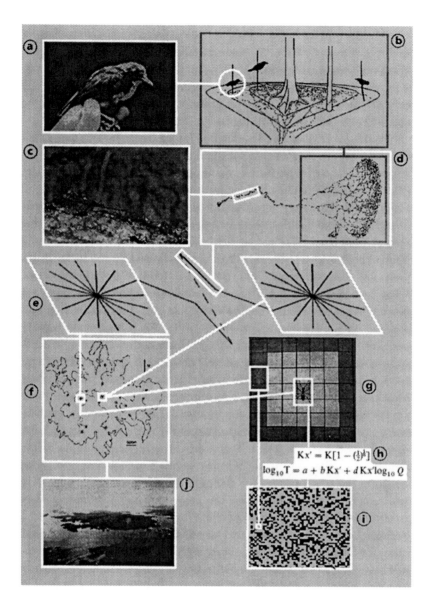

$$Kx' = K[1 - (\tfrac{1}{2})^k]$$
$$\log_{10}T = a + b\,Kx' + d\,Kx'\log_{10}Q$$

tramping the trails of Barro Colorado Island to find a single colony (Franks, 1982b). Discovering a raid is, however, worth all the effort. The biggest raids sweep out an area 200 m long by 20 m wide in a single day (Franks, 1985). Such is the density of raiding army ants at the swarm front that the leaf litter through which they raid is almost completely engulfed in the viscous flow of their dark writhing bodies (Figure 9.1(b),(d)).

The colonies live to a 35-day rhythm. For 20 days they raid out from a single 'statary' bivouac site like the spokes of an immense wheel. For the following 15 days they become nomadic and raid every day and emigrate down that raid path to a new foraging area almost every night (Figure 9.1(e)). These alternating nomadic and statary phases continue relentlessly throughout the whole life of each colony (Schneirla, 1971). They keep precisely to 35 days because their activity cycles are associated with the timing of brood production. In the middle of the 20-day statary period the single queen lays 100 000 eggs. Their embryonic development takes 10 days. When these eggs synchronously hatch into larvae the colony becomes nomadic. So like a good Napoleonic army, the ant army marches on its (social) stomach; the immense single cohort of larvae. The massive food requirements of these larvae explain why the colonies raid every day in the nomadic phase and must find a completely new area to raid every day. These larvae, because they are the same age, terminate their growth and spin their pupal cases in synchrony. This triggers the next statary phase. Pupal development takes 20 days so that the pupae hatch into new workers, ready to march into the next nomadic phase, just as the next cohort of eggs hatch triggering that nomadic phase. The beauty of 35-day activity cycles is that during the statary phase the colony has no brood to feed. The colony can remain in the same bivouac site during the time the queen is vulnerably swollen with eggs and the colony can take a holiday from raiding every other day throughout the statary period.

Colony growth is slow, because colonies lose large numbers of workers through raiding casualties every 35 days (Franks & Partridge, 1993). Approximately 1/3 of the colonies, i.e. the biggest ones, become reproductive each year (Franks, 1985). They still maintain a 35-day cycle but instead of rearing just workers they also rear about 4000 males and six queens (Franks, 1985). Such colonies reproduce in a process of binary fission, the new queens competing with one another and the old queen to head the daughter colonies (Franks & Hölldobler, 1987). Outbreeding is probably the rule because males fly off to mate with queens in other colonies. The queens are never winged and hence colonies cannot invade new islands of habitat that are isolated by water or inhospitable terrain. However, because the males are strong flyers, isolated populations of colonies are not necessarily genetically isolated and therefore inbreeding in otherwise moderately isolated small populations is probably of little importance. Queens may live 6 years and during this time they will lay about 6 million eggs and walk 63 kilometres between bivouac sites (Franks, 1989).

The local impact of raids is immense and because raided areas are slow to recover (Franks, 1982a; Partridge *et al.*, 1996), their impact on the tropical rain-forest leaf litter fauna overall is massive, even though army ant colonies are rare. Franks & Bossert (1983) estimated that about half of the area of Barro Colorado Island is still recovering from an *Eciton burchelli* army ant raid at any one time. The social insect prey, mostly other ants, being less mobile than the other arthropods the army ants eat, are slowest to recover (Franks & Bossert, 1983). Some of these prey ant species appear to be competitively dominant and other ant species can take advantage of their absence, or low density, in recent raided areas. In this way army ants maintain diversity (Franks & Bossert, 1983). In addition, overall throughout Central and South America about 50 species of bird rely on the army ants' raids, to flush their prey out of the leaf litter (Figure 9.1(a),(b)) and would probably go extinct in their absence (Willis & Oniki, 1978). Countless species of invertebrate 'camp followers' of the army ants would also become extinct following the demise of these wonderful Huns and Tartars of the insect world (Franks, 1989; Hölldobler & Wilson, 1990).

In sum, army ant colonies are the Bengal tigers of the insect world. They are rare and magnificent in their own right, and their ecological impact as keystone top predators in the tropical rain forest means they should occupy the first rank as subjects for conservation.

THE ROLE OF MATHEMATICS IN CONSERVATION BIOLOGY

Conservation biology is about understanding how an ecosystem works, how species within it interact, and how they depend on, one another. This means that when one species is threatened, for example through habitat loss, we, as conservation biologists, may be able to step in to minimize the overall damage. Mathematical models can help provide guidelines for such damage limitation. They can make qualitative and quantitative predictions, both in the long and short term, as to how an ecosystem will behave. We

Box 9.1 A simple metapopulation model

Suppose we have a series of patches that can be in one of two states, either occupied or unoccupied by a certain species. Suppose that this species colonizes patches at a constant rate c and that occupied patches become empty (that is the population on this patch dies) at a constant rate m. Let us write p to represent the proportion of occupied patches. Then the rate at which empty patches are colonized is given by $cp(1-p)$, since there are p patches from which the colonization can occur and $1-p$ patches available for colonization. Occupied patches become empty at a rate mp, because each occupied patch has death rate m. Thus we have the following equation describing how the meta-population changes over time

$$\frac{dp}{dt} = cp(1-p) - mp$$

This model is due to Levins (1969). We are interested in the model's steady states, since this is precisely when the metapopulation's size will be constant. That is, we wish to solve the system with $dp/dt = 0$. Writing p^* as the solution at the steady state,

$$p^* = \begin{cases} 1 - \frac{m}{c} & \text{if } c > m \\ 0 & \text{if } c < m \end{cases}$$

We see that provided the colonization rate exceeds the mortality rate, persistence is guaranteed and that the species is rare if m is nearly as high as c. It is possible to extend this model to include predator–prey interactions or multi-species competition simply by increasing the number of equations (Sabelis et al., 1991; Tilman et al., 1994). It is also possible to extend it to a system as complex as the army ants by includ-ing size classes of colonies and recovery rates of raided patches in the model (Britton et al., 1996). It turns out that these models describe the behaviour of real populations reasonably well (Hanski et al., 1994; Holyoak & Lawler, 1996).

need long-term predictions; it is not enough to know that a species in a reserve will exist for the next 5 or 10 years, we need to know if it is likely to exist in several hundred years time. Mathematics is the only way to extrapo-late into the distant future with any hope of predictive power.

The modelling also serves a second purpose, namely it can suggest new experiments to be conducted in the field. This will yield more data and better insights, which in turn can be used to refine the model, which will lead to better experiments and analyses and so on. So, as we repeat such cycles of investigation, we will obtain better knowledge of the system, better models and so better conservation strategies.

Army ants would appear at first to be an incredibly hard system to model. However, because so much is known about their behaviour, the modelling is not as difficult as one might first imagine. This knowledge enables us to judge which features of the enormously complex biological system we can neglect and which we must include. That is, it helps us to determine the essential requirements for army ant persistence.

A mathematical model for army ant dynamics

As Alan Turing said in 1952, a model is 'a simplification and an idealization, and consequently a falsification'. The aim of mathematical modelling is *not* to obtain a system of equations that reflects a natural phenomenon as closely as possible, but to obtain one that is sufficiently complex to reflect the essential features of the phenomenon while being sufficiently simple to allow important general principles to emerge. For example, it is often *better* to use a Newtonian model of dynamics than a relativistic one. The art of mathematical modelling lies in deciding what can safely be omitted.

Since the biology of army ants is so well known, such omission becomes a process of careful abstraction. The habitat consists of a series of patches. During the nomadic phase, a colony will travel through many different patches and so, provided the numbers of good and bad patches are the same, the nomadic phase, for the purposes of modelling, can be considered to make a minimal contribution to changes in colony size. Also, since raids during the nomadic phase only briefly pass through an area, rather than occupy it for a prolonged time, the recovery of patches from nomadic raids become negligible compared to recovery from statary raids. We conclude that the nomadic phase can safely be omitted and we are left considering purely the statary phase. Thus we consider a patch-occupancy model (Caswell & Etter, 1993).

Hence, some features of army ant dynamics that we consider essential are as follows:

- The habitat is a mosaic of patches in different stages of ecological succession.

- A colony becomes larger if it spends its statary phase on a good patch, smaller if it spends it on a bad patch.
- A colony undergoes fission if it becomes sufficiently large.
- A colony dies if it becomes too small.
- A colony dies if its queen dies.

We model the patches as squares whose size is equal to the area typically exploited by a colony in a statary phase, and whose status (good or bad) depends on how long it is since the patch was last exploited. At each time step, equal to the duration of one activity cycle, each colony moves at random to a square at a distance typical of the species (Figure 9.1(g)). It then increases or decreases in size by one size class, depending on whether the patch is good or bad, and may also split or die. The process is repeated indefinitely. The size of the patches, the time for recovery of an exploited patch, the distance moved in a nomadic phase, and the number of ants in a size class are parameters obtained from the biological data (Franks, 1989). For more details see Britton *et al.* (1996).

The model as described is ideally suited to computer simulation and this reveals complex spatial patterns of extinction and re-colonization of habitat patches, with the total number of colonies oscillating about a fixed level. However, it does not lead to expressions for the quantities of interest in terms of the parameters of the system, and we perform some mathematical analysis on a simplified model to achieve this. The first step is to apply the ideas of metapopulation theory outlined in Box 9.1. This gives a deterministic steady state about which the simulation model oscillates, but does not consider the stochastic variations that will eventually drive the population to extinction.

To do this, we think of the population as a gambler, with capital equal to the number of colonies it contains. It wins a toss every time a colony divides, and loses one every time a colony dies. But there is an upper limit to its capital, since there can never be more colonies than there are patches on the island. Hence the population is doomed to extinction, as stated in Box 9.2, and the only question is how long we might expect it to survive before this doom overtakes it. This time, in number of tosses, is determined as outlined in Box 9.2, for a game against an infinitely rich opponent. It is the expected time to extinction of the population. To convert it to a time in years we need to know how frequently, on average, a coin is tossed, i.e. how frequently colonies divide or die. To do this we consider a colony newly formed by fission. It may be thought of as a second kind of gambler entering a new

Box 9.2 Gambler's ruin problems

The simplest problem of this sort is the following. Two gamblers start a coin-tossing game with a pool of £N in cash, Janet having £n and John £$(N-n)$. At each toss of the coin Janet wins one pound from John if the coin falls heads, and John wins one pound from Janet if the coin falls tails. The probability that the coin falls heads is p, tails is $q = 1 - p$. The game ends when one of the gamblers loses all their money. Let the probability that Janet wins the whole game be P_n. Then by considering what happens after one toss it can be seen that

$$P_n = p P_{n+1} + q P_{n-1}$$

and it is clear that $P_0 = 0$, $P_N = 1$. (Janet has lost the game if she has no money, but has won if she has all the money.) This is a simple set of equations to solve for P_n, and the solution is given by

$$P_n = \frac{(q/p)^n - 1}{(q/p)^N - 1} \text{ if } p \neq q, \ P_n = \frac{n}{N} \text{ if } p = q = \tfrac{1}{2}$$

Now let the expected time (number of tosses) that the game takes be given by T_n. Then, by a similar argument,

$$T_n = p T_{n+1} + q T_{n-1} + 1,$$

with $T_0 = 0$, $T_N = 0$. (The game finishes immediately if Janet has no money or all the money.) Again these equations can be solved for T_n, and the solution is given by

$$T_n = \frac{n}{q-p} - \frac{N}{q-p} \frac{(q/p)^n - 1}{(q/p)^N - 1} \text{ if } p \neq q, \ T_n = n(N-n) \text{ if } p = q = \tfrac{1}{2}.$$

We have assumed that p is constant. If this is not so, as in the army ant application, the problems may not have solutions that can be written in such simple forms, but the systems of equations can be solved on a computer.

A variation occurs if Janet again starts with £n, but now plays against a bank with unlimited funds. Then the game can never end in a win for Janet, and must either go on indefinitely or end in her becoming bankrupt. If there is a fixed ceiling to her fortune, the probability that the game ends in her bankruptcy is 1.

game with capital equal to the number of ants it contains. It gambles with these ants, a size class at a time, each time it exploits a new patch, winning the toss if the patch is good and losing it if the patch is bad. Eventually it wins the whole game, by becoming large enough to divide, or loses it, by dying. The expected duration of the game is determined as in Box 9.2, for a game against an opponent with finite funds, and leads to a determination of the expected time between tosses for the population game. The expected time to population extinction can then be calculated in years (Figure 9.1(h)).

Habitat destruction and percolation theory

We shall now consider the problems that occur if some parts of the army ants' habitat are not easily accessible from other parts. The tropical rain forests are being destroyed at an increasingly alarming rate (Hartshorn, 1995; Whitmore, 1997). With habitat loss being the main cause of species extinction (Lawton & May, 1995), this is a major concern. Much of this destruction occurs through logging and as conservation biologists we wish to determine harvesting strategies that will have minimal impact on the rain-forest ecosystem. One such proposed strategy has been to randomly fell strips of woodland comparable in area to the patches used in our mathematical model (Hartshorn, 1989). This attempts to mimic naturally occurring tree-fall gaps and should allow for quicker recovery of the forest (Hubbell & Foster, 1992). Here we will test how such management strategies may affect army ants and, from this, will be able to consider several key conservational issues.

Due to the complex mathematics that would be involved, we have no choice but to resort to simulation on a computer. We consider an island containing 2500 patches, that is a reserve 9 km × 9 km. Then to observe the effects of habitat destruction we randomly remove some of these patches from the lattice. Since *Eciton burchelli* will not venture into unshaded areas for a prolonged time (Willis & Oniki, 1978; N. R. Franks personal observations), these removed patches are truly uninhabitable.

Simulations were conducted for various numbers of removed patches with the average number of colonies being recorded only once the system had settled down (Figure 9.2). We observed persistence to be impossible once 45% of the habitat had been randomly removed.

There is a branch of mathematics, known as percolation theory (see Box 9.3), which deals with how lattices behave under the random removal of sites (Figure 9.1(i)). Using standard results from percolation theory (Stauf-

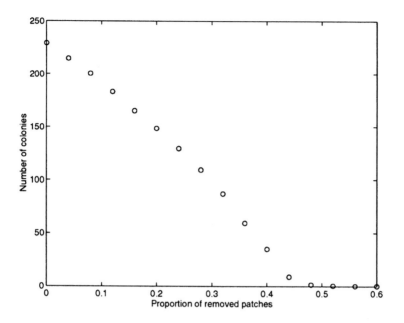

Figure 9.2 The average number of colonies as a function of the proportion of removed patches. Twenty simulations were conducted in each case for various numbers of removed patches from a 50 × 50 lattice. The average number of colonies was recorded on each run only after the system had settled down to a quasi-steady state. (Quasi-steady since the only true steady state is extinction.) Notice that there is a transition from persistence to extinction when the proportion of removed patches reaches 0.45.

fer, 1985), we know that our lattice becomes truly fragmented only when about 60% of the habitat has been randomly removed. We discovered army ant persistence to be impossible when just 45% of the habitat patches were removed, but this occurs when the habitat is still connected. These results are true irrespective of island size. Compare this to the situation where 45% of the habitat is removed, but what remains is left as a large square – this should support on the order of 100 colonies for the island size considered here (Britton *et al.*, 1996). Thus, even if a large habitat is connected persistence cannot be guaranteed.

The reason for this behaviour is surprisingly simple. If we randomly remove about 40–45% of the habitat, the remaining habitat still consists of just one connected cluster (Figure 9.3). This cluster takes the form of lots of small habitat pockets connected to one another by narrow corridors (Figure 9.1(i)). Each of these pockets is too small to support even one colony for a

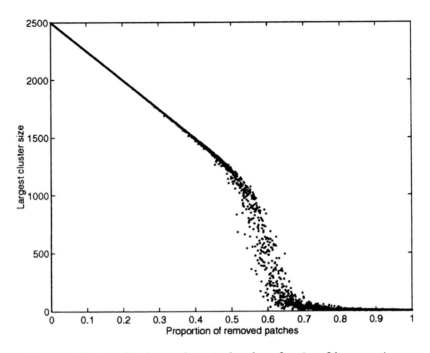

Figure 9.3 The size of the largest cluster is plotted as a function of the proportion of patches removed for a 50 × 50 lattice. Each data point is the average of five runs. Here patches are neighbours of one another in the sense of nearest and next-nearest neighbours. Notice that, when less than 60% of the patches are removed, the lattice consists of just one connected cluster with possibly a few isolated patches. When more than 60% are removed, we obtain many small clusters. This suggests that the critical percolation value occurs when about 60% of the patches are removed. In fact, the critical percolation value in this instance occurs when 59.28% of the patches are removed. For other examples see Stauffer (1985).

considerable time period (Partridge *et al.*, 1996) and so any colony in such a pocket will need to leave quickly. Since the corridors are small (hence hard to find) and the ants do not remember how they arrived in the pocket in the first place, it is pure luck if they escape. However, if the colony is lucky and escapes from this pocket it will find itself in another pocket, facing the same problem. Again, luck is its only hope for survival. At some point, however, the colony's luck is going to run out and it will die.

This result is remarkable. The forest does not even have to be fully fragmented to cause major problems, it just has to be sufficiently difficult to get from one area to another. This has important implications for conservation.

Box 9.3 Percolation theory

Imagine we have an infinite square grid, all the patches coloured white. Let's take a coin which has a probability p of coming up heads and probability $q = 1 - p$ of coming up tails. For each square in our lattice we toss the coin; if it comes up heads (with probability p) the square remains white whereas if it comes up tails (with probability q) the square turns black (Figure 9.1(i)). We now ask whether it is possible to get from any one white square to any other white square via some white 'path'. A path refers to how one is allowed to travel around the lattice; we may be allowed to travel only to the nearest neighbours or more complicated paths may be possible. Percolation theory is concerned with how p affects one's ability to get from any one white square to any other white square via a path of white squares.

We can rephrase this slightly in terms of clusters and cluster sizes. Two white squares are said to be in the same cluster if there exists a white path between them. The size of a cluster is simply the number of white squares it contains. Now percolation theory is concerned with how p affects the average cluster size (see Figure 9.3). It turns out that there is a critical percolation value p_c above which the mean cluster size is infinite and below which it is finite (Stauffer, 1985).

For our purpose we define a path with the nearest and next-nearest neighbour as stepping stones, that is a colony can move straight to any of the eight surrounding patches during the course of its migration. In Figure 9.3 we show that $p_c \sim 0.4$ in this instance.

IMPLICATIONS FOR CONSERVATION

Harvesting strategies

Many sustainable forest harvesting strategies involve randomly felling strips of forest comparable in area to the square patches used in the model (Hartshorn, 1989). We have seen that this approach may have severe consequences if the felling is too widespread, since breakages in the forest caused by harvesting combined with the natural breakages such as rivers and large tree-fall gaps, will begin to fragment the forest. This suggests that a better strategy would be to concentrate harvesting in a large square pocket of rain forest and systematically move into other areas in order to lessen the fragmentation. Admittedly this does not take into consideration forest recovery, since small strips of forest will recover a lot quicker than large square clumps. Clearly, forest recovery will play an important role, since after just 8 to 10 years parts of the environment have recovered sufficiently in a forest fragmentation project near Manaus in Brazil for army ant inhabitance (Stouffer & Bierregaard, 1995). However, since random felling increases the risk of species extinction, forest recovery rates will play a less important role in determining conservation strategies.

Clearly, as conservation biologists, we need to spend a great deal more time and effort in trying to understand the effects of heterogeneous habitats since the results can be counter-intuitive.

Habitat corridors

It is widely believed that the establishment of habitat corridors between reserves will reduce a species' extinction risk (Hobbs, 1992). This seems obvious – by increasing the area of a reserve we increase the number of individuals it contains and so increase the persistence time. We have seen previously, however, that even a very large reserve with poor internal connections is not sufficient for species persistence. Thus, if the corridors are too small they will not allow movement of individuals between habitat pockets so that the corridor will not play a role in extending persistence times. We conclude from this that corridors have to be wide; how wide depends on the biology of the species under consideration. To determine this width we need to understand how an individual finds a corridor, whether it will enter the corridor and then its subsequent behaviour in the corridor.

With regard to army ants, corridors seem to be found by chance, so the entrance to the corridor ought to be reasonably wide. Once in the corridor, given how colonies move during the nomadic phase (with each successive raid and emigration taking much the same compass bearing), a narrow corridor would channel movement and speed up corridor passage (Tishendorf & Wissel, 1997). So an ideal corridor would have wide entrances but may well have a narrower strip of forest in the middle. This shape of corridor may well also suit the antbirds, which are quite happy to follow raids of army ants wherever they may lead (Willis, 1967; Stouffer & Bierregaard, 1995). However, most habitat corridors are species specific (Hobbs, 1992) and whilst the ants and ant-following birds might be content with such corridors, other associates may not. This is a real cause for concern. We must understand how species in an ecosystem rely on one another before any effective conservation strategy can be implemented.

Suitable reserves

When planning a reserve, most important figures to be taken into consideration are expected times to extinction. These depend very strongly on the number of individuals a reserve can support (MacArthur & Wilson, 1967). For example, a reserve so large that army ants would be expected to persist for 10 000 years would need to contain at least seven colonies at equilibrium (Partridge *et al.*, 1996). However, this figure only considers the ant colonies themselves and not any of the associated species, nor does it allow for catastrophes (see below). Certainly antbirds require a much larger reserve in which to function, since on Barro Colorado Island, Panama (Figure 9.1(j)), which is seven times this size, about ten bird species (including one or two ant-following birds) were becoming extinct every decade (Willis & Oniki, 1978) some 50 years after the reserve was isolated from the surrounding mainland.

Clearly, the relationship between army ants and their associates, in particular the ant-following birds, requires further investigation. Mathematics may well be the ideal tool for such studies.

Role of catastrophes

Large reserves supporting many colonies will have long persistence times. For such reserves, the most likely cause of extinction will be catastrophes (Mangel & Tier, 1994). These catastrophes may take several forms, e.g. disease, forest fires or El-Niño induced events. With El-Niño events becoming ever more frequent (Wuethrich, 1995), we need to understand how catastrophes influence the population dynamics of species in order to prevent extinction. One thing is for certain, catastrophes can dramatically reduce expected times to extinction and so any conservation plan must include their effects.

Concluding thoughts

If army ants became extinct many other species would also disappear. Conversely, if large viable populations of army ants can be conserved many other species will also be conserved. (Note however that Barro Colorado Island has viable army ant populations.) Conservation of army ants and their followers will require large tracts of continuous rain forest or much larger tracts of forest fragments linked by broad corridors. It would be a distinct mistake to consider army ants as especially poor dispersers and hence particularly vulnerable to habitat patchiness. The ant-following birds, for example, will not cross more than a few metres of water (Willis & Oniki, 1978) and various tree species are known to be poor dispersers (Laurance & Bierregaard, 1997). Preserving what remains of the tropical rain forest will require expert studies both in field work and mathematical modelling in addition to a major change in the way we treat the natural world.

ACKNOWLEDGEMENTS

G.P.B. would like to thank the Engineering and Physical Sciences Research Council for a studentship. N.R.F. wishes to thank the Smithsonian Tropical Research Institute, Panama, for their gracious provision of facilities on Barro Colorado Island.

PART IV

Sexual selection, threats and population viability

Sexual selection and conservation

ANDERS PAPE MØLLER

INTRODUCTION

Sexual selection arises from competition among individuals for mates and ultimately for fertilization of gametes. The exaggeration of secondary sexual characters and other displays imposes costs on individuals of the chosen sex, and thereby causes a reduction in mean fitness of individuals of a population. The costs of sexual selection and demographic stochasticity arising from mating preferences and the effects of small population sizes have important consequences for the probability of extinction. Conservation strategies implemented by determination of minimum viable population sizes and sizes of nature reserves necessary to ensure sustainable populations for a given period of time have to take sexual selection into account as an important factor determining the risk of extinction.

Sexual selection is defined as the evolutionary process arising from competition among individuals for access to mates and their gametes (Andersson, 1994). The outcomes of sexual selection are the evolution of costly and apparently wasteful secondary sexual characters and other displays. Competition among individuals of the same sex, usually males, has given rise to large male body size, but also the evolution of weaponry such as horns in beetles and antelopes, antlers in deer, and spurs in pheasants, while choice by females of attractive males has resulted in the evolution of pheromones in insects, elaborate calls in insects, frogs, birds and mammals, extravagant feathers in birds, and bright coloration in many different groups of animals. The origin and maintenance of such extravagant display is hotly disputed, with traits either being reliable indicators of direct fitness benefits, indirect viability benefits, arbitrarily attractive, or attractive because of so-called pre-existing biases (review in Andersson, 1994). Common to all these models of sexual selection is that extreme ornamentation, that is generated by female mate preferences or male–male competition, is beneficial through its effect on enhanced mating success of certain individuals, but is otherwise detrimental to the mean fitness of individuals in a population. This has important implications for the probability of extinction and hence for conservation.

This review considers four aspects of conservation and sexual selection: (1) the fitness costs arising from sexual selection and their consequences; (2) Cope's rule and sexual selection; (3) introduction success and sexual selection; (4) speciation, extinction, and sexual selection. The review finishes by introducing some practical conservation consequences of sexual selection and a discussion of future prospects of conservation biology and sexual selection.

COSTS OF SEXUAL SELECTION

All models of sexual selection predict that the evolution of extravagant displays displace males from their survival optimum. For example, according to the Fisherian model of arbitrary attractiveness the male trait continues to become exaggerated until the benefits in terms of attractiveness are balanced by the costs of the sexual display (Fisher, 1930; Pomiankowski et al., 1991). Similarly for viability indicator models of sexual selection the male trait becomes exaggerated in a condition-dependent fashion so that benefits in terms of mating benefits are balanced by the costs of display (Heywood, 1989; Iwasa et al., 1991). Hence all models of sexual selection predict costs of sexual display that have to be paid for by reductions in other components of fitness.

Although there is an immense literature on sexual selection, the empirical findings concerning costs of sexual displays are relatively limited. This does not necessarily imply that such costs are unimportant, but rather that this particular subject of sexual selection has received relatively little attention. The mere size of many secondary sexual characters makes them potentially costly to carry, and this may affect foraging ability and escape from predators. Experimental manipulation of tail length has been shown to affect flight ability and foraging success in birds (Møller, 1989; Evans & Thomas, 1992; Møller et al., 1995). High levels of activity and heavy burdens impose metabolic costs on individuals, and such costs are partially compensated by increased ability of oxygen transportation. Hematocrit is a measure of the relative amount of red blood cells in the blood, and high hematocrit values are directly related to high activity levels. Tail length in males of the sexually dimorphic barn swallow *Hirundo rustica* is directly related to hematocrit, both for natural tail length and for experimentally manipulated tails (Saino *et al.*, 1997b,c).

Costs of locomotion, but also mere increased attractiveness of males with exaggerated traits, increases the risks of predation (Endler, 1980; review in Magnhagen, 1991). Although predation costs of signalling have been reported in several invertebrates and vertebrates (Magnhagen, 1991), there is only a single study on who gets eaten and who survives. Male barn swallows with long and symmetrical tails were more likely to survive than males with short and asymmetrical tails (Møller & Nielsen, 1997).

Since production and maintenance of secondary sexual characters and the production of sexual displays competes for limited amounts of energy with other activities, sexual selection may result in trade-offs with other factors affecting fitness. A particularly important trade-off concerns the use of energy for sexual display and anti-parasite defences such as immune function (Folstad & Karter, 1992; Møller & Saino, 1994). Experimental exaggeration of secondary sexual characters reduces responses to a standard challenge test of the immune system, although the response is positively associated with the magnitude of the sexual display (Saino & Møller, 1996). Therefore, increased risk of parasitism is a common cost of sexual display in a diverse array of invertebrates and vertebrates (Cade, 1975; review in Clayton, 1991; Møller *et al.*, 1999).

If sexual displays are costly, we should expect more exaggerated displays to reduce survival prospects while reduced displays should increase the survival rate. A series of experiments in three populations of barn swallows have demonstrated that survival prospects of males indeed are inversely related to experimental tail length (Møller & de Lope, 1994; Saino *et al.*, 1997a). These experiments also show that the survival cost of a long tail is condition dependent; males with naturally long tails are better able to cope with tail elongation than males with short tails, while tail shortening particularly improves the survival prospects of naturally short-tailed males (Møller & de Lope, 1994).

The effect of male ornamentation on survival prospects have also been demonstrated using a comparative approach. The survival rate of male birds is reduced in the presence of extravagant coloration compared to that of females, even when males do not provide any parental care as in ducks (Promislow *et al.*, 1992, 1994; Owens & Bennett, 1994). Hence sexual differences in appearance covary with sex differences in survival.

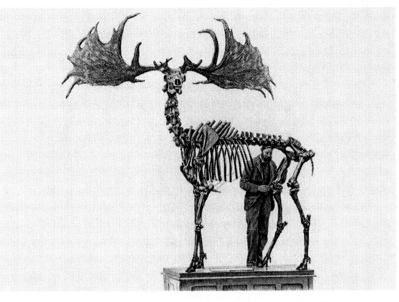

Figure 10.1 Sexually dimorphic species have a higher probability of extinction than monomorphic species. This may be because the costs associated with sexually selected traits make them more vulnerable. The Irish elk, *Megaloceros giganteus*, had the most extreme sexual dimorphism of any known cervid and became extinct about 11 000 BP (the Zoological Society of London).

Given that males incur substantial fitness costs of their sexual displays, species with particularly extravagant displays should suffer from a reduction in mean fitness compared to less sexually selected species. A reduction in mean population fitness caused by sexual selection will result in a significant selective load that under a range of circumstances may increase the risk of local or global extinction (Tanaka, 1996). During periods of environmental hardship, when costs of sexual displays are likely to be particularly high, we should expect mean population fitness to be particularly reduced. This is also the period when reproductive rates reach a minimum. Hence extinction may follow during extensive periods of adverse conditions.

Sexual selection can be considered an extreme case of specialization, with male ornamentation being a specialized means of mate attraction (Parsons, 1995). Extreme specialization generates species that are particularly good at doing particular tasks, but such extreme specialization may also impose a risk in terms of extinction (Figure 10.1). Studies of the causes of extinction have considered specialization to be one such important factor (Jablonski, 1987; Parsons, 1993; Hoffmann & Parsons, 1997).

COPE'S RULE AND SEXUAL SELECTION

A commonly observed evolutionary trend in the fossil record of mammals and other taxa is that the body size continuously increases within lineages, and the risk of extinction increases concomitantly with body size (Cope, 1896; Eisenberg, 1981). These trends are referred to as Cope's rule. Recently, a mechanism has been proposed for the increasing body size in lineages represented by Cope's rule (McLain, 1993). Because one mechanism of sexual selection is competition among males for access to females, males of larger body size are generally favoured over small-sized males in competition for mates (review in Andersson, 1994). Thus the trend for increased body size in lineages could be generated by sexual selection.

The second part of Cope's rule concerns the increased risk of extinction with increasing body size, which could simply be a consequence of large-sized species having lower population sizes and hence greater risk of extinction owing to demographic stochasticity. Alternatively, the increased risk of extinction could be a result of the increased costs of sexual selection within lineages. McLain (1993) tested these predictions by using information on body size of different mammalian lineages and the longevity of these lineages. There was a clear tendency for increasing body size within a lineage in most of the taxa investigated, and the mean longevity of small taxa within lineages was considerably larger than the mean of large-sized taxa (Figure 10.2).

Recently, a very large data set on trends in body size and extinction in molluscs has been analysed by Jablonski (1997). There was no general trend for an increase in body size, and it was concluded that Cope's rule does not apply. Gould (1997) argued in a 'News and views' article that Cope's rule, therefore, is not a rule. This conclusion may be premature for at least two different reasons. First, molluscs may be different from many other organisms by almost exclusively being hermaphrodites. Sexual selection is particularly weak in hermaphrodites because the costs of sexual selection are paid immediately, while the costs of sexual selection in species with separate sexes are not paid by females. Hence, if sexual selection is playing an important role in generating the patterns characteristic of Cope's rule, we should expect to see no or weak effects for molluscs, as was

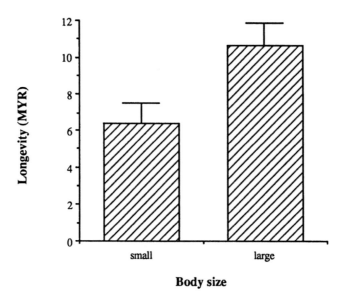

Figure 10.2 Mean (SE) longevity in millions of years in genera of small and large body sized mammals belonging to five different families. Adapted from McLain (1993).

indeed the case. Second, although large-sized males generally are favoured by sexual selection, there are many taxa where small males are at a selective advantage (review in Andersson, 1994). We should only expect Cope's rule to apply in lineages, where there is a selective advantage for large males, if sexual selection is the mechanism generating Cope's rule.

SEXUAL SELECTION AND INTRODUCTION SUCCESS

If sexual selection has important consequences for the extinction probability of species, we could predict the introduction of species to novel sites to disfavour species that are currently subject to intense sexual selection. This appears to be the case.

Bird species have repeatedly been introduced to oceanic islands such as those of the Pacific Ocean. In an analysis of the success of introductions of birds to Tahiti and Oahu (Hawaiian Archipelago), McLain et al. (1995) found that sexually dichromatic species were considerably less likely to become successfully established than monochromatic species (Figure 10.3).

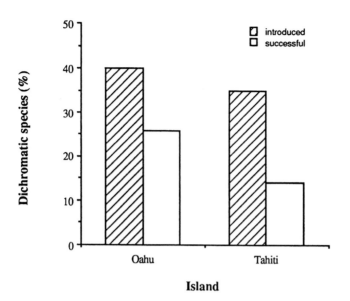

Figure 10.3 Sexual dichromatism among introduced and successfully established bird species on Oahu and Tahiti. Adapted after McLain *et al.* (1995).

Bright male coloration in sexually dichromatic species is known to be sub-ject to current mate preferences, while that is apparently not the case for coloration in monochromatic species (Andersson, 1994). Furthermore, the degree of male brightness compared to that of females is known to give rise to a reduction in male survival prospects (Promislow *et al.*, 1992, 1994; Owens & Bennett, 1994). Although the study by McLain *et al.* (1995) con-trolled for a number of potentially confounding variables, the size of the inoculum was not controlled. A subsequent analysis of the determinants of introduction success for birds released on New Zealand showed sexual dichromatism was again an important determinant of success, even after controlling for the number of individuals introduced (Sorci *et al.*, 1998).

What causes this differential introduction success of sexually mono- and dichromatic species? At least two possibilities exist. First, the costs of sexual selection may render dichromatic species less likely to succeed, as described above. Second, demographic stochasticity may be particularly im-portant in species subject to intense sexual selection and therefore fre-quently result in extinction. Demographic stochasticity arises as a consequence of the dictatorship of small numbers. In a small population all individuals may, for stochastic reasons, end up being males or females, and the population is then doomed to extinction (Goodman, 1987). In a similar vein, all surviving males of a particular species that were released on New Zealand may have been unattractive, or the surviving females may have had particularly strong mate preferences for extravagantly adorned males. Such random effects may lead to increased risks of extinction.

The data set on the introduction success of birds released on New Zealand was used to test several models of demographic stochasticity, using a standard computer programme for calculating the population demographic consequences of the behaviour of individuals (Legendre & Clobert, 1995; Ferrière et al., 1996). Demographic stochasticity arises from the fact that individuals survive and reproduce in a random manner, and demographic parameters like survival, fecundity, probability of mating and sex ratio have average values at the population level, but are realized randomly at the individual level. For example, chance realization of demography may result in small populations sometimes only being composed of males, or a season with average low fecundity being followed by a high mortality rate, and this may result in extinction for random reasons. Demographic stochasticity has recently been suggested to play an important role in population extinction (Gabriel & Burger, 1992; Mode, 1995; Kokko & Ebenhard, 1996).

We used a standard demography programme to test for the effects of demographic stochasticity under a range of different conditions. Using the population sizes of birds released on New Zealand, we were able to demonstrate using realistic survival, fecundity and sex ratio parameters that the patterns of introduction success could to a large extent be accounted for by demographic stochasticity (Legendre et al., 1999). We also varied the mating system in an attempt to investigate its effects on the probability of extinction. A one-sex model resulted in a much lower extinction probability than a two-sex model, simply because populations sometimes are composed entirely of individuals of one sex owing to random variation in sex ratio (Legendre et al., 1999). Interestingly, the probability of extinction was considerably higher in a strictly monogamous than in a polygynous mating system. The reason for this finding is that because of random events, populations more often consist of unmated, non-reproducing individuals in a monogamous than in a polygynous mating system (Legendre et al., 1999). This situation is even more extreme in a lekking mating system, with very low probabilities of extinction, because one or a few males may copulate with the entire population of females (Legendre et al., 1999). Obviously, this may have other negative effects such as inbreeding that were not treated in the simulations. The consequence of female choosiness on the probability of extinction was severe. If, for example, 10% of the females only mate with males above a certain threshold, and if such males are unavailable these females will not reproduce, then the probability of extinction for demographic reasons is extremely high (Legendre *et al.*, 1999). This situation with a 10% fraction of the females being choosy is far from extreme, because many studies have shown much greater proportions of non-reproducing females, if males and females are allocated randomly to each other. Demographic stochasticity may therefore have important consequences for the probability of extinction directly related to the mating system and female choosiness.

SPECIATION, EXTINCTION AND SEXUAL SELECTION

Finally, there is an indirect argument why sexual selection should increase the risk of extinction. If sexual selection disfavours particular taxa by increased risks of extinction, we should expect oppositely directed forces to counterbalance this trend. One such force is the role of sexual selection in speciation. If sexual selection continuously gave rise to the generation of new species, but simultaneously removed such species by extinction, then we should not be able to detect any change in the proportion of extravagantly ornamented species.

Sexual selection has been predicted to give rise to speciation since Darwin's days (review in Andersson, 1994). The theoretical arguments suggest that intense sexual selection may give rise to rapid divergence among populations. Such divergence may eventually result in taxa that are reproductively isolated. Theoretical arguments based on both indicator mechanisms and arbitrary attractiveness models of sexual selection have been suggested to promote speciation (Fisher, 1930; Lande, 1981; West-Eberhard, 1983; Schluter & Price, 1993). Although several case studies have been presented as providing evidence for sexual selection and speciation being related, we can only make inferences about the role of sexual selection in speciation, if a large number of statistically independent evolutionary events of intense sexual selection gives rise to diversification more often than predicted by chance.

The empirical evidence for speciation being associated with sexual selection derives from three studies of birds. A comparative study of species richness in sister tribes of birds differing in the frequency of sexual dichromatism revealed a consistently larger species richness in the more dich-romatic tribes (Barraclough *et al.*, 1995). A second comparative study inves-tigated species richness in pairs of bird taxa with and without a lekking mating system, which is presumed to give rise to particularly intense sexual selection (Mitra *et al.*, 1996). There were generally more species in the lekking taxa, although the result was on the borderline of significance de-pending on the phylogeny used. The third comparative study was based on statistically independent evolutionary events of extravagant feather orna-ments in birds (Møller & Cuervo, 1998). Pairs of closely related genera with and without species with feather ornaments differed in species richness, with the genera with ornamented species having consistently more species. Furthermore, the difference in species richness between pairs of genera was positively related to the proportion of ornamented species. The in-crease in species richness caused by the presence of extravagant ornamen-tation was also associated with social mating system: the increase in species richness was consistently greater in lekking species as compared to monog-amous species, as predicted by Mitra *et al.* (1996). Finally, ornamented taxa consistently had more subspecies than non-ornamented sister taxa, which suggests that sexual selection indeed is associated with divergence among populations. Thus the available studies suggest that speciation in fact is, promoted by intense sexual selection.

CONSERVATION CONSEQUENCES OF SEXUAL SELECTION

The previous paragraphs have briefly reviewed the evidence concerning sexual selection, extinction and conservation. Which are the practical con-servation consequences of sexual selection? We may immediately draw two consequences.

First, the results concerning extinction of introduced species of birds on oceanic islands suggest that the minimum viable population size for in-tensely sexually selected species is considerably larger than that for less intensely sexually selected species. This implies that nature reserves de-signed for the protection of particularly threatened species need to be larger for intensely sexually selected species.

Second, many species are now so threatened that major parts of popula-tions are kept in zoos or other protected breeding sites. Such captive breed-ing is maintained with the long-term aim of reintroduction and release of species. Such reintroductions must consist of a larger number of individuals, if intensely sexually selected species are involved, in order to reduce the risk of introduction failure.

FUTURE PROSPECTS

We currently have extensive information on the occurrence and strength of sexual selection in many different organisms. However, we know relatively little about the costs of sexual selection, and the consequences for fitness at the population level. We have also only the most meagre understanding of the conservation status of species in relation to sexual selection. The studies reviewed here provide some clues to the importance of sexual selection in conservation. However, a better understanding of these results depends on an understanding of the mechanisms generating such differences in conservation status. For example, how important is parasite-driven sexual selection, and what are the consequences for fitness costs of sexual selection in relation to conservation?

There are still important problems that particularly need attention concerning the causes of extinction. Recent studies have demonstrated that extinctions and conservation status are not randomly distributed among species of birds, implying that particular taxa run an elevated risk (Bennett & Owens, 1997). Furthermore, body size and clutch size are associated with increased risks of being threatened (Bennett & Owens, 1997). Both these effects might arise as a consequence of sexual selection. More recent studies have indicated that sexual dichromatism, which reflects sexual selection, and body size indeed are associated with elevated risks of extinction in birds (A. P. Møller, unpublished data). We need more comparative studies of the causes of extinction and the correlates of threatened conservation status in different animals. Such studies might also benefit from analyses of the success of introductions of species that have already gone extinct.

ACKNOWLEDGEMENTS

My research was supported by a grant from the Danish Natural Science Research Council.

11

Dispersal patterns, social organization and population viability

SARAH DURANT

INTRODUCTION

While dispersal patterns and social organization undoubtedly exert an influence upon population dynamics and genetics and hence are likely to affect population viability, there have been few attempts to estimate the impact of these behaviours. In this chapter simulation modelling is used to address this gap in our knowledge. First a general population viability analysis (PVA) is outlined, secondly the possible impact of differences in behaviour patterns on a PVA is discussed, and finally simulation modelling is used to show how such differences might impact populations. This chapter focuses on mammalian systems.

Population viability analysis

All populations are subject to fluctuations in numbers that result from random variation in their demography and environment. Even in a constant environment, rates of reproduction and mortality vary within and between individuals. This variation among individuals is termed demographic stochasticity (Gilpin & Soulé, 1986). Demographic stochasticity may cause a small population to go extinct because of a chance sequence of high mortality, low fecundity or skewed sex ratios. It is most important for small populations and decreases in significance as population size increases. Environmental stochasticity refers to variation which is not intrinsic, but results from changes in the external environment (Shaffer & Samson, 1985; Gilpin & Soulé, 1986; Goodman, 1987). Examples of sources of environmental stochasticity include epizootics and droughts. Generally environmental stochasticity removes a fixed proportion of the population regardless of population size (Goodman, 1987) and therefore its impact stays constant as population size increases. The concept of a minimum viable population (MVP) was originally devised to quantify the minimum size of a population that is viable under one or both of these stochasticies. It has been defined as the smallest isolated population with a 95% chance of persistence over 100 years (Shaffer, 1983).

Small populations may also be afflicted by genetic problems. These fall into two broad categories: inbreeding and loss of genetic variance (Miller, 1979; Gilpin & Soulé, 1986; Caughley & Gunn, 1996). Inbreeding occurs when individuals mate with their close relatives, and is most likely to occur in small and isolated populations, when dispersal is limited. Matings between relatives lead to a decrease in the heterozygosity of offspring and an increased likelihood that homozygous lethal recessives will be expressed (Vrijenhoek, 1985; Ledig, 1986; Ralls et al., 1986). This has been shown to result in high juvenile mortality and poor reproduction across a wide range of captive (Bouman, 1977; Shoemaker, A. H., 1982; Templeton & Read, 1984; Templeton, 1987; Ralls et al., 1988) and wild populations (Greenwood & Harvey, 1978; Packer, 1979; Jimenez et al., 1994; Keller et al., 1994).

In small populations genetic variance is also lost as a consequence of the random sampling of genes between generations (Wright, 1931). This phenomenon is known as genetic drift and leads to decreases in measures of genetic variance, such as allelic diversity and heterozygosity (Miller, 1979). There are few data on the long-term effects of low levels of genetic variance, but theory suggests that it may reduce a population's ability to respond to natural selection (Dobzhansky & Wallace, 1953; Franklin, 1980) and increase susceptibility to disease (O'Brien & Evermann, 1988). Once genetic variance is reduced it is not easily restored and can only be regained by mutation or immigration. Mutation rates for mammals are estimated to be very low, ranging between 10^{-6} to 10^{-4} per gene per generation (Hedrick, 1983; Strickberger, 1985). Therefore it has been generally assumed that the best policy when conserving a species is the preservation of maximum genetic variation (Benirschke, 1977). The recommended aim for captive and wild populations is the preservation of 90% of the original quantitative genetic variation over 200 years (Foose et al., 1986; Ralls & Ballou, 1986; Soulé et al., 1986). This figure has become accepted for estimating the size of a genetic MVP.

The relative importance of demographic versus genetic considerations within conservation biology has been the topic of much debate within the scientific literature (Lande, 1988a; Caro & Laurenson, 1994; Caughley,

1994; Lacy, 1997). Genetic concerns have been long supported by the established field of population genetics, whilst demographic concerns have attracted more recent interest. Therefore the higher relative abundance of research addressing genetic problems within conservation biology might appear to exaggerate its importance. There is currently only a single well-documented example where genetic considerations have been shown to be at least partly responsible for the extinction of populations, that of the Glanville fritillary butterfly in Finland (Saccheri et al., 1998). The example of the Florida panther is a further demonstration that genetic problems can drive populations to very low levels, from which they may not be able to recover without management intervention (Roelke et al., 1993). Genetics and demography are interrelated, and the interactions between these two factors have been little explored and yet may have great potential to impact populations; it is therefore advisable for both to be taken into account within any conservation plan.

Population viability analysis (PVA) is a process of examining the viability of a particular population taking into account all the various risk factors (demographic and genetic) that may affect that population. PVAs are generally conducted by computer simulation using various purpose-built computer programs such as Vortex (Lacy, 1987), Gapps (Harris et al., 1986) and RAMAS (Ferson & Akçakaya, 1990). However these models generally neglect the impact of different breeding systems and dispersal behaviour upon factors causing extinction. These behavioural factors are likely to affect strongly both population dynamics and genetics, and hence extinction risk.

Behaviour and population viability

In general, wild populations are not homogeneous, but form distinct breeding units that may be separated geographically, socially or even ecologically. These sub-populations are unlikely to be completely isolated (unless the intervening terrain between sub-populations cannot be crossed) and individuals will disperse from their natal group to join other groups or to found new groups. These movements will inevitably affect the dynamics and genetics of populations. In several species, particularly small mammals, dispersal helps regulate populations (Myers & Krebs, 1971), because in these systems dispersers may suffer heavy mortalities and, even if they survive, they may be unable to establish themselves within a new breeding unit. In small populations there is also a chance that dispersers leave behind individuals of the opposite gender without a mate or move to new locations where there may be no other individuals with which to reproduce. In such a situation individuals lose opportunities to reproduce and dispersal effectively acts as an extra mortality on the population. However dispersal may also help to avoid local extinctions, such as those caused by over-predation (Hilborn, 1975), and may aid recolonization. In addition, dispersal, by mixing a meta-population, decreases genetic differentiation between breeding units. Frequent local extinctions and recolonizations reinforce this effect (Slatkin, 1987).

Within each sub-population breeding systems may additionally affect both the persistence and the genetics of populations. There are three main categories of mating system: monogamy, polygyny and polyandry (Ridley, 1986). Monogamy is rare in mammals, and polygyny, combined with some degree of polyandry, is typical. Generally, under polygyny, a proportion of the males in the population do not breed, because only the strongest and most competitive males have access to females, leading to a high variance in male reproductive success (Clutton-Brock, 1988). For example, the polygyny of northern elephant seals can lead to one male mating with as many as 90 females (Le Boeuf & Reiter, 1988). Such a breeding system strongly affects genetic diversity, because very few males contribute to the gene pool (e.g. Hill, 1979; Lande, 1987). In addition, larger group sizes may increase interference, and hence lower foraging rates of individuals in the group. This in turn can lead to density-dependent mortality and can hence affect population regulation (Sutherland, 1996a).

As yet there have been very few studies investigating the role social structure and dispersal exert upon the persistence of real populations. One important exception is a study conducted by Forney & Gilpin (1989) on two species of *Drosophila*. By comparing three experimental population structures this study demonstrated that panmictic populations persisted longer than subdivided populations. However one species showed a higher propensity to migrate, and persisted well when migration was possible, while the other did not migrate and hence did not benefit from dispersal opportunities. This experiment neatly demonstrated the effect that population subdivision, migration and the species-specific propensity to migrate have upon persistence.

There is a common argument that spatial heterogeneity and hence population patchiness enhances population persistence (Wiens, 1985). This argument rests on the concept of the *'spreading of risk'* (den Boer, 1968), which postulates that extinction is less likely to occur in several sub-populations at once than in a single panmictic population. Such a hypothesis can only be valid when extinction factors operate independently between sub-populations because it depends on some sub-populations escaping detrimental environmental conditions that might wipe out or greatly reduce other sub-populations. Demographic stochasticity, because it depends on variation between individuals, could act in this way, but because of its inverse relationship with population size is unlikely to be strong enough to result in subdivision benefits. Certain forms of environmental stochasticity, which is uncorrelated between sub-populations, are more likely to exert such an effect.

Computer simulation models have been used to investigate the consequences of meta-population structure on extinction and genetic diversity. By modelling a single polymorphic locus, Boecklen (1986) predicted that subdivision would decrease heterozygosity in the short term but would increase it in the longer term, provided migration occurred between sub-populations. However, this model assumed that sub-populations remained at a constant size, and so its results may not apply to sub-populations that fluctuate in size. Stelter *et al.* (1997), in a model of meta-population structure for grasshoppers, showed that the frequency of catastrophes, in the form of flooding, had a critical impact on the persistence of populations. Vucetich *et al.* (1997) demonstrated the impact of reproductive suppression within wolf societies upon chance of extinction for this species. However, despite their importance for the development of species-specific conservation strategies, examples of detailed models that incorporate extinction dynamics, behaviour and meta-population structure are rare.

The remainder of this chapter addresses three hypotheses that are generated from the expectation that population structure affects the viability of populations. First, that population structure affects persistence. Second, that population structure affects the genetics of populations. Finally, that populations can benefit by subdivision through the spreading of risk. This will be carried out by making use of computer simulation models that are examined through four case studies. These studies are restricted to four species with very different social organizations and demographies: the Mediterranean monk seal, where migration is random within meta-populations; the cheetah, which has very high juvenile mortality and thus has a very different demography to the other species; the mountain gorilla, which has organized breeding groups with non-random migration; and the roan antelope, which has organized stable female harems each with a single male.

Table 11.1
Demographic parameters of species covered in the text (overall multiplicative growth rate for all four species equals 1)

	Monk seal	Cheetah	Gorilla	Roan
Age of first reproduction	5^a	2^c	9^d	3^g
Longevity	30^a	12^c	35^d	20^f
Annual production of offspring	0.500^b	5.226^c	0.148^e	0.400^g
Juvenile survival	0.844^b	—	0.977^f	—
Year 1	—	0.093^{c*}	—	0.835^g
Year 2	—	0.832^c	—	0.580^g
Year 3	—	—	—	0.923^{g*}
Adult survival	0.900^b	0.874^c	0.960^f	0.940^g

[a](Kenyon, 1981); [b](Durant & Harwood, 1992); [c](Kelly et al., 1998); [d](Harcourt et al., 1980); [e](Harcourt et al., 1981); [f](Durant, 1998a); [g](Beudels et al., 1992); * estimates were adjusted slightly from those measured in order to give a growth rate of 1, for comparison across all species.

THE MODELS

The models used in this chapter are developed from one described in detail elsewhere (Beudels et al., 1992; Durant & Harwood, 1992; Durant & Mace, 1994; Durant, 1998a). In the basic model the number of births and deaths were assumed to follow a binomial distribution, where the probability of giving birth or dying is equal to the corresponding mean demographic rates estimated from field studies. If the population reaches a certain level, termed the population ceiling, then reproduction is suppressed in order to prevent numbers exceeding this level. The parameters included in this model are mean annual survival for each yearly age class, mean reproductive rates, the age of first reproduction, longevity and the population ceiling. The demographic rates for the different species in this study were collected from the literature and are presented in Table 11.1.

Environmental stochasticity was incorporated as a series of discrete disasters. These disasters reduce the mean survival or birth rate to a low value during the year in which they occur. The number of survivors or offspring can then be calculated from a binomial distribution, where the mean rate of survival or birth in a disaster year is substituted for that in a disaster-free year. Therefore, numbers of survivors and births are still distributed binomially in a disaster year but their means are lower than in a disaster-free year. It was assumed that there was a constant probability that a disaster would occur in any particular time interval resulting in an exponential distribution of times between disasters. Once a single disaster occurred in a year, birth or survival rates could not be further reduced by additional disasters.

Two types of disaster were simulated. First, *frequent reproductive* disasters, which act on the birth rate and not the survival rate, and are therefore generally weaker in effect (Durant & Harwood, 1992), but are likely to occur fairly frequently within the population. These disasters occurred with a mean frequency of one every 5 years and reduced the birth rate to zero when they occurred. Second *infrequent survival* disasters, which were assumed to occur at a mean frequency of once every 50 years, and reduce survival across all age classes by 50%. Severe epizootics, such as phocine distemper or rinderpest, may follow such a course (Harwood & Hall, 1990). When disasters were simulated the survival or birth rate during disaster-free years were adjusted so that the long-term mean survival or birth rate was the same as that under only demographic stochasticity (M. J. Kelly & S. M. Durant, unpublished data).

The model monitors genetic diversity by tracking a single locus in each individual in the population. At the start of every simulation each individual was allocated two alleles out of a possible ten, using the approach of Boecklen (1986). It was assumed there was no selection acting on the locus under investigation. Results were obtained from 1000 simulations. Each sub-population was initialized at a stable age distribution, and alleles were allocated in equal proportions at random to individuals. When different levels of subdivision were simulated the population ceiling was set on each sub-population so that the overall meta-population would not exceed that of an equivalent panmictic population.

The random migration model

In this model it was assumed that migration within subdivided populations was random and all individuals had the same probability of migrating in any particular year. This probability is termed the *migration probability*. Whether or not an individual migrated was established by sampling from a uniform distribution on the interval (0,1) using a random-number generator. If the random number sampled was less than the migration probability then the individual attempted to migrate, otherwise it remained. The number of emigrations from each sub-population therefore had a binomial distribution, which approximated to the Poisson for low migration probabilities. The destination of a migrating individual was chosen randomly with equal probability from all sub-populations, including those with no resident individuals. If the destination sub-population was at its ceiling level then immigration was not allowed unless a sub-population with space could not be found within 50 attempts. Thus it was possible, although rare, for a sub-population to exceed its ceiling level because of immigration, however there would be no reproduction until it dropped below this level. This model was used to simulate the dynamics of monk seal and cheetah populations.

The Mediterranean monk seal

The monk seal population in the Mediterranean is highly subdivided, because it is scattered around numerous islands and rocky coastlines (Marchessaux, 1989). There are few data on individual movements in this species, however it is known that individual seals are capable of moving over distances of many kilometres. It is therefore likely that individual monk seals move between sub-populations, but the rates of movement are unknown. There is no apparent social cohesion within population centres and so it is likely that movement patterns are random, and the population structure fits that of the random migration model.

The main centre of the monk seal distribution is in the Aegean sea in Greek and Turkish waters, with a population estimated at between 200 and 300 individuals (Sergeant et al., 1978; Marchessaux, 1989). This population is subdivided into small sub-populations which probably contain fewer than 20 seals (Marchessaux, 1989). For this study the highest estimate of 300 seals was chosen as the initial population size, and the most likely population structure is assumed to be that of a meta-population made up of 30 sub-populations.

The cheetah

Compared to the other species investigated in this chapter, cheetahs have unusually high levels of juvenile mortality (Table 11.1); over 90% of cubs die before independence (Laurenson, 1994). When cheetahs occur across large stretches of protected habitat, such as the Serengeti National Park in Tanzania, the populations are largely panmictic (Caro & Durant, 1995). However, cheetahs are increasingly becoming confined to small national parks or large game ranches managed by 'cheetah friendly' ranchers. In these circumstances the population can form a meta-population structure, where small populations cluster around scattered patches of favourable habitat separated by areas of unfavourable habitat. They may be able to move across these unfavourable patches without incident, but they would be unable to survive within them over the long term.

For direct comparison with the other species investigated here an initial population size of 300 cheetahs was used for simulation. The population structure modelled was that of a meta-population made up of 15 sub-populations, each containing 20 individuals, a size that contains five adult cheetahs under a stable age structure. This might be the number of cheetahs that could be supported by an area of around 200 km², or 20 000 ha, the size of a large game ranch.

The mountain gorilla model

The mountain gorilla model was constructed in order to simulate the more complex nature of the social structure of this species. A mountain gorilla population is organized into social groups that contain adult males and females and their offspring (Schaller, 1963; Weber & Vedder, 1983). Some adult males are also solitary (Schaller, 1963; Caro, 1976; Weber & Vedder, 1983). Movement is common between groups, however only sub-adults and adults migrate (Harcourt et al., 1981). The youngest individual observed to migrate was 6.5 years old (Harcourt et al., 1976). Females only leave a breeding group to join bachelor adult males or to join another breeding group, and are never solitary (Harcourt et al., 1976). Males may also migrate into breeding groups, but they can also become solitary (Caro, 1976; Harcourt et al., 1981). Migration is therefore not random but dependent upon age and sex, and was simulated by adapting the random model by placing the following additional conditions upon dispersal: (1) individuals had to be 7 years old before they could migrate; (2) a female could only migrate to join another breeding group (with a resident adult male), or a solitary adult male; (3) an adult male did not migrate if he was the only adult male in his group; (4) a group which has lost all adult males because of mortality could either merge with another group, or be joined by a solitary adult male; (5) if the population was near its ceiling level and a migrating male could not find space in any group, then he died; (6) if all the adults in a group died, leaving only juveniles below 6 years of age, then these juveniles were assumed to be unable to survive on their own.

The mountain gorilla population in the Virunga mountains numbers around 293 individuals (Aveling & Aveling, 1989). This population is organized into 29 groups with an estimated 11 solitary males (Aveling & Aveling, 1989). Group size ranges from three to 21 individuals (Harcourt *et al.*, 1981; Weber & Vedder, 1983). Therefore the initial population size of 300 individuals divided into 30 breeding groups, used in the monk seal simulations, is also appropriate for the Virunga mountain gorilla population.

The roan antelope model

The roan antelope model simulates a harem social structure with female philopatry (Beudels *et al.*, 1992). As with gorillas, populations of roan antelope are separated into breeding groups, and individuals do not migrate randomly. However roan differ from gorillas in that females do not migrate between groups. Instead, all females tend to remain within their natal group for life, forming a cohesive unit (Joubert, 1974). Each group of females or herd is held by a single adult male (Joubert, 1974). Female offspring remain within the herd for life but males are driven out by the resident male when they are aged around 2.5 years, before they reach sexual maturity (Joubert, 1974). These males then remain in bachelor herds until they reach 5–6 years of age, when they separate and become solitary (Joubert, 1974). It has been hypothesized that if the number of females within a herd rises above a certain level then the herd becomes vulnerable to fragmentation by these solitary adult males (Starfield & Beloch, 1986). This theory is supported by the rarity of herds containing more than 12 adult females (Smithers, 1983). The following assumptions were made in order to model this social organization: (1) only one male within any herd could reproduce; (2) a group of females could only be dominated intact by a new bull upon the death of the old one; (3) if the number of breeding females was above a certain size, termed the *split level*, then, provided there was at least one solitary adult male, and the number of herds was below maximum capacity, the adult females split into two equal-sized groups. One of these groups, along with the young animals in the herd, remained with the resident bull, whilst the other formed a new herd with a new male chosen at random; (4) if a resident bull died, then his herd was immediately taken over by one of the solitary adult males chosen at random; (5) every adult male without a herd had an equal probability of taking over part of a fragmented herd or an intact herd upon the death of its dominant bull, regardless of age or natal herd; (6) if the number of adult females in any herd exceeded a ceiling level, then reproduction ceased.

Note that for this model, the migration rate is irrelevant, because migra-

Table 11.2

Percentage persistence over 750 years for populations of monk seals, cheetahs, gorillas and roan antelope under different levels of subdivision

Species	Number of subdivisions				
	1	5	10	15	30
Monk seal	87.5	73.8	52.2	30.1	0.0
Cheetah	27.2	13.8	2.2	0.2	0.0
Gorilla	98.4	97.5	97.1	97.3	94.0
Roan	—	83.9	81.0	80.5	61.6

The annual migration probability in all models excepting that of the roan antelope was 0.1 and populations were subjected to only demographic stochasticity.

tion is determined demographically by the rate at which herds grew large enough to split. The model should give a good reflection of the number of females in the population, as well as the number of herds, but will not accurately depict the number of bachelor herds or the distribution of males. Genetically, the population can be modelled accurately by keeping track of only those individuals who reproduce, whilst the demographic status of the population is principally reflected by the number of females in the population, rather than the number of males. However, this model was not directly comparable to the gorilla and monk seal models because ceilings on the herd numbers were measured in terms of the number of adult females rather than the total number of individuals. This problem was surmounted by setting the ceiling on numbers of adult females in a roan herd to a level which was equivalent to the number of adult females in a herd with a stable age structure that is at its ceiling. This process ensured that the maximum number of breeding females in each herd of roan was comparable to that in groups of monk seals, cheetahs and gorillas. The split level of the herd was set at half the population ceiling.

For comparative purposes populations were initialized at 300 individuals. This is within the range of the estimated size of the population in the Kruger Park in South Africa at the end of the 1960s (Joubert, 1974), but is larger than the population in the Parc Nationale d'Akagera in Rwanda (Beudels et al., 1992), from which many of the demographic parameters were obtained (Table 11.1). Populations were examined under varying levels of subdivision. As with monk seals and gorillas, subdivision into 30 herds most accurately reflected the true population structure. Under this level of subdivision the maximum number of adult females was 14 and herds could split when they contained more than seven adult females, a figure in agree-

Table 11.3

Percentage persistence over 750 years for populations of monk seals, cheetahs, and gorillas under different migration rates

Species	Panmictic	Migration rate		
		0.0	0.01	0.1
Monk seal	87.5	7.4	0.0	0.0
Cheetah	27.2	0.0	0.0	0.0
Gorilla	98.4	72.8	92.2	94.0

Populations were subdivided into 30, except for the cheetah, which was subdivided into 15 sub-populations. Populations were subjected to only demographic stochasticity.

ment with actual reported herd sizes of between five and 12 adult females (Smithers, 1983).

POPULATION STRUCTURE AND PERSISTENCE

When monk seal populations were subjected to only demographic stochasticity, persistence decreased with increasing levels of subdivision (Table 11.2). For example, a panmictic population had a chance of persistence of 88% over 750 years, but this was reduced to zero for a population subdivided into 30. This greater vulnerability to extinction for subdivided populations was a consequence of the random nature of migration in the monk seal model. As subdivision was increased, there was a greater likelihood of a monk seal migrating to a locality containing no resident individuals of the opposite gender, where it would be unable to breed. In effect, this acted as an extra mortality on the population.

The simulations above were conducted under a high migration probability of 0.1. When this rate was decreased in a meta-population made up of 30 sub-populations, persistence increased (Table 11.3). Therefore, perhaps rather surprisingly, a meta-population with no migration between sub-populations showed markedly higher persistence than one with migration. Even a low migration probability of 0.01 reduced persistence from levels with no migration. This was again a consequence of random migration acting in effect as an extra mortality on meta-populations. This effect outweighed the benefits that were gained from recolonization under high migration rates.

Persistence of cheetah populations was even lower than that of monk seals (Table 11.2), because of the high juvenile mortality and shorter life span of this species (Table 11.1). When populations were subdivided into 15 sub-populations, and the migration rate was decreased, persistence increased (Table 11.3). In fact, persistence showed a relationship with both the level of subdivision and a change in the migration rate similar to the monk seal. However, overall levels of persistence were low, and all subdivided populations had a less than 15% chance of persisting for the duration of simulation.

Persistence of populations of gorillas was higher than that of panmictic monk seal or cheetah populations (Tables 11.2 and 11.3). This was in part a consequence of a longer generation time for this species, but was also a consequence of a lower variance in the growth rate. The latter results from the high mean survival rates and low mean birth rate for gorillas. These parameters form the probability parameters of a binomial distribution that is used to compute births and death. The variance of a binomial distribution is a maximum when the probability parameter is 0.5, and falls to a minimum as this probability decreases to 0.0 or increases to 1.0. Therefore, because the survival rates and birth rates for the gorilla are overall much further from 0.5 than for the monk seal or cheetah, variance in the growth rate is reduced. Persistence of gorilla populations was not greatly affected by population subdivision (Table 11.2). Persistence was also unaffected when the migration probability was decreased to 0.01 in a population subdivided into 30 gorilla groups. However, persistence decreased when no migration occurred. This contrasts with the random migration model where persistence showed a marked increase as the rate of migration declined.

Persistence of meta-populations of roan antelope was generally higher than that of monk seals and cheetahs, but lower than that of gorillas (Table 11.2). As with gorillas, persistence was unaffected by subdivision of the population except under the highest level of subdivision into 30 herds, where it was reduced. In the gorilla and roan models the breeding capability of the population was not reduced by migration, because a migrating female only immigrated into a group where she could reproduce. Therefore migration did not act as an extra mortality on the population as was the case in the random migration model. These populations thus retained the benefits of migration, in that areas could be recolonized by new groups, but had few of the disadvantages.

Table 11.4

Percentage loss of allelic diversity over 750 years for populations of monk seals, cheetahs gorillas and roan antelope under different levels of subdivision

Species	Number of subdivisions				
	I	5	IO	15	30
Monk seal	47.2	42.6	37.5	35.0	100.0
Cheetah	77.3	77.1	70.9	100.0	100.0
Gorilla	21.9	23.0	24.3	19.3	25.1
Roan	—	87.1	81.5	78.0	72.1

The annual migration probability in all models excepting that of the roan antelope was 0.1 and populations were subjected to only demographic stochasticity.

Conclusions

Subdivision of monk seal and cheetah populations markedly decreased persistence. Increasing the migration rate between sub-populations also decreased persistence. However, subdivision of gorilla and roan populations did not significantly affect persistence except under the highest level of subdivision, and then only in the case of the roan antelope. Changes in the migration rate in the gorilla model also did not affect persistence, unless no migration could occur.

POPULATION STRUCTURE AND GENETICS

As monk seal populations became more subdivided, slightly less allelic diversity was lost (Table 11.4). For example, a population divided into 15 subpopulations lost a mean 35% of allelic diversity over the duration of simulation, whereas a panmictic population lost only 47% of alleles. However, losses of allelic diversity were much higher in populations with low migration rates (Table 11.5). Loss of allelic diversity from cheetah populations was much greater than that from similar monk seal populations (Table 11.4). However, as with the monk seal populations, whilst subdivision did not greatly affect losses in allelic diversity, a decline in the migration rate to zero or 0.01 had a more marked effect (Table 11.5).

Losses of allelic diversity from gorilla populations were much lower than from monk seal or cheetah populations (Tables 11.4 and 11.5). For

Table 11.5
Percentage loss in allelic diversity over 750 years for populations of monk
seals, cheetahs and gorillas under different migration rates

Species	Panmictic	Migration rate		
		0.0	0.01	0.1
Monk seal	87.5	88.5	100.0	100.0
Cheetah	77.3	100.0	100.0	100.0
Gorilla	21.9	75.7	49.4	21.9

Populations were subdivided into 30 except for the cheetah, which was subdivided into
15 sub-populations. Populations were subjected to only demographic stochasticity.

example, over the duration of simulation only 22% of allelic diversity was
lost from a panmictic population of gorillas compared with 47% and 77%
from similar populations of monk seals and cheetahs respectively. This was
a consequence of a longer generation time and lower variability in individual contributions of offspring in gorillas, and could be predicted by an
analysis of inbreeding effective size (Durant, 1998a). Losses of allelic diversity were not greatly affected by subdivision of the population, however
losses were significantly greater in a population subdivided into 15 groups
than at all other levels of subdivision (Table 11.4). Allelic diversity was lost
more rapidly as the migration rate decreased, with the highest rates of loss
from populations with no migration (Table 11.5), nonetheless overall rates
of loss were still much lower than those of monk seals and cheetahs.

Losses of allelic diversity were much greater from roan populations
than from gorilla, monk seals or cheetah populations of similar structure
(Table 11.4). This loss was also strongly affected by the level of subdivision,
and increased as the number of herds decreased. This was not surprising,
because in the roan model each extra herd meant an extra breeding male,
and genetic diversity is known to be strongly affected by the relative numbers of breeding males and females (Crow & Kimura, 1970). Even in the
best-case scenario for this species, obtained from a meta-population of 30
herds, as much as 72% of allelic diversity was lost over 750 years.

Conclusions

Overall, both subdivision and an increase in the migration rate decreased
losses of genetic diversity, provided sub-population size was not too small.

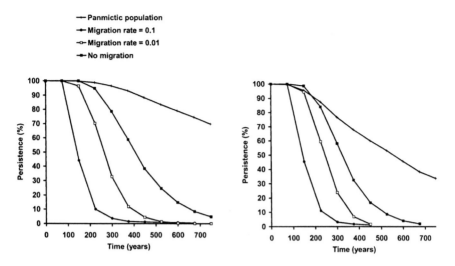

Figure 11.1 Percentage persistence over time for panmictic populations and meta-populations of Mediterranean monk seals under different migration probabilities. Populations were subjected to (a) frequent reproductive disasters and (b) infrequent survival disasters, which occurred independently across all sub-populations. Meta-populations were made up of 30 sub-populations. Initial population size was 300 individuals, divided equally across sub-populations, and the overall meta-population ceiling was 1200, also divided equally across sub-populations.

Subdivision had a particularly marked effect on roan populations. Here subdivision increased the number of males contributing to the gene pool.

ENVIRONMENTAL STOCHASTICITY AND THE SPREADING OF RISK

Monk seals

Unless migration rates were high, persistence of Mediterranean monk seals was reduced both by frequent reproductive and infrequent survival disasters, although the latter type of disaster had the greatest effect (Figure 11.1). There was no spreading of risk benefit for frequent reproductive disasters (Figure 11.1a), but there was some benefit for populations subjected to infrequent survival disasters (Figure 11.1b), but only in the short term and only if no migration occurred. In these circumstances persistence over the first 150 years was slightly higher than in a panmictic population.

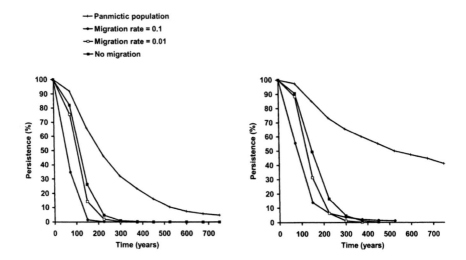

Figure 11.2 Percentage persistence over time for panmictic populations and meta-populations of cheetahs under different migration probabilities. Populations were subjected to (a) frequent reproductive disasters and (b) infrequent survival disasters, which occurred independently across all sub-populations. Meta-populations were made up of 15 sub-populations. Initial population size was 300 individuals, divided equally across sub-populations, and the overall meta-population ceiling was 1200, also divided equally across sub-populations.

Cheetahs

Persistence of cheetah populations was more strongly affected by frequent reproductive disasters than by infrequent survival disasters, in contrast to the pattern seen in monk seal populations (Figure 11.2). For example, when populations were subjected to frequent reproductive disasters persistence over 150 years was 16%, 16%, 3% and 66% for a population with migration rate of zero, 0.01, 0.1 and a panmictic population respectively compared with 46%, 31%, 15% and 85% for the same population structures subjected to infrequent survival disasters. This indicates that the demography of this species, with its high juvenile mortality and relatively short life span, makes it more vulnerable to disasters that affect reproduction than the monk seal. However, there were no apparent benefits from the spreading of risk for both types of disaster. Subdivided populations persisted for shorter times than panmictic populations and persistence decreased as the migration rate increased.

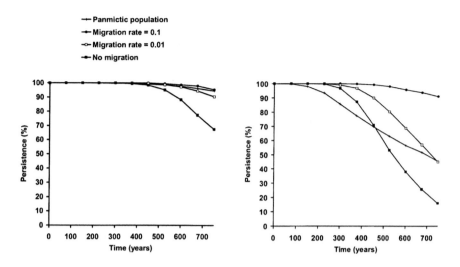

Figure 11.3 Percentage persistence over time for panmictic populations and meta-populations of gorillas under different migration probabilities. Populations were subjected to (a) frequent reproductive disasters and (b) infrequent survival disasters, which occurred independently across all sub-populations. Meta-populations were made up of 30 sub-populations. Initial population size was 300 individuals, divided equally across sub-populations, and the overall meta-population ceiling was 1200, also divided equally across sub-populations.

Gorillas

As with monk seals, frequent reproductive disasters had a very weak effect on the persistence of populations of gorillas to the extent that persistence was reduced significantly from its levels under demographic stochasticity only in a meta-population with no migration (Figure 11.3a). Persistence remained fairly high, above 60% over 750 years, and was highest for a panmictic population and, as under only demographic stochasticity, persistence did not differ significantly between populations with high and low migration rates. Infrequent survival disasters exerted a much stronger effect upon persistence (Figure 11.3b). This type of disaster also completely overturned the relationship between the migration rate and persistence hitherto observed. Under these disasters, provided migration could occur, persistence of subdivided populations was markedly higher than panmictic populations, and populations with a high migration rate showed the highest chance of long-term persistence. These differences were very marked. For example, populations with migration rates of 0.1, 0.01, and zero and a panmictic population had respective chances of persistence of 100%, 97%, 87% and 78% over 375 years, whilst over 600 years these chances were 96%, 69%, 38% and 56%. Only the meta-population with no migration had a lower persistence than a panmictic population, and then only over a longer time interval.

Roan antelope

As was the case with the monk seal and gorilla models, frequent reproductive disasters exerted less of an effect on roan antelope meta-populations than infrequent survival disasters (Figure 11.4). For example, the chance of persistence over 750 years was 74% and 57% for a population subdivided into five herds under frequent reproductive and infrequent survival disasters respectively, compared with chances of 84% under only demographic stochasticity. These differences between the two types of disaster became less marked under higher levels of subdivision, indicating that some protection from extinction was provided by subdivision. For example, persistence of a population subjected to infrequent survival disasters was 57%, much the same as that of 58% for one subjected to frequent reproductive disasters and not much lower than the 63% level recorded under only demographic stochasticity. Nonetheless there was no net benefit from the spreading of risk.

Conclusion

There was no long-term spreading of risk benefit for three of the four species examined here: monk seals, cheetah and roan antelope. However, in the case of the gorilla there were benefits if disasters were sufficiently strong in their effect. In these circumstances persistence of subdivided populations was much higher than persistence of panmictic populations, provided limited migration could occur, implying that there can be demographic advantages in segregating a population in this situation. In addition, although there were no net gains in persistence, the losses from subdividing roan antelope populations were low.

IMPLICATIONS FOR CONSERVATION

The results described in this chapter reflect only a small subsection of the possible variety of behaviours existing in real populations. Many other be-

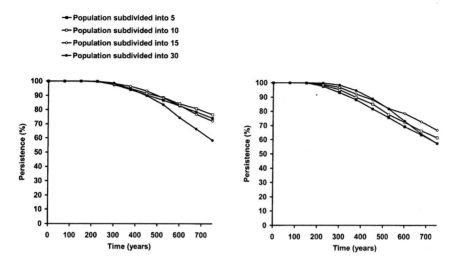

Figure 11.4 Percentage persistence over time for meta-populations of roan antelope under different levels of subdivision. Populations were subjected to (a) frequent reproductive disasters and (b) infrequent survival disasters, which occurred independently across all sub-populations. The initial population contained 150 females divided equally across herds. The overall meta-population ceiling was 420 adult females, which was also divided equally across herds. The herd split level was half the herd ceiling.

haviours, not examined here, also have the potential to affect persistence through demographic or environmental stochasticity or loss of genetic diversity in unpredictable ways (Figure 11.5). For example, in a number of species sociality leads to reproductive suppression within breeding groups [e.g. dwarf mongooses (Creel & Waser, 1994) and wild dogs (Creel et al., 1997)]. This can lead to an increase in demographic stochasticity, because reproduction in these circumstances depends on the productivity of the alpha pair rather than the productivity of a whole group or pack. It will also increase the rate of loss of genetic diversity because only one pair will contribute to the gene pool. In other species sociality and territoriality can lead to infanticide, where immigrating males will kill offspring sired by predecessor males [e.g. lions (Pusey & Packer, 1993), gorillas (Harcourt et al., 1981)], factors that affect losses of genetic diversity and that lead to environmentally stochastic effects, because the proportion of infanticides in a population is likely to stay constant, and may even increase, as population size increases. Allee effects, where stochasticity is increased owing to difficulties individuals may have in finding mates, can occur through many

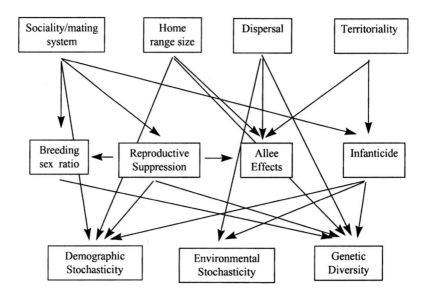

Figure 11.5 Some links between different behaviours and population persistence through their effects on stochasticity in population growth and losses of genetic diversity.

different behavioural means, such as through large home range size, where individuals may occur at low density and may have difficulties finding each other. Territoriality may also increase the likelihood that individuals are not able to breed if they cannot find an empty territory (Lande, 1987). This will be exacerbated if combined with a low dispersal distance, because then individuals may not discover unoccupied territories if they are more than a small distance from their natal home range. All these factors may interact and affect persistence and genetic diversity in complex ways.

Nonetheless the results described demonstrate that the aspects of behaviour examined in this chapter can affect persistence (Table 11.6). Species that migrate randomly show a decline in persistence with an increase in population subdivision or migration rate, whereas species with non-random migration showed no such relationship with persistence. In fact, the crucial factor that determined whether subdivided populations persisted better than panmictic populations, aside from the strength of environmental stochasticity, was the cost to individuals of recolonizing extinct sub-populations. Where these costs were virtually zero, as for gorillas, then populations benefited from subdivision. However if these costs were in-

Table 11.6

Summary table depicting the effects of subdivision and dispersal on persistence and genetic diversity

Model	Subdivision and dispersal		Spreading of risk
	Persistence	Genetic diversity	
Monk seal	−		−
Cheetah	−		−
Gorilla			+
Roan		+	

creased, as for monk seals and cheetahs, then populations were unlikely to benefit from subdivision in the long term. Mortalities arising from the act of migration itself were neglected from these models, and therefore it is likely that these models underestimate costs of migration.

Both social structure and dispersal behaviour also affected the loss of genetic diversity from populations. Previous work has suggested that subdivision reaps genetic benefits, even if the migration rate is extremely low (Boecklen, 1986; Lacy, 1987). However the simulations presented here suggest that genetic benefits are only gained in certain circumstances: when migration rates are high and when population subdivision increases breeding opportunities for individuals. Over the same time span used in this study, Boecklen (1986), using simulation models, found that population subdivision decreased losses in allelic diversity. The results presented here suggest that losses of allelic diversity are only lower in subdivided populations if the migration rate is much higher than that used in Boecklen's (1986) simulations (excepting roan antelope). However Boecklen (1986) did not consider sub-population extinction and recolonization in his models, which are likely to reduce genetic differentiation of sub-populations and overall diversity (Slatkin, 1977). This study shows that very often the population structure that best ensured persistence was not the structure which best decreased losses of genetic diversity. For example, a decrease in the migration rate for monk seal populations increased persistence whereas it tended also to increase losses of genetic diversity.

There is a common argument that spatial heterogeneity and hence population patchiness enhances population persistence (Wiens, 1985). This argument rests on the concept of the 'spreading of risk' (den Boer, 1968), which postulates that extinction is less likely to occur in several sub-populations at once than in a single panmictic population. However, the simulations show that this is only the case under strong environmental stochasticity that occurs independently between sub-populations. Even here, persistence was only higher than in a panmictic population if migration costs were very low. However, in the particular set of circumstances where there were benefits, these benefits were large and resulted in substantial increases in persistence above levels in panmictic populations.

A sequence of independent disasters could only drive a single sub-population to extinction at any one time and hence was unlikely to affect the entire population. If a single sub-population was eliminated it could be recolonized by a sub-population which had escaped the effects of disasters. If disasters were strong enough and subdivision *per se* did not greatly reduce persistence, subdivision could enable a meta-population to persist better than a panmictic population. Whilst infrequent survival disasters were strongest in their effect on persistence of monk seal, gorilla and roan antelope populations, frequent reproductive disasters were strongest in their effect on cheetah populations. This implies that the extremely high cub mortality in this species makes it more vulnerable to further reductions in recruitment. Nonetheless, even under these disasters cheetah populations showed no spreading of risk benefit.

Models of population dynamics inevitably make a number of simplifications, and the models here are no exception. One important feature that might affect the results would be stronger density-dependent effects upon the growth rate. In the models examined here, density dependence acts only weakly through a cessation in reproduction when the population grows to its ceiling level; if instead growth rates increased as population size declined then this would affect both the impact of catastrophes and the response to dispersal (Thomas & Singer, 1996). Density dependence may not be important for two of the species examined here, monk seals and cheetahs, because for these species predation and/or persecution may depress population density below levels where density-dependent effects may become important (Marchessaux, 1989; Laurenson, 1995). However it may have important effect on the dynamics of roan and gorilla populations.

In addition, only the roan antelope model makes an allowance for differences in the variance in reproductive success between males and females. The results show that these differences have a large impact on losses of allelic diversity. The losses within the roan antelope model may be unrealistically high, because real roan herds can be taken over by rival males (Joubert, 1974). Gorilla groups, which are dominated by a single silverback male (Harcourt *et al.*, 1980), are likely to have similar differences in reproductive success between males and females. However, monk seals and cheetahs, which are more mobile, may show less extreme differences, but even here it is likely that males will have a higher variance in reproductive success than females, and this will be reflected in higher losses of allelic diversity than that portrayed. The loss of genetic diversity in the roan antelope model could be seen as a maximum, whilst the loss within the other species is a minimum. For populations of the size examined in this study mutation is unlikely to have a significant impact on losses of diversity.

Disasters in real populations are unlikely to be entirely independent but instead have a varying degree of correlation between sub-populations. For example, epizootics will be correlated to an extent that depends upon the transmission rate between sub-populations. Monk seals could be vulnerable to phocine distemper (Osterhaus et al., 1997) and roan antelope are vulnerable to anthrax epizootics (Joubert, 1974). Both these epizootics fall within the category of infrequent survival disasters. In addition, monk seals are vulnerable to collapses of the caves in which they breed (Maigret, 1986); cheetahs suffer large cub mortalities because of predation (Laurenson, 1994); roan antelope are vulnerable to a delayed onset of the rainy season, which can lead to high yearling mortality (Beudels et al., 1992); whilst gorillas are likely to suffer mass infanticides on contact with unrelated adult males (Harcourt et al., 1980). Where disasters are climatically induced, there is likely to be a high correlation between sub-populations, whilst more local disasters, such as cave collapses and infanticides, are more likely to be independent.

One major assumption of all the models used in this chapter is that an individual can migrate to any other sub-population with an equal probability: the so-called *island model* of migration (Latter, 1973). This is undoubtedly not the case for monk seals, cheetahs and gorillas but may be appropriate for roan antelope. Monk seal sub-populations are widely dispersed (Marchessaux, 1989), and it is physically easier for a migrating individual to immigrate into a neighbouring locality than to a distant one. A similar situation is likely to arise with cheetahs. Gorilla females are noted to migrate to neighbouring groups, often during inter-group interactions (Harcourt et al., 1976). Males may disperse across wider areas if they are solitary, however there is some evidence that even here they remain within their natal home range (Caro, 1976; Harcourt et al., 1976). In the case of roan antelope only males migrate, and these males spend long periods from reaching sexual maturity to holding a breeding herd, during which they may roam extensively (Beudels et al., 1992). They are therefore less likely to show a preference for neighbouring groups. Migration to only neighbouring groups will in effect lower the migration rate and therefore increase persistence in subdivided populations of monk seals and cheetahs, and decrease persistence in gorillas. Such an alternative *stepping stone* model (Kimura, 1953) is likely to increase losses of genetic diversity, but for the migration rates investigated and the maximum of six steps from one sub-population to another in a meta-population of 30 sub-populations, the effects are not likely to be great (Slatkin, 1985).

My models demonstrated that a population of 300 monk seals did not form a demographic MVP, with a 90% chance of persistence over 100 years, if the level of subdivision was high or if the migration rate was high, even when subjected to only demographic stochasticity. However populations of cheetahs fared even worse. Only panmictic populations subjected to demographic stochasticity alone satisfied the conditions of an MVP. All subdivided populations and all populations subjected to disasters were non-viable. Populations of 300 gorillas or roan antelope were demographically viable under all types of stochasticities and all population structures. Genetic viability, defined as a less than 10% loss of original quantitative genetic variation over 200 years, can be assessed using allelic diversity. Under this definition populations of gorillas were genetically viable, however populations of monk seals were only genetically viable if there was no migration between sub-populations or if the level of subdivision was below 30. Populations of cheetahs and roan antelope were never genetically viable. Rates of loss of genetic diversity were nowhere low enough to allow mutation rates to restore genetic diversity.

Overall, therefore, the Virunga gorilla population appears to be both demographically and genetically viable whatever the migration rate. The Kruger park roan population is demographically viable but is not genetically viable. The viability of the Aegean monk seal population depends on the migration rate. Over a 6 year period, Johnson *et al.* (1982) resighted a small proportion of (5%) of 850 tagged Hawaiian monk seals at atolls other than the site of birth. This suggests that the migration probability for this species is in the region of 0.01. If this also applies to the Mediterranean species then the monk seal population is neither demographically nor genetically viable. A cheetah population of similar size is never genetically viable and only demographically viable if it is not subjected to environmental stochasticity and it is panmictic. For this species particular effort should be made to ensure large areas which can support larger population sizes are set aside for their conservation. These observed differences in viability be-

tween the different species are largely owing to differences in social structure and dispersal.

CONCLUSIONS

This study tested three hypotheses that result from the expectation that
dispersal patterns and social organization may affect the viability of populations: (1) population structure affects persistence; (2) population structure
affects the genetics of populations; (3) populations can benefit by subdivision through 'spreading of risk' (den Boer, 1968). This was carried out by
using detailed stochastic computer simulation models, examined through
four case studies of species with very different social organizations and
demographies: the Mediterranean monk seal (Monachus monachus), mountain gorilla (Gorilla gorilla beringei), roan antelope (Hippotragus equinus) and
cheetah (Acinonyx jubatus). The crucial factor that determined whether subdivided populations persisted better than panmictic populations, aside
from the strength and independence of environmental stochasticity, was
the cost to individuals of recolonizing extinct sub-populations. Where these
costs were virtually zero, as for gorillas, then populations benefited from
subdivision. However if these costs were high, as for monk seals and cheetahs, then populations were unlikely to benefit from subdivision in the long
term. Genetic benefits from subdivision were also only gained in certain
circumstances: when migration rates were high and when population subdivision increased breeding opportunities for individuals. Frequently the
population structure that best ensured persistence was not the structure
which optimized the maintenance of genetic diversity. In addition, limited
support was found for the spreading of risk theory for population subdivision. This only occurred under strong environmental stochasticity, which
occurred independently between sub-populations.

ACKNOWLEDGEMENTS

This research was funded by the Natural Environmental Research Council.
I thank John Harwood and Rosaline Beudels for their support, encourage-
ment and discussions. I thank Tim Caro and Marcella Kelly for access to
records of cheetahs prior to 1991. The manuscript was much improved by
comments from Théresa Jones and Georgina Mace.

12

Incorporating behaviour in predictive models for conservation

RICHARD A. PETTIFOR, KEN J. NORRIS & J. MARCUS ROWCLIFFE

INTRODUCTION

E. M. Nicholson (1963), in the foreword of a book dedicated to the conservation of waterfowl, noted that 'The practice of conservation today is partly an art and only partly a science, but its basis must be firmly assured by the ecological sciences'. The context within which conservation biologists work has changed considerably over the past 35 years: scientifically, a far more rigorous approach to conservation biology has developed, based on the theory of population dynamics, incorporating evolutionary and behavioural ecological principals (e.g. Soulé, 1987; Burgman *et al.*, 1993; Sutherland, 1996a; Hanski & Gilpin, 1997; Tuljapurkar & Caswell, 1997). Politically, too, the environment within which conservation biologists now operate is very different to that of the 1960s: much of our work is currently informed by the requirements of national and international legislation and directives. On the debit side, conflict between wildlife and human interests continue, whilst habitat change (often, but not always, deleterious) is widespread. Perhaps the biggest challenge facing conservation biologists today is in being able to understand the response of wildlife to anthropogenic change, and to predict the consequences of such change. This chapter is therefore concerned with various modelling approaches that conservation biologists may use in their attempts to understand the dynamics of populations, and in particular how the incorporation of behaviour into these models can enable more robust predictions to be made as to the likely response of populations to novel circumstances.

Describing the current behaviour of a population and diagnosing declines

The first step in defining any conservation objective is to describe the problem, based on objective criteria. Simple snapshots of population abundance over two discrete time-periods have been used to document declines in common breeding bird populations, resulting in research specifically targeting these populations in order to understand the reasons for the declines [see Fuller *et al.*, 1995; e.g. skylark *Alauda arvensis* (Wilson *et al.*, 1997)]. Gibbons *et al.* (1996) used historical and more recent surveys to score whether or not there had been a decline in breeding numbers for each bird species in the United Kingdom. Where counts have been carried out over a number of years, then trends in population size can be assessed, either over time or at various spatial scales with respect to the global population (see Pettifor, 1997). Long-term monitoring has led to the development of 'alert limits', whereby the health of populations is regularly assessed (Greenwood *et al.*, 1995; Kirby & Bell, 1996; Pettifor, 1997) – when 'alarm bells' ring, proactive conservation action can then be initiated, both in terms of research and policy. Once conservationists have statistically established a decline in abundance for a particular population/species, it may also be necessary to diagnose the cause of the decline before remedial action can be undertaken. This is not the only option – conservationists could instead simply generate measures to increase fecundity or reduce mortality (Green, 1995). However, in the case of many critically endangered species it is essential first to arrest the decline in abundance, before designing any additional measures to improve the species' conservation status. Diagnosis involves understanding two processes: (1) the external factor(s) causing the decline; (2) the demographic mechanism(s) through which the decline occurs. Green & Hirons (1991) describe a 'comparative' approach to diagnosis, in which factors that are considered likely to affect the species under investigation are compared between different populations of the species. Populations are defined either in time or space. The application of these approaches is reviewed by Green (1995).

Statistical models, coupled with an ecological understanding, can initiate and inform conservation action. However, they are not predictive *per se*, even when the slope of a regression line indicates imminent extinction – at small population sizes stochastic events (genetic, demographic and environmental) make predictions extremely difficult. At high population densities (as might occur when habitat is lost, forcing animals onto ever more restricted areas), previously observed trends in population growth rate may not hold – even when the animals are able to settle on an already occupied area, it is quite likely that the vital demographic rates might change (e.g. age at first breeding might increase, mean fecundity decline and/or survival probabilities decrease), dramatically changing lambda. Even longish

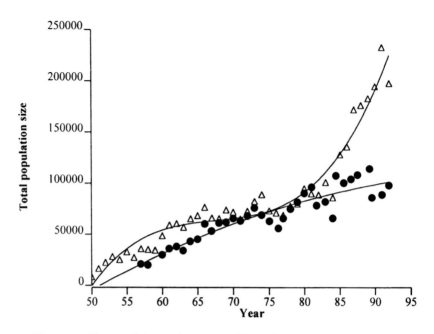

Figure 12.1 The growth in population size of the Icelandic/Greenlandic pink-footed goose *Anser brachyrhynchus* (triangles, solid line) and Icelandic greylag goose *A. anser* (circles). From Pettifor *et al.*, 1997.

runs of data are not necessarily sufficient to prevent quite erroneous conclusions being drawn regarding the future trajectory of a population. For example, Figure 12.1 illustrates the growth in size of two populations of geese – there would have been no statistical basis in the early 1980s to predict the subsequent disparity in the trajectories of the two populations. Below, we review modelling approaches that are more predictive than the statistical ones outlined above, particularly emphasizing the need to incorporate a behavioural-based framework into our understanding of population processes.

PREDICTIVE MODELS

Population-based models

The majority of models used as predictive tools for conservation fall into the category that may be termed population-based. These rely on estimates of population growth rates or demographic rates, which can be used to simulate population growth, and explore the implications of changing specific parameters. Where changes in certain parameters can be equated to specific changes in management, these models may therefore be used predictively.

A wide range of model structures have been used, depending on the specific questions addressed and the availability of data. In the simplest case, logistic or exponential growth rate models may be used (e.g. Ginzburg et al., 1982; Bayliss, 1989; Fogarty et al., 1991; Burgman et al., 1992). Alternatively, population models may be based on estimated demographic rates, allowing the effects of changes in either survival or productivity to be explored (e.g. Duncan, 1978; Wanless et al., 1996). Often, population-average demographic rates are used, although where data are available on age- or stage-specific rates, these may be incorporated into structured population models (e.g. Houllier & Lebreton, 1986; Crouse et al., 1987; Caswell, 1989; Price & Kelly, 1994). Such models may be deterministic, exploring which aspects of the life history have the greatest influence on population growth rate (e.g. McCarthy et al., 1995), or stochastic, allowing the models to be used as a form of risk assessment (Burgman et al., 1993). Population-based models often assume a single isolated population, but they may also be spatially structured, either taking a patch-occupancy metapopulation approach (Levins, 1969; Hanski, 1991; Hanski & Thomas, 1994; Hanski & Gilpin, 1997), or using a more complex structure of migration within a landscape of varying habitat (e.g. Lande, 1988b; Akçakaya et al., 1995; Lindenmayer & Lacy, 1995; McCullough, 1996).

The problem with predictions from such models is with application to novel circumstances. First, it is difficult to obtain the necessary data for demographic correlations, especially density dependence. Such correlations incorporate an unknown degree of estimation error, and it is therefore difficult or impossible to assess the true level of variability in the population (Bayliss, 1989). Also, population correlations rarely cover a sufficiently broad range of environmental or population variables. In particular, in populations that are strongly regulated, density is often relatively stable over time, so population behaviour at more extreme sizes will rarely be known (Goss-Custard, 1993). Further, the necessary assumption, that the conditions under which correlational data were collected will apply to the situation for which predictions are required, is unlikely to apply. Thus an age-structured stochastic matrix model of the Svalbard barnacle goose (*Branta leucopsis*) population, based on detailed statistical analyses of demographic data obtained from a 20-year study of individually ringed birds up until 1990, indicated an equilibrium population size of around 12 000 birds (Rowcliffe *et al.*, 1995) – the current population is in excess of 23 000! This discrepancy between observed and predicted arose largely because we over-estimated the strength of a density-dependent decline in fecundity (Pettifor *et al.*, 1999), and because of recent changes in behaviour, with the birds founding new breeding colonies that have grown more rapidly than previously established colonies (R. A. Pettifor & J. M. Black, unpublished data). Despite these limitations, population models have an important role to play in the exploration of data and the generation of new hypotheses (Taylor, 1995; Harcourt, 1996; Starfield, 1997). However, their utility as predictive tools is highly restricted. Behavioural-based models can potentially avoid these problems.

BEHAVIOURAL-BASED MODELS

As conservation biologists, one of our most exacting tasks is in being able to predict the response of populations to novel circumstances. How can the confidence in the predictive abilities of population ecologists be improved? Here we concur fully with Goss-Custard (e.g. Goss-Custard *et al.*, 1995a) and Sutherland (e.g. Sutherland, 1996a) that an individual-based, behavioural approach is the way forward. The reason for this is that just as genotype and phenotype interact to determine the observed response of an individual, so individuals and conspecifics interact resulting in the observed dynamics of a population. The underlying rationale is that because the decision-rules governing individual behaviour have been honed over evolutionary time through natural selection (Maynard Smith, 1982), then an understanding of the behaviour of individuals, and the incorporation of these rules into a modelling framework, can enhance the confidence placed in the model predictions (see Goss-Custard & Sutherland, 1997).

Theory and a worked example

Individual pairs of great tits *Parus major* commonly lay between five and 12 eggs, with the larger clutches resulting in higher recruitment. For a long time the question has been why the smaller clutches have not been weeded out of the population, so that all birds lay the most productive brood-size, and a number of hypotheses have been advanced to explain the discrepancy between the mean and the most productive clutch-sizes (e.g. Charnov & Krebs, 1974; Boyce & Perrins, 1987; Nur, 1987). However, as soon as one considers that individual birds are simply attempting to maximize their own fitness, and that the observed variation in clutch-size merely reflects intrinsic and/or extrinsic differences (e.g. in age, territory quality, etc.) between individuals, then this apparent conundrum no longer appears so interesting and the answer is obvious, namely individual optimization, expressed through phenotypic plasticity (Högstedt, 1980; Pettifor et al., 1988; Pettifor, 1993a,b). By the same token, observed population behaviour is simply a reflection of the constituent individuals attempting to maximize their fitness.

It is therefore through understanding the behaviour of individuals that we can begin to understand populations, and predict their potential trajectories in response to anthropogenic change. Fretwell & Lucas (1970) are attributed with providing the key theoretical breakthrough, although a number of ecologists were converging on the same framework at a similar time (e.g. Brown, 1969; Orians, 1969; Parker, 1970). The paradigm shift was that they attempted to explain the distribution of animals across resources of differing quality in terms of a game-theoretic distribution. Evolutionary stable strategies and the Ideal Free Distribution (IFD) are cornerstones of behavioural ecology. The importance of game-theoretic models is that they provide a predicted distribution against which the observed can be compared. Thus the IFD requires that individuals are free to distribute themselves without cost over the resource in such a way that they maximize their gain function (i.e. they ideally have perfect knowledge of resource availability and conspecific competitive behaviour), and that each individual is equivalent in competitive ability, foraging efficiency, or some other suite of characters. However, as simple observations show with respect to the intake rate of shorebirds, individuals are never 'ideal', and therefore models have been developed to account, for example, for differing competitive abilities (Sutherland & Parker, 1985; Parker & Sutherland, 1986) and perceptual limits (Bernstein et al., 1988, 1991). However, these additional refinements can lead to extra complexity, as shown by Sutherland (1996a, Figure 3.3). Despite these complexities, game-theoretic models have been used to great effect in predicting the distributions of animals in relation to resource availability when tested against observed distributions.

Goss-Custard and co-workers (Goss-Custard et al., 1995a,b,c) have probably been most successful to date in using individual-based, game-theoretic models structured around the actual physiology and behaviour of the animals not only to match the theoretical distribution of predators across patches of differing prey densities, but also in exploring future trends in numbers of predators in response to changes in prey availability (see Figure 12.2). Fundamental to these models of oystercatchers (*Haematopus ostralegus*) feeding on an array of mussel beds are three behavioural responses: the interference function, the depletion rate, and the dominance structure of the flocks. Based on over 20 years of detailed observations in the field, Goss-Custard *et al.* were able to obtain the shape and relevant parameter values for the local density-dependent functions. The model is game-theoretic in that each simulated individual is allocated a patch out of the 12 available mussel beds, dependent on its own dominance rank and the decisions made by all the other competitors, also dependent on their own ranks. Depletion occurs on these patches in a manner quantified in the field through both the predation by the oystercatchers and natural losses (storms, etc.). Thus each daily iteration of the model allocates each individual bird, defined by its specific foraging efficiency and susceptibility to interference, to the patch where it will obtain the highest intake rate, taking into account the decisions of all other birds present. The model is physiologically structured in that daily food and metabolic requirements are incorporated. Gut-processing time, assimilation efficiency and fat deposition are included from empirically derived studies such that daily food consumption and energy reserves are tracked through the course of the winter.

The three characteristics of the model – individuals-based, game-theoretic and physiologically-structured – allow simulations to be run in order to examine the consequences of novel changes to the habitat upon the local population. At a theoretical level, insights into the key parameters influencing the dynamics of this local population can also be explored. However, the changes in demography at this local scale cannot be translated into the consequences for the global population (although see Goss-Custard *et al.*, 1995d,e). Thus the approach pioneered by Goss-Custard is unique in its ability to predict the consequences of habitat or demographic changes not previously observed in the field (and is therefore crucial for site protection), but is limited in its relevance to the conservation of the global population. However, attempts are now being made to adopt this approach in modelling global populations throughout the annual cycle (see Population models below).

Figure 12.2 Schematic illustration of the decision processes feeding into Goss-Custard *et al.*'s (1995a,b,c) model of overwintering oystercatchers. Each individual assesses where it will maximize its intake rate based on both resource availability and its own susceptibility to interference owing to the presence of other individuals already present on the different patches. Taken from Goss-Custard, West & Sutherland (1996c) with permission.

Depletion models

Conservationists often need to understand how a particular proposal might affect the number of animals that can be supported within a given area over a given time period. Such proposals might include human activities that could be detrimental to an animal population; alternatively, they might include activities designed to enhance the suitability of an area for a particular population. In either case, a framework is required to predict how proposed changes might affect animal abundance.

In principle, such a framework could be very simple. Consider a hypothetical animal population occupying an area of habitat during the non-breeding period. The habitat contains food resources of a single type, but the density of these resources varies between patches within the habitat. If animals feed in the patch(es) with the highest resource density, then it is possible to use the model of Sutherland & Anderson (1993) to calculate the maximum number of animals the habitat could support, termed the maxi-

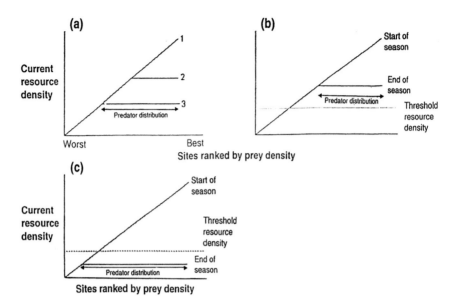

Figure 12.3 Graphical models of resource depletion. Predators are assumed to occupy the patch(es) containing the highest resource density. (a) Shows the distribution of predators across a series of patches (ranked from best to worst) over a period of time (e.g. winter). The predators initially congregate in the best patch at time 1. As resource density in the best patch declines due to depletion, the predators become dispersed between a greater range of patches at time 2. Further resource depletion increases the range of patches occupied by the predators at the end of the time period (time 3). This basic model can be used to estimate whether a given population size is sustainable by the available resources by assuming predators need to occupy a patch containing a threshold resource density in order to survive. (b) Shows that even though by the end of the season (e.g. winter) predators are dispersed over a greater range of patches than at the start, predator abundance is not sufficient to deplete resources below the threshold. This population size is sustainable by the available resources. (c) Shows that resource depletion by the end of the season has reduced resources densities below the threshold. This population size is unsustainable.

mum sustainable population. Figure 12.3(a) shows how their model works. We could use such a model to predict the impact of habitat changes on the number of animals the area could support, and so decide whether a proposal was likely to have an impact on animal abundance. For example, if a development proposal involved the loss of 15% of the habitat, we could predict whether the remaining habitat could support the population over a given time period (Figure 12.3(b), (c)).

Depending on the predictions of the model, we might conclude that the proposal was likely to have a negligible impact (Figure 12.3(b)), or have a detrimental impact on abundance (Figure 12.3(c)). Such models are often termed 'depletion' models because the number of animals an area can support over a given time period is determined by the net rate at which resources are consumed by the animal population, i.e. the total consumption rate minus the rate at which resources are replenished due to growth or reproduction (Sutherland & Anderson, 1993; Sutherland, 1996a).

The dynamics of resource depletion have been studied in a wide range of animal taxa (recently reviewed by Dolman & Sutherland, 1995b). However, most depletion models to date based upon the modelling framework of Sutherland & Anderson (1993) have been used to describe relatively simple natural systems, involving in particular grazing wildfowl (swans, ducks and geese) (e.g. Sutherland & Allport, 1994; Gill *et al.*, 1996; Percival *et al.*, 1996). These models have addressed a range of conservation issues, including the impact of food availability, habitat management, human disturbance and interspecific competition for resources on maximum sustainable population sizes. This type of simple behaviour-based model has, therefore, been practically valuable in influencing decisions about how particular habitats should be managed, especially at a local level.

Ideal free models

Depletion models are essentially based on the process of scramble competition for resources. However, in many natural situations competition can involve contests. This type of competition is termed 'interference' competition, as the presence of competitors reduces or interferes with an animal's ability to exploit resources. Various types of interference competition have been described in animal populations, ranging from overt aggression in which resources are stolen from other animals (Brockmann & Barnard, 1979), to passive interference, which includes, for example, disturbance of prey by competitors (e.g. Goss-Custard, 1980). In such cases, behaviour-based models are founded on the 'ideal free' distribution described by Fretwell & Lucas (1970).

Consider an animal population occupying an area of habitat consisting of a good and poor-quality patch of food resources. For simplicity we will assume quality is determined by resource density. The intake rate of resources by a solitary animal in relation to resource density is described by the 'functional response' (Figure 12.4(a)). We assume that animals are stressed for feeding time, and so attempt to feed in the patch giving them the highest intake rate (i.e. are rate maximizers). Animals initially occupy patch A. As the density of animals in patch A increases, intake rates decline due to interference (Figure 12.4(b)). Eventually, intake rates in patch A are depressed by interference to such an extent that patch B offers equivalent intake rates, so animals begin to occupy both patches. If we assume that animals need to achieve a threshold intake rate to survive (see also Sutherland & Parker, 1985; Goss-Custard et al., 1995a,b,c), then it is easy to see how such a model could be used to determine whether a given population size could be sustained by the available resources (Figure 12.4(c),(d)). In a similar way to depletion models, therefore, ideal free models can be used by conservationists as the basis for assessing the impact of various proposals on animal abundance. In order to construct an ideal free model for a particular system, it is clearly important to estimate the model's parameters. The critical components are the functional response, the effect of interference on intake rates, the energy requirements of the animals, and their individual differences in susceptibility to interference.

Functional responses have been described for a number of species (e.g. Sutherland, 1996a, Figure 3.7). To describe the functional response it is necessary to quantify the density of food resources actually available to individuals in the population. It is worth noting that resource density can sometimes be a poor measure of food availability (e.g. Zwarts & Wanink, 1993). Provided animals always attempt to maximize their intake rate (as this is assumed to maximize fitness), then describing the functional response is sufficient for a model. However, there is some evidence that animals do not always attempt to maximize their intake rate. For example, Norris & Johnstone (1998) showed that oystercatchers feeding on cockles (*Cerastoderma edule*) rarely fed at the maximum possible rate during a particular winter. Furthermore, Swennen et al. (1989) showed that if the feeding time of captive oystercatchers was reduced, they increased their intake rate. These data suggest that some birds might experience additional costs associated with feeding at the maximum possible rate. Such costs might include, for example, damaging their bills, an increased risk of parasitism or predation, or becoming too fat. Where such costs exist, predicting intake rates will be difficult, because they will not only depend on the benefits of obtaining food at a particular rate, but also on the costs of maintaining a given intake rate. This is a complex optimization problem, i.e. determining which intake rate maximizes fitness for a given set of environmental circumstances and behavioural decision rules. However, there is a theoretical framework for addressing this issue (see McNamara et al., 1994), which could be applied

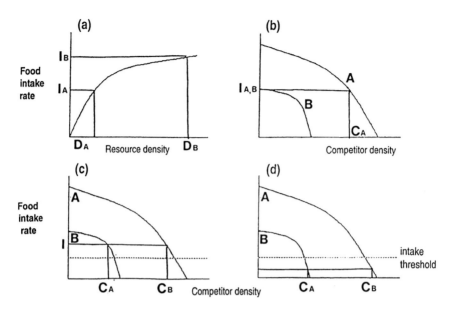

Figure 12.4 Ideal free models. (a) and (b) illustrate aspects of a basic ideal free model; (c) and (d) illustrate how the model can be used to determine whether a given population size is sustainable by the available resources, by assuming predators need to achieve a threshold intake rate to survive. The model describes a simple habitat consisting of two patches (A and B) that differ in resource density. For simplicity, we assume no prey depletion, and all predators are equally susceptible to interference. (a) Shows the functional response, and gives the resources densities in A (D_A) and B (D_B), and the respective intake rates predators could achieve by feeding alone in either patch (I_A and I_B respectively). (b) Shows the influence of competitor density on intake rates in each patch (interference). C_A shows the competitor density in patch A at which intake rates ($I_{A,B}$) are equal in each patch, so predators start to occupy patch B as well as patch A. (c) Shows that the abundance of predators is not sufficiently high for competitor densities in the two patches (C_A and C_B) to depress intake rates (I) below the threshold. This population can be sustained by the available resources. (d) Shows that predator abundance is sufficiently high for competitor densities in the two patches (C_A and C_B) to reach levels that depress intake rates (I) below the threshold. This population cannot be sustained by the available resources.

to conservation problems. The challenge will be in estimating the fitness costs and benefits involved.

In birds, interference competition has been described in a number of different families (e.g. Carpenter *et al.*, 1993; Kotrschal *et al.*, 1993; Bautista *et al.*, 1995; Ens & Cayford, 1996). Interference has been incorporated into ideal free models in two main ways (see van der Meer & Ens, 1997). First, some workers have simply described how intake rates change as competitor density increases (Sutherland & Koene, 1982; Ens & Goss-Custard, 1984; Goss-Custard & Durell, 1987; Dolman, 1995). This approach implicitly assumes that resource density and the distribution of competitive abilities within the population remains unchanged over the observed range of competitor densities (Milinski & Parker, 1991). However, this assumption is likely to be violated in many natural situations, the consequence of which is hard to predict. Second, recent studies have constructed simulation models of the behavioural process of interference (e.g. Ruxton *et al.*, 1992; Holmgren, 1995; Stillman *et al.*, 1997). It is also possible to incorporate individual differences in susceptibility to interference, which is common in natural populations (see Sutherland, 1996a for a review), within this modelling framework.

To predict maximum sustainable population sizes using ideal free models, it is important to estimate the intake rate animals must achieve to fulfil their energy requirements. When feeding for a given time period, this threshold intake rate can be viewed as the mean intake rate the animal needs to achieve during the feeding period. It is determined by the energy value of food (minus the energy required to warm the food to body temperature in the gut), the net costs of thermoregulation and behavioural activities (net cost refers to the fact that certain activities generate energy as heat), and the costs and benefits of storing any surplus energy in the form of fat. While there are good empirical estimates of the energy value of food (e.g. Zwarts *et al.*, 1996) and the costs of thermoregulation (e.g. Wiersma & Piersma, 1994), there are little empirical data on the energetics of particular behavioural activites, such as searching for food, handling food, fighting with competitors, etc. for free-living animals. However, the use of double-labelled water on wild and captive animals is now providing estimates of these various energetic costs.

The ideal free framework described above does not consider prey depletion, although we have already seen how important prey depletion is in determining animal abundance. It is relatively simple to incorporate prey depletion within ideal free models (see for example Sutherland & Dolman, 1994; Dolman & Sutherland, 1995b). Consider a non-breeding period (e.g. winter) divided into discrete time intervals. An ideal free model can be used to determine the distribution of animals between patches, depending on resource density and the effect of interference on intake rates, at time interval t. Next, depletion occurs over a specified time period (= feeding time) within time interval t, and resource densities are reduced. At time t_{+1}, the

ideal free model recalculates the distribution of animals between the feeding patches, depletion occurs again, and so on until the end of the non-
breeding period. If a threshold intake rate for survival is assumed, such a
model could be used to predict the maximum sustainable population size
over any number of time intervals.

Social structure models

Social structure can play a central part in the dynamics of animal populations. For example, it has been suggested that territoriality interacting with
kin groupings may cause cyclic population dynamics in microtine rodents
and red grouse (Krebs, 1985; Lambin & Krebs, 1993; Watson et al., 1994),
and a plausible population model based on this hypothesis has been developed (Mountford et al., 1990). It is thus important to consider the incorporation of social structure into predictive models of populations in many
cases. This is particularly so when the model is used to explore the effects of
management perturbations, because these may influence the population
indirectly through their effects on social structure, as well as directly
through their effects on survival or fecundity (e.g. Prins et al., 1994; White
& Harris 1995; Swinton et al., 1997).

Considering the fate of small populations, a number of studies have
incorporated social structure into population viability analyses (Lande,
1988b; Durant & Mace, 1994; McCarthy et al., 1994; Young & Isbell, 1994;
Green et al., 1996; Vucetich et al., 1997). These models consider such factors as social grouping, territoriality, and the effective availability of mates,
providing a framework by which key elements of social structure can be
incorporated into essentially demographic models.

Models of this kind are generally not explicitly behaviour-based, because they treat social structure as a black box, not allowing the behavioural
strategies of individuals to change in response to changing conditions. This
approach can be effective when considering very small populations; however, if we want to predict the responses of larger populations, we need a
model that is more fully behaviour-based. In many cases, this will require
that social structure is taken into account. For example, Sutherland (1996a)
presents a simple model in which territoriality in breeding oystercatchers
determines the overall breeding output of the population. The model is
based on the case of breeding oystercatchers in which there are two types of
territory with sharply differing profitabilities (Ens et al., 1992). Individuals
entering the breeding population thus choose to settle on a good territory if

one is available. However, if no good territories are free, newcomers must decide whether to breed immediately on a sub-optimal territory, or wait without breeding for a good territory to become available (Ens *et al.*, 1995). The optimal strategy for an individual can be calculated in terms of expected lifetime reproductive success, and depends on the expected annual survival rate of individuals and the size of the population. This model can thus be used to define a density-dependent response in the average reproductive output of the population, and hence to define an expected equilibrium population size. The predicted effects on population size of changing mortality rate can therefore be explored.

Social interactions are particularly important in species that form dense aggregations. Flocking in geese has important implications for variance in foraging performance between individuals, and any behaviour-based model of goose populations therefore needs to take account of this. When feeding on an abundant, uniform food such as pasture, geese are constrained to feed in flocks by the perceived risk of predation. This can be quantified in the form of a negative relationship between flock size and time spent vigilant; in large flocks, individuals spend little time vigilant, and thus spend more time feeding (Lazarus, 1978; Inglis & Lazarus, 1981). However, the benefits of flocking are countered by the effects of local depletion, because some members of tightly packed flocks are unable to avoid previously depleted patches (Black *et al.*, 1992). There are thus inequalities in foraging performance within flocks, and these are determined by the social structure of dominance hierarchies. In the case of barnacle geese *Branta leucopsis*, the most dominant individuals feed near the edges of flocks, where there is little localized depletion (Black *et al.*, 1992), whilst dominance itself is related to family status; large families are dominant over small families, which are dominant over pairs without offspring, which are dominant over single birds (Black & Owen, 1989). These are the essential behavioural characteristics that determine the relative foraging performance of individuals, which itself can determine the demographic performance of the population (Rowcliffe *et al.*, 1999). Empirical relationships determined for geese foraging in winter between flock size and time feeding, and between available biomass and intake rate, can be used to drive a model of their flocking behaviour and the intake rates of individuals, with individuals choosing where to feed on the basis of these relationships. A behaviour-based population model for geese must therefore take account of them. Determining the predicted intake rate achieved by all individual geese in the population over a winter season thus allows the consequences for average demographic rates to be predicted. The outcome will clearly be dependent on the initial size of the population, and on the availability of resources, and this model therefore allows the strength and timing of density dependence in demographic rates to be explored under a range of altered conditions.

Spatial scale

The above models attempt to understand patterns of animal dispersion, based on the assumption that travel is 'free' (i.e. is not costly). In some instances travel costs may be inconsequential (e.g. an oystercatcher moving between two adjacent mussel beds), but in other cases can be considerable (e.g. geese may disperse up to 35 km from their roost sites – in mid-winter, when foraging is already limited by short days, such travel is expensive both in terms of time and energy costs). Many animals range over much vaster areas, requiring considerable investment in terms of time and energy. In addition, the actual timing of migration and which staging sites to use are critical decisions in the annual cycle of many birds. Understanding the evolutionary basis of the decision rules used by animals is of interest not only to the theoretical biologist or behavioural ecologist (McNamara & Houston, 1986), but is of major import to site managers, policy makers, and conservation biologists: in terms of site protection and designation, what is the optimal distribution of sites within a flyway network between wintering ground and breeding area?

Three approaches have been taken in an attempt to answer this last question. The first is a theoretical exploration of the consequences of habitat loss using a novel approach (Sutherland & Dolman, 1994; Dolman & Sutherland, 1995b; Sutherland, 1996a): interference and depletion on the wintering (staging) sites result in density-dependent survival probabilities, whilst breeding success is also density-dependent. The evolutionary stable migration strategy is then calculated for a set of given parameters (including phenotypically determined competitive abilities, plus, more importantly in this instance, a variable energetic cost for reaching each site). Once the ESS is known for a given suite of parameter values, those of interest are changed (in this case the distribution of available wintering sites) and the ESS recalculated, thereby establishing the effects of losing a given site through habitat loss on both the global population and on individual site usage. The second approach is currently being developed by ourselves and co-workers (including A. Lang & A. I. Houston), where simulated individuals choose both departure rules (date and reserve levels) and arrival sites, based on simple optimization criteria, empirically determined rules, physiological costs and intake rates (see section below). The third makes use of stochastic dynamic programming (McNamara & Houston, 1986; Mangel & Clark, 1988), where state variables reflect differences between individuals and the environment, such that the optimum decision sequence is determined (see Weber *et al.*, 1997). This approach has now been successfully applied to describing the migration strategies of western sandpipers (*Calidris mauri*) moving between their wintering sites in southern USA, and breeding in western Alaska, making use of a number of staging sites along the western coast of North America (C. W. Clark & R. W. Butler, unpublished data). Their inputs into the model include a breeding window based on an empirically derived relationship between reproductive success, energy reserves and date of arrival on the breeding grounds, various environmental characteristics of each staging site, physiological costs and constraints of the birds, variable wind speeds and differential predation risk between sites. The model predictions matched the observed data with respect to arrival times and departure dates for each of the staging sites, and the authors were then able to examine the consequences of hypothetical loss of various staging sites on the population growth trajectory for this species.

Thus the scale at which we attempt to make predictions is partly dependent on the questions that need to be answered. However, even at a site-based level, the costs of movement between sites can be considerable. These costs, though, can easily be incorporated both in energetic terms and as time constraints. At larger spatial scales, the same approach can be adopted within the framework of game-theoretic distribution models. In addition, stochastic dynamic programming can be a useful alternative modelling approach allowing one to examine optimal patterns in site choice and timing of movement. In the following section, we combine the modelling of decision rules with the more traditional game theoretic models to examine individual behaviour feeding through to global population trajectories.

Population models

Currently, no explicit empirically derived, behaviour-based models of entire populations exist. The depletion models of grazing waterfowl to date are local, and ignore the consequences of birds being displaced to other sites.

Goss-Custard *et al.* (1995d,e) have developed an oystercatcher model that operates at a population level, but is not strictly behaviour based. However, Goss-Custard and ourselves are now developing such explicit population-wide, individual, behavioural, and game-theoretic-based models of two goose populations – the Svalbard barnacle goose and the dark-bellied brent goose (*Branta bernicla*).

Our own approach is outlined in Figure 12.5, which also describes the annual cycle of the Svalbard barnacle goose. Each of the stages in the annual cycle (over-wintering, spring-staging and breeding) is coded separately so that the effects of changing individual parameters in differing combinations can be explored both within and between seasons, and locally or globally. The input to and output from each stage are specified in a fixed format (a stage-structured population matrix) in which individuals are characterized by (1) age, (2) abdominal profile (fat reserves), (3) family size, (4) tradition (i.e. site occupied in each season), (5) quality (a variable that can be used to explore various individual differences in behaviour, particularly feeding efficiency and competitive ability) and (6) time of transfer to the next stage of the yearly cycle. The inputs to the model are a description of the characteristics of each site (including area) and an initial population structure. The outputs are the numbers of geese that use each site, total population size, and other attributes (e.g. age distribution) of the population. Within each season, models are specific to the particular area and phenology: thus in winter the model output is driven by the flock model (Rowcliffe *et al.*, 1999 and unpublished data) described above (see Social structure models), whilst on the spring-staging grounds, birds are distributed between agricultural fields (where they feed in flocks), and managed and unmanaged islands used for sheep grazing (where the geese feed in pairs, defending territories, thus following a despotic distribution based on their dominance). As on the wintering grounds, the spring-staging areas differ in the extent and quality of food available, resulting in considerable differences between pairs in both the rate of energy acquisition and the maximal reserve levels attainable (Lang *et al.*, 1999 and unpublished data). An optimization model is used to determine the departure dates of the birds from the spring-staging areas to the breeding grounds, whilst the allocation of birds to specific feeding grounds in spring or breeding colonies in summer is based on empirically derived rules.

It is hoped that such individual-based, whole-population models will allow both ecological questions to be explored [e.g. effects on λ of changes in demography, including population structure; sensitivity analyses and

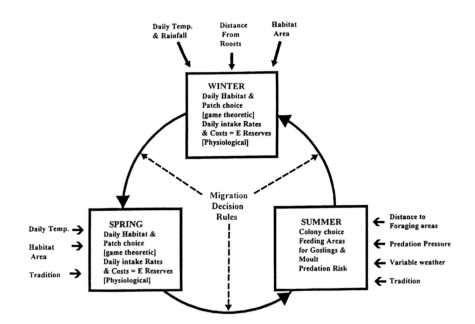

Figure 12.5 Schematic figure of our individual-based approach in modelling the annual cycle of the Svalbard barnacle goose population. Birds over-winter on the Solway Firth, United Kingdom, arriving in late September and departing late April/early May for the spring-staging areas in Helgeland, an archipelago off the coast of Norway. Dominance is a central feature of any goose's life throughout the annual cycle – in winter it determines its position within the flock, and consequently its intake rate (see Social structure models). The model is updated daily with respect to the intake rate of each individual stage, each stage having a certain mean and variation determined empirically. Their daily distribution within or between habitats is determined using a game-theoretic framework (see Social structure models). Birds leave the Solway for their spring-staging grounds on a variable schedule determined via an optimization model that examines the trade-off between reserves and date, and also determined by suitable weather conditions for migration. Birds again choose between habitat types (agricultural land where they feed in flocks, and managed or unmanaged islands, where they feed as pairs in a despotic manner). After attempting to put on further reserves, they depart for the breeding grounds on Svalbard. Breeding success is established partly in a deterministic manner (an empirically derived trade-off between clutch-size and lay-date) and partly stochastically (good/bad years, e.g. date of snow-melt, predation pressure by foxes, etc.). The birds also undergo a full moult in summer, when they are rendered flightless, and this and the vulnerability of their young to predation constrains them to 'safe' areas in which they can forage. The birds then need to make the return journey to the Solway, having put on sufficient reserves to complete the migration before inclement arctic weather sets in again on Svalbard.

exploration of the shape of some of the parameter functions (cf. van der Meer & Ens, 1997)], as well as those of more immediate interest to conservation biologists. These latter would include assessments of changes in habitat management practice (e.g. timing, extent and intensity of fertilizer applications and sheep and cattle grazing on the fields used for over-wintering by the geese; the re-introduction of sheep grazing on currently unmanaged islands in Helgeland) and direct loss of habitat (e.g. conversion of permanent pasture to either intensive agriculture or set aside) and other anthropogenic change (e.g. decline in extent of merse owing to sea-level rises).

Gaps in behaviour-based models

Most of the current game-theoretic models using empirically derived data are taxonomically biased [largely waterfowl – swans, geese, ducks and waders (shorebirds)]. This more reflects the ease of measurement of foraging behaviour (especially intake rates) and prey distribution and depletion (the prey tending to be sessile), rather than that the predator–prey system is somehow 'simple' (it is not!). However, even though the species currently studied tend to be the more tractable, there remain significant problems in developing behaviour-based models of entire populations: (1) defining the extent of available resources; (2) understanding constraints, such as knowledge animals have about the extent, quality and location of resources, and also genetic constraints, e.g. migration programming (see Sutherland, 1996a); (3) incorporating both demographic and environmental stochasticity, which are often ecologically important.

For small, closed populations, or studies at a local scale with clearly defined boundaries (e.g. an habitat island – see Sutherland & Allport, 1994), specifying the habitat or other resource available is not problematic. At larger scales, or, for example, where the local population is subject to high but variable immigration and emigration rates, then clearly defining resource availability and animal densities becomes more difficult. This becomes even more of a problem when we have no objective basis for defining constraints in, say, foraging behaviour. This applies not only to perceptual limits, but also in understanding the basis for patch selection. Thus in geese, 'tradition' plays a major role in the individual's choice of foraging area or breeding site. Some species, and individuals within populations, may show much greater exploratory behaviour than others. In some cases it may be possible to quantify patch choice or habitat avoidance in terms of perceived predation risk. In other instances, 'switching' may be sudden and currently unpredictable. However, the advantage of the behavioural-based approaches discussed above is that sensitivity analyses can be carried out, allowing 'what if' scenarios at least to be explored.

WHY USE BEHAVIOUR-BASED MODELS IN CONSERVATION?

Behaviour-based models are probably the most robust way of producing genuinely predictive models of animal populations, providing a reliable basis for forecasting what might happen if environmental conditions change. Predicting future changes in abundance from current observations of demographic rates and how these are affected by population density is fraught with problems. First, it can be extremely difficult to measure density-dependence in natural populations, especially if population density is relatively stable over time (Goss-Custard, 1993). Second, even if a density-dependent survival or fecundity function can be estimated from observed data, it is unlikely to be directly applicable to future environmental conditions, unless the prevailing conditions are very similar to those experienced by the population that provided the observational data. Behaviour-based models can potentially overcome both of these problems. They overcome the first problem because behaviour-based models can easily be used to predict density-dependence in fecundity or survival outside the observed range of population densities. Behaviour-based models overcome the second problem because animals in the population are allowed to respond flexibly to environmental change by adjusting their behaviour to maximize their fitness, and the model then simply uses the demographic consequences of these decisions to predict how abundance might change by generating new density-dependent functions. Therefore, behaviour-based models are of enormous potential value to conservationists who need some means of predicting the likely future impact of current plans. However, behaviour-based models can be complex and time consuming to construct. This is a problem for conservationists, who often require answers to problems such as, 'would bird numbers decline in this area if X happens?' over relatively short time scales. Can this potential conflict be resolved?

Behaviour-based models should simply be viewed as a potential tool to aid conservationists in assessing the impact of environmental change. This seems obvious, but is, nevertheless, extremely important. If the conservation issue concerns how to manage a particular habitat within a relatively small area for part of a more widespread animal population, then the type of model required would be 'local', and predictions are likely to be required within a relatively short timescale. Behaviour-based models have been successfully applied to such problems. To date, these applications have primarily concerned the management of grassland habitats for grazing wildfowl. However, there are other issues that could be approached using 'local' models, particularly the potential conflict between wintering wading birds and commercial shell-fisheries, or the effects of agri-environment measures designed to provide food for seed-eating birds on the abundance of seed-eating birds during the winter, for example. Behaviour-based models describing the dynamics of entire populations, during the breeding and non-breeding period, require considerably more information, and are, therefore, likely to take a relatively long time period to construct. Such 'global' models are likely to be of most value in addressing environmental change that affects entire populations, rather than local populations. A current example might be the impact of sea-level rise on populations of wading birds wintering on the east Atlantic seaboard. They are also valuable in exploring how changes in demography at a local level can affect the global population. Such questions are critical to conservation biologists and ecologists in general. It is also worth noting that 'global' models, once constructed, can be applied to 'local' problems. Indeed, they provide a distinct advantage over local models in that they can potentially consider the impact on abundance caused by birds being displaced from local populations to other areas as a result of environmental change. However, it is unlikely that a 'global' model is the best option if the conservation issue is local, if predictions are required relatively quickly, and if no current global model exists.

Conservation biology will come-of-age as a science when it is able to predict the consequences of environmental change on animal abundance. Indeed, conservation biology has been described as the acid-test of ecology – human-induced perturbations of the system under study are all potentially experimental manipulations which may be used to test the predictive abilities of ecological theory. We strongly believe that behaviour-based models provide an important development in increasing the confidence which can be placed in such predictions to novel circumstances.

ACKNOWLEDGEMENTS

Thanks to John Goss-Custard for first introducing us to behavioural, individual-based models, and to Tim Coulson for comments on an earlier manuscript, and to Jeff Black for discussion regarding geese in general and Svalbard barnacle geese in particular. Much of the thinking in this paper was carried out whilst holding NERC grants on this subject, and the Wildfowl & Wetlands Trust supported our study of the Svalbard barnacle goose population on their Caerlaverock reserve, as well as providing access to their long-term study data. The Institute of Zoology also core-funded two of us whilst the manuscript was in preparation.

13

Controversy over behaviour and genetics in cheetah conservation

TIM CARO

INTRODUCTION

Over the last 10 years there has been a debate within the field of conservation biology over the relative importance of genetic and environmental factors in determining the fate of small populations (see Caughley, 1994; O'Brien, 1994a; Frankham, 1995a; Caughley & Gunn, 1996; Hedrick et al., 1996; Clinchy & Krebs, 1997; Young & Harcourt, 1997. On the one hand, population genetics theory suggests that small populations will suffer from inbreeding depression and loss of heterozygosity while, on the other, ecologists say that populations rarely remain at small population size for a long enough period of time to suffer from these genetic problems. Either they will recover or soon go extinct for environmental reasons. Here I want to contrast genetic and environmental problems faced by small populations using a single case study, the cheetah (*Acinonyx jubatus*), and try to assess the importance of both factors. The cheetah is an endangered species that lives at low densities in sub-Saharan Africa (Nowell & Jackson, 1996). It once inhabited southern and central Asia but was exterminated by hunters (Divyabhanusinh, 1995), although a small population survives in Iran. The species holds a central place in conservation biology as it has been regarded as perhaps the classic example of a species being endangered because of its lack of genetic variation.

In this chapter, I shall go through this unresolved controversy about the importance of genetics and environmental factors influencing the fate of this species by first outlining the genetic problems faced by cheetahs; then examining some of the misgivings over the genetic findings; next reviewing behavioural and demographic discoveries in the wild; and then discussing the captive data in regard to juvenile mortality and disease. Finally, I shall attempt to assess the circumstances under which lack of genetic variation

Table 13.2
Summary of the genetic argument for reduced survival of cheetahs

Finding	Prediction	Evidence
Lack of genetic variation across loci	Increased juvenile mortality	Juvenile mortality high in captivity
Sperm abnormalities	Poor breeding performance	Reproduces poorly in captivity
Lack of variation at the MHC	Compromised immune system	Case study of disease susceptibility

polymorphic group of tightly linked loci in vertebrates. It encodes for cell surface antigens that are responsible for cell-mediated rejection of allogenic skin grafts. Because of the extraordinary polymorphism, grafts from unrelated donors are recognized as being foreign and are rejected within a few days. The cheetah findings again suggested that the cheetah genome lacked variability (see also Sanjayan & Crooks, 1996; Sanjayan *et al.*, 1996).

In theory, genetic monomorphism may result in reduced juvenile survival (e.g. Ralls *et al.*, 1988). This is because any deleterious recessive alleles will not be masked by dominant more benign alleles if loci are not heterozygous. In addition, monomorphic populations are likely to be susceptible to disease. This is because if a disease successfully overcomes the immune system of one individual, it will overcome every member the population because they are genetically identical (e.g. Falconer, 1989).

O'Brien *et al.* (1985) looked for these and other effects. Examining data on juvenile mortality in captive mammals, they pointed out that cheetahs were at the high end of the distribution with only reindeer (*Rangifer tarandus*) and dik-dik (*Madoqua kirki*) having higher non-inbred mortality (but see Loudon, 1985). Consequently, they reasoned that juvenile mortality was high in cheetahs probably as a result of genetic monomorphism. They also described an outbreak of feline infectious peritonitis (FIP) at the Wildlife Safari breeding colony. There, out of 42 cheetahs, 90% were infected with FIP and 18 died, while none of the 10 lions (*Panthera leo*) died of the disease. They argued the genetically compromised immune system prevented cheetahs from fighting off the challenge of FIP. The impression from this influential paper is that the cheetah is doomed because of the consequences of low genetic variation and this was taken up by the popular press (Cohn, 1986; Steinhart, 1992; Sunquist, 1992). Table 13.2 summarizes the genetic argument. The satisfying point was that theoretical predictions

Table 13.3
Comparison of genetic variation in carnivora and other mammals

	Mean ± SE (N)	
	Polymorphism	Heterozygosity
Carnivora	0.089 ± 0.088 (26)	0.028 ± 0.027 (26)
Mammalia	0.163 ± 0.129 (78)	0.042 ± 0.030 (81)

Adapted from Merola (1994).

were apparently confirmed by empirical observations, apart from the un-resolved issue of explaining the genetic bottleneck (Hedrick, 1987, 1996).

MISGIVINGS ABOUT CHEETAH GENETICS

Nevertheless, there are difficulties with the genetic argument. First, we know that not all cheetahs are genetically identical. For example, the so-called 'king cheetah' which has a blotched coat rather than a spotted one, was originally thought to be a different species principally confined to Bot-swana and Zimbabwe (Hills & Smithers, 1980; Bottriell, 1987). We now know, however, that the blotched coat pattern is simply controlled by a single recessive allele (van Aarde & van Dyk, 1986). Clearly there is genetic variation in at least some parts of the cheetah's geographic range. Second, fieldworkers can recognize individual cheetahs by the variable banding on their tails. For example, Caro and Durant (1991) scored the widths of the black and white bands on the tails of cheetahs, and calculated the degree of dissimilarity between them. They found that offspring resemble their mothers more than unrelated cheetahs. Although their data indicated that coat colour is affected by environmental influences *in utero*, family resem-blances are most attributable to shared genes between relatives. Again this suggests genetic differences between individuals.

A more formal questioning of the cheetah's homozygosity surfaced in 1994. Merola (1994) examined percentage polymorphism and heterozygos-ity among different classes of animals. She showed that mammals are at the low end of this continuum. Furthermore, among mammals, she showed that terrestrial carnivores have approximately half the levels of polymorphism and heterozygosity of other mammalian species (Table 13.3). Thus, if one compares cheetahs to other species or other mammals (as O'Brien *et al.*, 1985 did), one is bound to find relatively low levels of genetic variation. The appropriate comparison is to other terrestrial carnivores, where one finds that more than 30% of species exhibit levels of variability lower than the cheetah. This calls into question O'Brien *et al.*'s original finding that the cheetah is unique in its low level of genetic variation (but see O'Brien, 1994b; May, 1995). In addition, we now know that there is extensive variation in cheetahs for mitochondrial DNA, minisatellite and microsatellite DNA (Menotti-Raymond & O'Brien, 1993, 1995) and some variation in proteins at the MHC (Yuhki & O'Brien, 1990) although the authors argued that this variability is much lower than in other related species.

DATA FROM THE WILD

Behavioural observations

Thus far, this discussion of the cheetah's conservation predicament has been divorced from the field. Indeed, field research on cheetahs had, until the mid-1980s, been concerned principally with their ecology (Pienaar, 1969; Eaton, 1974) or behavioural ecology (Caro, 1994). For example, research that I began in 1980 in Serengeti showed that females live alone or with attendant cubs while males live alone or in small lifelong coalitions of two or three animals. Males live together not in order to hunt co-operatively, but to take over and defend small territories in which nomadic females collect. These and other studies had, nevertheless, indicated that cheetahs live at low population densities compared to other, similar sized carnivores even in protected areas, but no one knew the reason why (Schaller, 1972). None of the obvious reasons fitted: prey in the 15–60 kg range was usually abundant in the National Parks (Laurenson, 1995), and disease outbreaks had not been documented (Caro, 1994).

In order to investigate why cheetahs lived at low densities, Karen Laurenson started a project in which she radio-collared 20 female cheetahs. With the help of a light plane, she was able to locate females from the air twice a week, and then return to them on the ground and check their reproductive condition. By regularly sighting individual females, she could determine when they were heavily pregnant. Then, as soon as their signal came from the same place on subsequent flights, she could pinpoint the exact location of the mother's lair (Laurenson, 1993). Now she could observe the mother leave and enter the lair on hunting expeditions, and while the

Table 13.4
Mortality of cheetah litters and cubs

	Number in lair	Number emerging from lair	Number alive at 4 months	Number alive at 14–18 months
Litters	36	10	5–6	3–4
	(100.0)	(27.8)	(13.9–16.7)	(8.3–11.1)
Cubs	125	36	10–12	5–7
	(100.0)	(28.8)	(8.0–9.6)	(4.0–5.6)

Percentages in brackets.
Adapted from Laurenson (1994).

mother was away, she entered the lair to weigh, sex and count the cubs. She first found that if a female lost her cubs prematurely, she came back into oestrus and became pregant again within 3 weeks. It seemed that neither females nor males were reproductively impaired, despite the congenital sperm abnormalities in wild cheetahs (Laurenson *et al.*, 1992).

Next, through behavioural observations, she discovered that cheetah cub mortality was extremely high. Most of the cubs that were born never left their lairs. Of 125 cubs born, only 5–7 survived to independence at 14–18 months of age (Table 13.4) (Laurenson, 1994). From a combination of direct observations and piecing evidence together, Laurenson determined that the principal source of cub mortality was predation by lions. Lions would see the cheetah mother sit up in the marsh in which her cubs were hidden, and then make straight towards it. The cheetah would try to defend her cubs by growling and rushing at the lion, but a lioness weighs approximately four times as much as a female cheetah and the latter could not drive it off. On gaining access to the cubs, the lion would quickly kill them with a bite to the back of the head, although rarely consume them. In total, lions killed 73.2% of cheetah cubs born (Table 13.5). Although cubs died in other ways, none caused such high losses. In particular, it is worth noting that only 4.1% of cubs could have died from being non-viable, perhaps for genetic reasons. These behavioural observations (see also Durant, 1998b) contradict the idea that genetic factors are responsible for high juvenile mortality as theory predicts. Instead they show that an ecological factor, predation, is responsible for high cub mortality in the wild (Figure 13.1). A recent population viability analysis has confirmed the importance of juvenile mortality in influencing population growth rate (Kelly & Durant, 1999).

Of course, there are other explanations for high rates of predation on

Table 13.5
Causes and extent of cheetah litter and cub mortality

Cause of mortality	Litters	Cubs
Predation	10 (62.5)	35.5 (73.2)
Abandonment	2 (12.5)	4 (8.2)
Fire	1 (6.3)	4 (8.2)
Exposure	1 (6.3)	3 (6.2)
Non-viable cubs	2 (12.5)	2 (4.1)
Unknown	23	69.5–71.5
Total dead	39	118–120

Percentages in brackets.
Adapted from Laurenson (1994).

Figure 13.1 Cheetah, *Acinonyx jubatus*, cub less than 4 months old. The high rate of reproductive failure in cheetahs has been attributed elsewhere to high levels of homozygosity. However, subsequent research has shown that the principal source of reproductive failure of cheetahs in the Serengeti is cub mortality in the first 4 months of life and that this is mainly owing to predation by lions (photograph by Tim Caro).

Table 13.6

Median number of tracks made around lairs, and number of times lairs were
entered and litters were handled in each mortality classification

Mortality classification	N	Number of tracks made per day	Number of lair inspections per day	Number of times cubs handled per day
Survived	10	0.15	0.05	0.05
Abandoned	2	0.15	0.07	0.06
Predated	9	0.19	0.03	0.03
Environment	2	0.33	0.14	0.08
Unknown	13	0.20	0.04	0.04
Kruskal–Wallis test		$H = 1.02$	$H = 4.50$	$H = 4.10$
P-value		$P > 0.1$	$P > 0.1$	$P > 0.1$

From Laurenson & Caro (1994).

cheetah cubs. It could be that Laurenson's presence attracted predators, so
this had to be checked (Laurenson & Caro, 1994). We found that there were
no significant differences between the number of tracks made by Laurenson's vehicle around lairs in which cubs survived, or were abandoned, or
were predated. Nor were there any significant differences in the number of
times per day that lairs were inspected or cubs were handled between litters
that met different fates (Table 13.6). Nevertheless, the appropriate comparison is between litters that were watched and those that were not, and we
could not test for this.

Instead, Sarah Durant (unpublished data) performed field experiments
in which she played roars of lions and whoops of spotted hyaenas (*Crocuta
crocuta*) to female cheetahs. She found that the females' reproductive success (measured by number of surviving cubs/year) was significantly associated with an increased chance of looking at the speaker and of rapidly
moving away from it, especially in older females. This suggests that females that are more assiduous in avoiding sympatric predators raise more
cubs per unit time.

Demographic data

Recently, Marcella Kelly has examined long-term demographic records of
cheetahs in the Serengeti spanning a 25-year period and can look at the
impact of lions over time (Kelly *et al.*, 1998). Her first data set consisted of
prey and predators seen within an estimated 1 km radius of cheetahs,
sighted on the Plains between 1981 and 1990. She compared a suite of

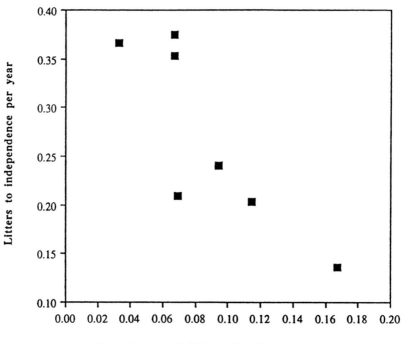

Figure 13.2 Litters raised to independence per year plotted against the proportion of sightings that lions were < 1 km away while cheetah mothers had cubs < 4 months old; $n = 7$ females, $r_s = -0.901$, $P < 0.05$ (from Kelly *et al.*, 1998).

different predator and prey variables to the reproductive output of females, restricting analyses to only those seven females for which we had complete records of reproductive careers. Results showed that lifetime reproductive success was significantly negatively correlated with the presence of lions within 1 km of the cheetah family containing cubs of less than 4 months of age (Figure 13.2). Measured over a mother cheetah's lifespan, the more often lions were present when she had young cubs, the lower her reproductive output. Note that these results were obtained from non-invasive observational data only and thereby circumvent O'Brien's (1994b) suggestion that Laurenson attracted predators to cheetah lairs (see also Laurenson *et al.*, 1995a,b).

Second, Kelly *et al.* (1998) compared cheetah reproductive parameters over two time periods, 1969–1979 when lion density was low on the Plains, and 1980–1994 when lion numbers nearly doubled. Table 13.7 shows that

Table 13.7
Average reproductive parameters of female cheetahs in times of low and high lion abundance on the Serengeti Plains

	1969–1979		1980–1994
Average number female lions/year:	26.5		42.5
Litter size at emergence (4 months)	2.9	**	2.1
	(23)		(58)
Litter size at independence	2.5	*	2.0
	(20)		(83)
Cubs to independence over whole life			
All females	2.1		1.6
	(22)		(87)
Breeders only	4.6	**	3.2
	(10)		(44)
Cubs to independence/year			
All females	0.42		0.36
	(20)		(84)
Breeders only	3.54	*	1.45
	(10)		(44)

Number of females in brackets.
$* \ P < 0.1$, $** \ P < 0.05$.
From Kelly *et al.* (1998).

average litter sizes at 4 months and at independence declined between the two periods. Furthermore, the average number of cubs reared per lifetime declined, and reproductive rate declined as lion density increased. It is unlikely that other environmental factors caused a decrease in cheetah reproductive success. Thomson's gazelle (*Gazella thomsoni*) numbers, the chief prey, remained stable; and we could find no effect of rainfall or even cheetah numbers on reproductive rates on an annual basis. As litter size in the lair was no different in the 1970s and 1980s (Laurenson, 1995), but litter size at 4 months of age and at independence declined, more cubs must have died between birth and leaving the lair in the 1980s. Since predation is the chief source of cub mortality, increased predation appears the most plausible candidate for the cheetah's declining reproductive success in recent years. Modelling similarly shows that the probability of the Serengeti cheetah population becoming extinct is strongly affected by lion abundance (Kelly & Durant, 1999).

Are lions a significant factor in areas other than Serengeti? Circumstantial data from farms in Namibia where lions have been exterminated show that litter sizes of 10-month-old cubs are almost double that in Serengeti; up to four cubs per litter reach independence there (McVittie, 1979). More

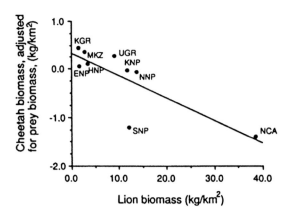

Figure 13.3 Cheetah biomass adjusted for prey biomass plotted against lion biomass across nine protected areas in Africa, $y = 0.34 - 0.046x$. ENP, Etosha National Park, Namibia; HNP, Hwange National Park, Zimbabwe; KGR, Kalahari Gemsbok National Park, R.S.A.; KNP, Kruger National Park, R.S.A.; MKZ, Mkomazi Game Reserve, Tanzania; NCA, Ngorongoro Conservation Area, Tanzania; NNP, Nairobi National Park, Kenya; SNP, Serengeti National Park, Tanzania; UGR, Umfolozi Game Reserve, R.S.A. (from Laurenson, 1995).

rigorously, Laurenson plotted cheetah biomass against lion biomass across nine different areas in Africa (Laurenson, 1995). Once she had corrected for prey biomass, she found that cheetah biomass was negatively correlated with lion biomass (Figure 13.3). This suggests that the Serengeti results hold up elsewhere.

In summary, six pieces of evidence suggest that lions are a crucial ecological factor lowering juvenile survival in the wild (Table 13.8).

THE CAPTIVE DATA

Demographic data on juvenile mortality

What about the data that showed high juvenile mortality in captive cheetahs (O'Brien *et al.*, 1985)? Since several problems seem to bear on these data too, Nadja Wielebnowski re-examined juvenile mortality in captivity. Using cheetah studbooks that give the pedigree histories and breeding success of all cheetahs in captivity in North American institutions, she discerned that juvenile mortality has declined with time (Wielebnowski, 1996). In particular it decreased as zoos became more accomplished at breeding

Table 13.8
Evidence that predators cause high juvenile mortality in wild cheetahs

Serengeti
1. Observations of lions killing cheetah cubs (Laurenson, 1994).
2. Enhanced response to experimental playbacks of predators in females with higher reproductive output (Durant, 1998b).
3. Correlations between presence of lions at cheetah sightings and cheetah reproductive output (Kelly *et al.*, 1998).
4. Differences in cheetah reproductive output between periods of high and low lion densities (Kelly *et al.*, 1998).

Namibia
5. Very large litter sizes near independence when lions are absent (McVittie, 1979).*

Across Africa
6. Negative correlation between cheetah biomass and lion biomass (Laurenson, 1995).

* Data obtained from farmer questionnaires and should be treated with caution

cheetahs in captivity and as more information was passed between them. She also found that juvenile mortality rates differed markedly across institutions. Some breeding facilities never bred a cheetah while others were forced to place a moratorium on breeding as so many were being produced. Both findings suggest that different management techniques must play a strong role in breeding success. Moreover, they speak against genetic factors being responsible for poor litter survival because such great variation would not be expected if homozygosity was the key to low cub survival.

In addition, Wielebnowski (1996) compared the causes of mortality in cheetah cubs sired by related and unrelated parents. Here she found that while extrinsic sources of mortality did not differ between the two groups, intrinsic sources did. Inbred cubs were more often stillborn than outbred cubs (Table 13.9). This suggests that cheetahs have sufficient variation at loci affecting juvenile survival to cause differences in juvenile mortality rates. This affirms that deleterious alleles are being expressed in inbred cubs, but not in all cubs, as was originally suggested.

In regard to poor breeding success in captivity, workers at the San Diego Zoo have noted that females can become pregnant after a single mating, indicating that sperm abnormalities may have little effect on reproduction (Lindburg *et al.*, 1993). In a final twist to the story, Dave Wildt himself conducted an enormous survey of the reproductive anatomy and

Table 13.9
Number of inbred and non-inbred cubs dying of different causes at five North American breeding facilities between 1970 and 1994

Causes of mortality	Non-inbred	Inbred
Extrinsic (fatal injury, husbandry, maternal problems)	31 (35.2)	6 (21.4)
Intrinsic (stillbirths, neonatal, premature, congenital)	33 (37.5)	19 (67.9)
Other (various infections, various other)	24 (27.3)	3 (10.7)
Unknown	15	1
Total	103	29

Adapted from Wielebnowski (1996).
Percentage of known causes are in brackets.

physiology of captive female and male cheetahs. He found no differences between proven breeders and non-breeders (Wildt *et al.*, 1993). These findings also independently point to the role of husbandry in determining whether breeding will be successful.

Disease

The other leg of O'Brien *et al.*'s (1985) argument was that cheetahs were particularly susceptible to disease as a result of their genetic monomorphism (O'Brien & Evermann, 1988). Evidence in favour of this hypothesis is that, first, 18 of the cheetahs but none of the lions died in the breeding facility in Oregon. Second, there is evidence of poor T-cell responses to feline herpesvirus in captive cheetahs (Miller-Edge & Worley, 1992) as well as increased susceptibility to feline leukemia virus (Briggs & Ott, 1986), FIP (Evermann *et al.*, 1988) and leukoencephalopathy. Evidence against is, first, that although many cheetahs died in Oregon, 24 survived, indicating a variability in response (Heeney *et al.*, 1990). Second, some wild cheetahs test seropositive to agents such as herpesvirus, feline coronavirus, immunodeficiency virus, and toxoplasmosis (see Caro & Laurenson, 1994) and some captive cheetahs produce antibodies to herpes and and calcivirus vaccinations (Spencer & Burroughs, 1991). All of these findings indicate a variability in immune response (see Evermann *et al.*, 1993) and show that no disease has yet circumvented the immune defenses of all cheetahs. Indeed, no disease outbreaks have been noticed in the wild, but it should be remembered that they have only been studied for the last 25 years and only at five major sites.

Table 13.10
Summary of findings highlighted in this chapter

Finding	Prediction	Evidence	Explanation
Homozygosity	—	—	Comparison groups may be inappropriate
Lack of genetic variation	Increased juvenile mortality	Wild	Yes but because of lions
		Captive	Yes but past husbandry
Sperm abnormalities	Poor breeding performance	Wild	Not supported
		Captive	No longer supported
Lack of MHC variation	Compromised immune system	Wild	No evidence
		Captive	Studies equivocal

THE EMERGING CONCENSUS

Table 13.10 pulls these various strands of evidence together and shows that although the predictions arising from low heterozygosity are apparently met, they are for non-genetic reasons. In short, what started off as a startling discovery, the cheetah's genetic uniformity, led to a series of predictions derived from genetic theory. These predictions are correct. However, it appears to have been too hasty to state that these genetic predictions were born out in the case of the cheetah. The cheetah does suffer problems but they appear to be of an ecological nature. There has been a history of poor husbandry in captivity, heavy predation occurs in protected areas, and outside protected areas information is accumulating that diverse anthropogenic factors are responsible for low numbers (Table 13.11). The only equivocal piece of the puzzle is whether cheetahs are unduly susceptible to disease. Behavioural and demographic data from the wild and from captivity have therefore been instrumental in overturning our view of the cheetah's conservation predicament.

OTHER SPECIES

Moving beyond cheetahs, however, there are now several studies from other species showing that reduced genetic variation has phenotypic consequences on reproduction and survival of individuals in wild populations just as it does in captive populations (Table 13.12). These effects certainly exist in the wild, but from a conservation perspective, the cheetah case study forces us to ask whether these manifestations are the most important factor controlling their populations (Caro & Laurenson, 1994). Indeed, in

Table 13.11
Probable causes of cheetah population declines in selected regions of Africa outside national parks

Cause	Area	Country	Source
Human population increase	Northern Region	Malawi	Gros (1996)
Direct killing by pastoralists	Kenya rangelands	Kenya	Gros (1998)
Direct killing by farmers	Namibian cattle ranches	Namibia	Morsbach (1986)
Overhunting of ungulate prey	Karamoja	Uganda	Gros (1997)

Table 13.12
Demonstrated effects of reduced genetic variation in wild populations of selected animals

Species	History of low heterozygosity	Individuals currently inbreeding?	Manifestations of low genetic variation	Source
European adder	Recent	Yes	Small litters Deformed young	Madsen et al. (1996)
Song sparrow	Occasional	Some	Differential loss in cold weather	Keller et al. (1994)
Sonoran topminnow	Recent	Yes	High mortality* Slow growth rate*	Quattro & Vrijenhoek (1989)
Florida panther	Recent	Yes	Testicular dysfunction	Roelke et al. (1993)
Ngorongoro lion	Very recent	Yes	Reduced yearling production	Packer et al. (1991)
White-footed mouse	Recent	Some (experiment)	Lower survival Male weight loss	Jiminez et al. (1994)
Cheetah	Long	No	Sperm abnormalities	Wildt et al. (1987)
Glanville fritillary butterfly	Recent	Yes	Low larval survival Low adult longevity Low egg-hatching rate	Saccheri et al. (1998)

* Sheffer et al. (1996) failed to replicate these findings.
European adder: *Vipera berus*; song sparrow: *Melospiza melodia*; Sonoran topminnow: *Poeciliopsis occidentalis*; Florida panther: *Felis concolor*; Ngorongoro lion: *Panthera leo*; white-fronted mouse: *Peromyscus leucopus*; cheetah: *Acinonyx jubatus*; Glanville fritiallary butterfly: *Melitaea cinxia*.

Table 13.13
Summary of the most important factors controlling populations of wild animals listed in Table 13.12

Species	Protected where study conducted?	Most important factor controlling population size in historical past
European adder	No	Agriculture
Song sparrow	No	Cold weather
Sonoran topminnow	No	Stream changes Introduced species
Florida panther	Yes	Hunting & habitat destruction
Ngorongoro lion	Yes	Ectoparasites, infanticide & possibily low heterozygosity
White-footed mouse	Not relevant	Not relevant
Cheetah	Yes	Predation
Glanvill fritillary butterfly	No	Human-induced environmental changes

See Table 13.12 for scientific names.

each of these examples, a recent history of low heterozygosity or current inbreeding appear less important than ecological factors in controlling population size in historical times (Table 13.13). Certainly reduced genetic variation may predispose a small population to extinction, but it needs sufficient time to act. It seems that other, more pressing environmental factors will come into play first.

The real question facing conservation biologists is whether a particular manifestation of low heterozygosity makes individuals additionally suspectible to environmental insult. It is difficult to see how low genetic variation could predispose cheetah cubs to escape stochastic predation events; but it is easier to see how song sparrows (*Melospiza melodia*) with high inbreeding coefficients might succumb to cold weather. Therefore, as conservation biologists, we need to know more about the interaction between the genetic landscape and environmental stresses in different species if we are going to make generalizations about the interactions between these factors that undoubtedly occur.

ACKNOWLEDGEMENTS

I thank Sarah Durant, Paule Gros, Marcella Kelly, Karen Laurenson and Nadja Wielebnowski for interesting discussions over the years, and Joel Berger, Phil Hedrick and Marcella Kelly for comments.

14

The role of animal behaviour in marine conservation

JOHN D. REYNOLDS & SIMON JENNINGS

INTRODUCTION

The marine environment has a much richer taxonomic diversity than the terrestrial world. Twenty-eight of the world's 33 phyla are found in marine habitats, compared with only 12 phyla in terrestrial habitats. Indeed, 21 phyla are wholly marine, whereas the terrestrial environment boasts only one phylum – Onychophora – that is not also found in marine habitats (May, 1994). Many of the species that comprise the broad taxonomic diversity in marine habitats remain poorly understood. For example, the first observations in 1996 of a living oarfish (*Regalecus glesne*), which can be up to 9 m long, seemed to lack the popular press one would expect for a large mammal.

Conservation of marine animals has received correspondingly little attention, with the exception of a few popular taxa such as cetaceans (Norse, 1993). A recent review of papers published during the first 9 years of the journal *Conservation Biology* (1987–1995) showed that of 742 papers that could be assigned to habitats, 5% were marine, 9% freshwater, 67% terrestrial, and the rest were general (Irish & Norse, 1996). There were no obvious trends in marine emphasis over this period. The International Union for the Conservation of Nature (IUCN) has recently stepped up its interest in marine fishes, with efforts to improve the criteria for assessing exploited species. This follows some highly controversial listings of exploited fishes in its latest *1996 Red List of Threatened Animals* (IUCN, 1996). For that list, only 10% of the world's 23 000 + fishes (fresh water and marine) had been assessed.

The lack of attention to conservation in aquatic habitats does not necessarily reflect a lack of threats to populations. An obvious potential problem is exploitation by fisheries (for food, power plants, fertilizer, the aquarium trade, and traditional medicines). In 1995, world fish production was 113 million tonnes and over 80% of this came from marine environments (Anon., 1997a). The fact that total catches are increasing much more slowly than total fishing effort suggests serious cause for concern about the conservation of many of the world's fish stocks (Anon., 1994). A recent review of the status of world fishery resources suggested that 57 of the 80 major stocks investigated were already fully or overexploited (Anon., 1994). Indeed, since at least 1950 there has been a shift toward lower trophic levels in global fisheries, especially in the Northern Hemisphere, lending further evidence to the consensus that the world's fisheries are not sustainable (Pauly et al., 1998). The reasons for these gloomy predictions are understandable: marine resources have often been regarded as common property, thereby suffering the tragedy of the commons (Hardin, 1968), with too many vessels chasing too few fish. Management is hampered by difficulties in obtaining accurate data on past trends in population dynamics and mortality, let alone the uncertainties in predicting the future against a back-drop of massive climatic effects for many species. Furthermore, management options must track continuing technological improvements in capture efficiency, often driven by economic incentives for increased participation and investment in the fishing industry (Anon., 1994, 1997b).

The direct effects of fishing are not, however, the only threat to many marine species. Large numbers of non-target species including invertebrates, fishes and marine mammals are killed accidentally during fishing operations (Alverson et al., 1994; Simmonds & Hutchinson, 1996; Jennings & Kaiser, 1998). Moreover, in some coral reef systems, fishing may induce ecosystem shifts between alternate stable states and lead to changes in the structure and composition of algal, coral, fish and invertebrate communities (Hughes, 1994). Additional threats to marine species include coastal development, coral mining, thermal and chemical pollution (Birkeland, 1997), as well as waste products from mariculture.

What role might animal behaviour play in marine conservation? One might expect that the history of fisheries research as a leading branch of applied ecology would have yielded some concrete answers. However, some of the biggest breakthroughs in fisheries theory have intentionally avoided the behaviour of individuals, to simplify analyses based on life tables or relationships between capture rates and exploitation effort (Russell, 1939; Schaefer, 1954; Beverton & Holt, 1957). Here, unfortunately, we have parallels with traditional studies of terrestrial population dynamics, though this situation is changing (Sutherland, 1996a; Clemmons & Buchholz, 1997).

To examine the role of behaviour in marine conservation, we first try to set the stage with some examples of declining populations, which can be compared with the terrestrial examples that dominate this book. We then consider how behaviour affects susceptibility to mortality, as well as the potential for population recovery. Our examples are drawn heavily from studies of commercial exploitation of fishes, because this has had the largest impact on animals in the marine environment. We will also consider the role of behaviour in non-targeted species such as seabirds, as well as effects of pollution, coastal development and other activities in the near-shore zone that affect many estuarine and migratory species.

THE PROBLEM: EXAMPLES OF DECLINING POPULATIONS

There are many examples of population declines in marine species that parallel those observed in their terrestrial counterparts. Consider three cases (Figure 14.1). Commercial fisheries were the main culprit in a 99% decline of the northern cod (*Gadus morhua*) stock off the Atlantic coast of Canada between 1962 and 1992 (Hutchings & Myers, 1994; Walters & Maguire, 1996) (Figure 14.1a). The commercial collapse of this stock put 40 000 people out of work. The abundance of adult (10 years and older) western Atlantic bluefin tuna (*Thunnus thynnus*) declined by the same percentage from 1970 to 1990, while under intensive study by scientists concerned about their populations (Figure 14.1b, ICCAT, 1991). Finally, a more familiar case is the blue whale (*Balaenoptera musculus*) (Figure 14.1c). The annual catch of these animals in the Antarctic declined by 97% from the winter seasons of 1930/31 to 1962/63, when restrictions came into effect (Small, 1971). This reduced catch was not the result of lower hunting effort, as the catch per day's hunting showed a similar decline, despite a trebling of the size of ships over that period, and a quadrupling of engine power.

In addition to these population declines, fishing has been responsible for the local extirpation of various grouper species in the tropical Indo-Pacific (Hudson & Mace, 1996) and 'common' skate *Raja batis* in the Irish Sea (Brander, 1981). However, it is widely believed that the probability of the extinction of marine species resulting from fishing remains low in comparison with well-documented effects of hunting on terrestrial animals (Steadman, 1995), owing in part to large distributions, and because the costs associated with pursuing marine species to extinction can be very high (Beverton, 1990).

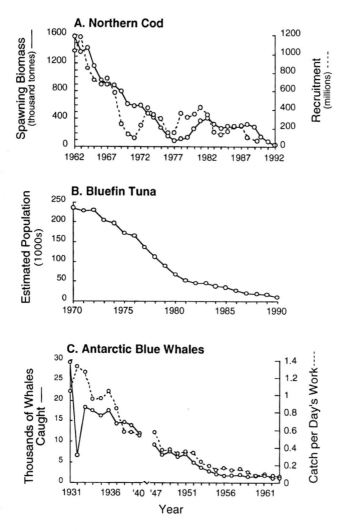

Figure 14.1 Changes in population sizes of three marine species: (a) northern cod stock (*Gadus morhua*) off eastern Canada (Bishop *et al.*, 1993, Hutchings & Myers, 1994). Spawners are 7 years and older, recruits are 3 years old; (b) western Atlantic bluefin tuna (*Thunnus thynnus*) (ICCAT, 1991); (c) blue whales (*Balaenoptera musculus*) in the Antarctic, including annual catches (solid line) and catch per hunter-day's work (broken line) (Small, 1971).

Are declines in marine species really conservation issues? Except for whales and dolphins (Simmonds & Hutchinson, 1996), these population declines have traditionally been confined to the realms of 'management failures'. However, what is the difference between a fish's 'management failure' and a whale's 'merciless persecution towards extinction'? Or, for that matter, the decline of tigers? The rates of decline have been strikingly similar for cod and blue whales, and indeed it was the recognition of this fact that allowed Atlantic cod to join the blue whale on the *1996 Red List of Threatened Animals* (IUCN 1996). The Atlantic cod's listing as 'Vulnerable' is less severe than the 'Endangered' status of blue whales, but on the basis of declines, it was well within the IUCN criterion of a 20% decline in 10 years for 'Vulnerable' status.

Perhaps the fish examples are different from whales or many of the more familiar terrestrial birds and mammals discussed elsewhere in this book because we think we can control the threats by stopping fishing. This would be news for the western Atlantic bluefin tuna, which continued to be exploited as its numbers declined because the value per fish reached as much as US$30 000 in Asian sashimi markets. Indeed, efforts to have the species listed on Appendix 1 of the CITES convention in 1992 proved too controversial, though agreements were reached to reduce quotas by 10% in 1992–93, and by 25% in 1994–95 (ICCAT, 1991). For a variety of understandable political, social, and economic reasons, the Hilborn–Walters Principle holds: 'The hardest thing to do in fisheries management is reduce fishing pressure' (Hilborn & Walters, 1992).

One might imagine that another difference between population declines in fishes and terrestrial vertebrates could be that the latter have slower intrinsic rates of natural increase, and are therefore less able to bounce back once threats to their numbers are removed. This is not an unequivocal rule. Some marine species have very low intrinsic rates of increase: those of marine mammals are well known (Beddington *et al.*, 1985) and many fishes have late maturity and low fecundity. Population growth rates of northern cod have been estimated between 0.09 and 0.17, which is comparable to many terrestrial vertebrates (Hutchings *et al.*, 1997). For example, the thornback ray *Raja clavata*, does not reach maturity until an average age of 10 years, and females produce only 50–150 eggs per year, with an incubation time of 4–5 months (Holden, 1971). Moreover, even species with 'fast' life histories, such as herring, may take decades or more for populations to recover after exploitation has ended because their reproductive success is so closely linked to environmental factors (Beverton,

1990; Bailey, 1991). It is a sobering thought that at the time of writing, 6 years after the 1992 closure of the northern cod fishery off eastern Canada, there is still no sign of recovery, and evidence of further declines in some areas.

In some coral reef areas there is also the worry that ecosystem changes associated with the decline of key species may prevent other species from ever recovering, irrespective of their potential rates of natural increase (Done, 1992; Hughes, 1994). Thus, declines in many populations of marine species suggest that both exploited and non-target marine species are not exempt from serious conservation problems, though so far fishes have proved comparatively robust against extinction.

Behaviour has two potential links to conservation, which should be kept distinct. First, behaviour may affect the vulnerability of individuals to being killed or having some aspect of their life cycles disrupted. This may occur, for example, owing to animals foraging in a way that makes them vulnerable to traps. Second, behaviour may play a key role in translating such individual mortality into population changes. This depends on density-dependent interactions between individuals and aspects of their life histories that affect intrinsic rates of natural increase (Jennings et al., 1998). Although we have been able to pull together many examples of the ways in which behaviour may affect individual mortality, deductions about population consequences have had to be much more speculative for some kinds of behaviour. Below we consider key behaviours relevant to conservation.

SHOALING

Shoaling behaviour has many functions, including foraging advantages and avoidance of predation when predators kill only one or a few individuals at a time (Pitcher & Parrish, 1993). However, this behaviour is exploited by fishers, who can surround entire schools with nets (Figure 14.2). This can yield profitable catches even when total population sizes are low (Pitcher, 1995; Mackinson et al., 1997). An example was provided by the Peruvian anchoveta (*Engraulis ringens*), which collapsed dramatically in the early 1970s (Pauly et al., 1987). In the decade from 1962 to 1971 the mean annual yield was 9.7 million tonnes, but from 1976 to 1985 this fell to 1.3 million tonnes. As the population declined, owing to a combination of commercial fishing pressure and the loss of ocean productivity during an El Niño event in 1972/73, the remaining fish formed shoals that continued to

Figure 14.2 Two-spot snapper (*Lutjanus biguttatus*) shoaling above a coral reef in Fiji. Their shoaling behaviour means that high catch rates with spears, fishing lines, and nets can be maintained even when their overall abundance is low (photograph by S. Jennings).

yield profitable catches. This further hastened the species' decline (Figure 14.3).

In traditional reef fisheries, fishers have long observed the spawning behaviour of their favoured target species in order to fish within spawning aggregations (Johannes, 1980) and the abundance of such species has often been reduced by intensive fishing (Hudson & Mace, 1996). When open sea spawning grounds can be trawled by large vessels, intensive fishing has led to dramatic population declines. For example, shoals of spawning herring *Clupea harengus* in the southern North Sea were easy targets for large trawlers, which maintained high catch rates even though the population size was dangerously low. By 1976 the herring stock had collapsed, but has since started to recover following closure of the fishery (Cushing, 1992).

Shoaling behaviour by cetaceans can be detrimental when individuals remain near others that have been wounded or killed. This has been noted for gray whales, *Eschrichtius robustus*, for example (Bogoslovskaya, 1986). It has also been noted that males are more likely to remain with females than vice

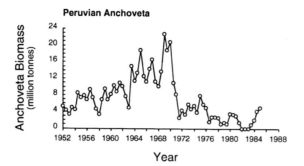

Peruvian Anchoveta

Figure 14.3 Population changes in the Peruvian anchoveta (*Engraulis ringens*) (Pauly & Palomares, 1989). The tendency of the fish to form shoals around food sources kept the fishery economically viable and exacerbated their decline.

versa. We speculate that this asymmetry might be the result of sexual selection. There is also evidence that whales in groups of three or fewer tend to swim more slowly than single individuals when pursued, because they maintain their synchronous swimming behaviour (Bogoslovskaya, 1986).

A practical concern for conservation is that high catchability caused by schooling behaviour may cause major difficulties for obtaining accurate stock assessments. This occurred in the Canadian cod, where scientific surveys indicated low population sizes, while commercial catches remained high (Walters & Maguire, 1996; Myers *et al.*, 1997). With hindsight, it is obvious that as the population declined, fish concentrated in favourable areas, and fishers tracked them. At the time, however, the uncertainty from conflicting signals between scientific and commercial surveys caused a disastrous delay in conservation measures.

SWIMMING BEHAVIOUR

For several species of fishes it has been shown that the ability to avoid nets depends on swimming speeds and escape responses (Wardle, 1993). Small fish are less able to outrun trawl nets than larger ones, because the ability to maintain aerobic swimming speeds is directly proportional to body length. Indeed, large species such as cod, haddock (*Melanogrammus aeglefinus*), saithe (*Pollachius virens*), and mackerel (*Scomber scombrus*) have been seen swimming for long periods of time in the mouth of trawl nets, having little trouble keeping up with the towing speed (Wardle, 1993). However, a recent experimental study has shown that even if fish manage to outlast the duration of the net tow, the stress afterward may still be fatal (Olla *et al.*, 1997). For example, in the case of juvenile walleye pollock (*Theragra chalcogramma*) a prolonged swim at 0.65 m/sec for 3 hours led to high mortality up to 6 days later (Olla *et al.*, 1997).

The behaviour of different species after they become exhausted has further implications for mortality and management. Haddock tend to rise upward as they drop back into the net, whereas cod stay low (Main & Sangster, 1982a). This means that nets can be designed with a horizontal separation and different mesh sizes above and below, so that different sizes of each species can be selected. Similarly, the bycatch of small fishes can be reduced by fisheries for Norwegian prawn (*Nephrops norvegicus*) by taking advantage of the fact that prawn tend to stay within 70 cm of the seabed, whereas with the exception of flatfishes, most young fish rise upward (Main & Sangster, 1982b). Attention to such details of behaviour could have important conservation implications, because shrimp and prawn trawls are made from very fine meshes and by-catches of fish and invertebrates may be five to ten times larger those of target species (Alverson *et al.*, 1994).

An understanding of differences in the behaviour of shrimps and by-catch species has provided the basis for net designs that reduce by-catch mortality. Sea turtle populations in many areas have been threatened by shrimp fisheries because they drown in nets. For example, it has been estimated that as many as 50 000 turtles are caught annually by prawn trawlers in the south-eastern United States, of which about 20% die (Henwood & Stuntz, 1987). Since 1990 it has been a legal requirement to fit turtle excluder devices (TEDs) to nets in the north-western Gulf of Mexico, and turtle mortality has fallen (Caillouet *et al.*, 1996). TEDs usually provide a large opening at the top of the net through which the turtles can escape. The loss of target shrimps is minimized because shrimp tend to stay low in the net. A study in Queensland, Australia showed a mean reduction in penaeid prawns owing to TEDs of 29%, but that this depended heavily on the site and season (Robins-Troeger, 1994). TEDs have also been shown to lead to significant reduction in by-catches of fish species (Rulifson *et al.*, 1992; Robins-Troeger, 1994).

FEEDING BEHAVIOUR

Feeding behaviour determines susceptibility to various methods of fishing, as well as vulnerability to being killed as a by-catch. Attention to the details of foraging behaviour can help us to understand which individuals are susceptible. For example, red king crabs (*Paralithodes camtschaticus*) make only a limited number of approaches to traps ('pots'), and then search for a route that gives access to the food over a very small area (Zhou & Shirley, 1997). The existing trap design has small entrances, and given the restricted searching behaviour of the crabs the probability of their entering the trap is relatively low. This inefficiency probably helps inadvertently to sustain the population.

Dominance behaviour by crustaceans while feeding can affect which individuals are caught. Experiments with traps that had been stocked with American lobsters (*Hommarus americanus*) have shown that the presence of lobsters deters crabs (*Cancer irroratus* and *C. borealis*) from entering traps (Richards *et al.*, 1983). Similar results have been found with interactions between European lobsters and *Cancer pagurus* (Addison, 1995). Thus, behavioural interactions between individuals can reduce levels of mortality. The implications of this at the population level deserve further study; strong density-dependence owing to competition for suitable habitat and shelter may mean that populations could absorb increased mortality of dominant individuals or species. On the other hand, studies of behaviour of European lobsters suggest that there may be an 'underworld' of individuals that never enter traps, and are therefore missed by censuses, and these may sustain the population (J. T. Addison, personal communication).

Feeding behaviour may be important for conservation owing to interactions between predators and prey. A combination of intensive fishing and a climatic shift to warmer temperatures led to the collapse of the Norwegian and Barents Sea capelin *Mallotus villosus* stock (Hamre, 1991; Blindheim & Skjoldal, 1993). This had profound effects on the distribution of the capelin predators such as cod (*Gadus morhua*), seals and seabirds (Figure 14.4). Some seals died of starvation, but others migrated from the Barents Sea to search for alternative food sources. This behavioural change led to a large proportion of the seal population dying when they were trapped in fishing nets (Haug *et al.*, 1991).

Scavenging species of fish, invertebrates, and seabirds may benefit from commercial exploitation. For example, fish species and invertebrates congregate in areas impacted by trawling disturbance and feed on organisms damaged or exposed by the passage of the net (Caddy, 1973; Kaiser & Spencer, 1994, 1996; Ramsay *et al.*, 1997). The behaviour and population biology of many seabird species have changed profoundly in response to the discarding of fishes at sea. In the North Sea, for example, almost 0.5 mil-

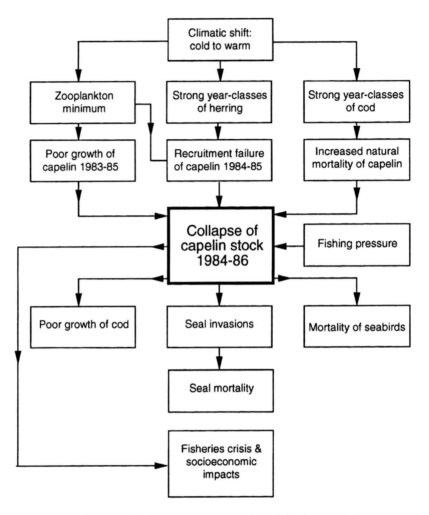

Figure 14.4 Relationships between the ecological events in the Barents Sea ecosystem from the climatic shift in the late 1970s to the departure of seals in the late 1980s (based on Blindheim & Skjoldal, 1993).

lion metric tonnes of fish, offal and benthic invertebrates are discarded annually (Camphuysen *et al.*, 1993). This provides enough food to maintain about 2.2 million seabirds. This additional food supply has had marked effects on bird behaviour, with many individuals feeding almost exclusively on fisheries discards. Discarding has been linked to the tenfold increase in the number of breeding seabirds in the North Sea from 1900 to 1990 (Furness, 1996).

The dependence of seabird populations on fisheries discards and offal production was shown by the effects of the closure of the bottom fishery in eastern Canada in 1992 owing to the collapse of northern cod stocks. Reproduction by black-legged kittiwakes (*Rissa tridactyla*) suffered from three interacting effects: loss of food from fisheries, delayed availability of an important prey item, capelin (*Mallotus villosus*) owing to anomalous cold water in the region, and intense egg predation by greater black-backed gulls (*Larus marinus*), which were also suffering from the first two problems (Regehr & Montevecchi, 1997). Egg predation led to low hatching success, while food shortages determined chick fledging success.

Dolphins often feed in association with tuna, to their detriment when purse seiners target the latter. Smith (1983) estimated that the population sizes of three dolphin species caught by tuna boats in the eastern tropical Pacific were reduced to 20%, 35–50% and 58–72% of pre-exploitation levels by 1979. Tuna are also caught using drift nets, lures (the pole and line fishery) and long lines. Pole and line fisheries are highly selective and there is little or no by-catch. However, the other methods lead to by-catches of sharks, seabirds, turtles and marine mammals. Deep sea drift netting was finally banned by the United Nations in 1992 but longline fishing is still permitted. Hooks are baited and set out on lines that may be over 100 km long. Before the hooks have sunk, they may attract seabirds, which become hooked and drown. It has been estimated that 44 000 albatrosses were killed each year in the southern oceans by Japanese tuna longlines (Brothers, 1991). A solution is to set the hooks out from below the water surface, so that they are beyond the reach of the birds.

MIGRATION AND DISPERSAL

Most commercially exploited fishes migrate, either entirely within the sea, or between the sea and fresh water. The latter species, termed diadromous, are particularly at risk from fishing, pollution in rivers and estuaries (which tends to be higher than in the open sea), and alteration to rivers and spawning grounds owing to reduced river flows, removal of riparian forests, and logging roads (McDowall, 1992). They also suffer from high catchability owing to their predictable movements and concentrations around river mouths and in rivers, and are susceptible to low-tech fishing methods such as nets that can scoop out fish from behind weirs. The main problem faced by species that migrate entirely within the sea is that stocks may enter the territorial waters of a number of countries, leading to political difficulties in bringing in effective conservation policies (Anon., 1997b).

Fishes undergo ontogenetic shifts in their migratory habits that determine their susceptibility to exploitation. For example, European bass, *Dicentrarchus labrax*, undertake short tidally related migrations in confined estuarine nursery areas until they reach 3–4 years of age and then start to migrate over hundreds of kilometres at maturity (Pawson *et al.*, 1987). The tidal migrations in nurseries make them vulnerable to pollution, barrage construction and habitat reclamation, and also make them highly accessible to fishers (Jennings, 1992). This behaviour, however, can also make conservation management easier because their confinement makes it easy to pinpoint sites that need protection. Indeed, many of the most important nursery areas for this species are now closed to fishing to protect young fish. As adults, the wide-ranging bass are much harder for fisheries to target and their mortality rates are lower.

Tidal migrations of many species, including bass, make them susceptible to power generators. Studies at the Annapolis Royal Barrage in the Bay of Fundy have shown that around 25% of young shad, *Alosa sapidissima*, in the water passing through the turbines may be killed (Dadswell *et al.*, 1986).

Recovery of diadromous populations often depends on restoration of rivers. This means removing or circumventing obstructions, for example, by using fish ladders and elevators to get around dams. For salmonids, freshwater spawning grounds can be enhanced or created by digging channels with suitable gravel and water flow. Artificial hatcheries are less desirable from a conservation point of view, because they can weaken the will of policy-makers to protect natural spawning habitats (Meffe, 1992). Furthermore, it is not clear what, exactly, is being 'conserved'. The 'population' relies on continual artificial propagation, and the replacement of sexual and natural selection on the spawning grounds with a hatchery environment causes genetic and phenotypic alteration of fish, including their behaviour (e.g. Fleming & Gross, 1989; Fleming, 1994; Fleming *et al.*, 1996; Olla *et al.*, 1994; Ruzzante, 1994). The resultant 'hatchery' phenotype of species, such as coho salmon (*Oncorhyncus kisutch*), includes reduced spawning coloration and smaller jaws that would normally be used for fighting.

Migratory behaviour determines the benefits of closed area management. Small closed areas on tropical reefs may provide high levels of protection for fishes that make short localized migrations (Holland *et al.*, 1993; Samoilys, 1997), but even large closed areas such as the mackerel (*Scomber scombrus*) 'box' of south-western Britain, which was over 200 km by 200 km, will provide only limited protection for a species that migrates over 1000 km and spends much of its life outside the protected area (Lockwood, 1988).

RESPONSE TO DISTURBANCE

The behaviour of fishes may change following fishing activity and this can affect catches and scientific assessments of the fishery. In areas where spears are used for fishing, the fishes rapidly become shy or wary and catch rates fall. Even disturbance by divers conducting fish counts for stock assessments may have a similar effects. Harmelin-Vivien *et al.* (1985) recorded apparent declines in fish abundance when consecutive visual surveys were conducted in the same area. Evidence from tropical reef fisheries also suggests that fishing disturbance may stimulate the redistribution of fish populations. When Sumilon Island Reserve in the Philippines was opened to fishing, Alcala & Russ (1990) attributed the significant increase in the abundance of parrotfishes (Scaridae) on the reef slope to their having been driven from normal reef habitats to deeper areas by the use of drive netting techniques in which fishers use weighted lines and palm fronds to scare fishes towards their nets. We still know relatively little about the effects of large commercial trawlers moving through spawning aggregations of fish, but there is evidence for cod, for example, that spawning may be disrupted for a considerable distance to each side of the net as the fish move away from the disturbance.

HABITAT CHOICE: REPRODUCTION AND PERMANENT RESIDENCE

Fish aggregation devices (FADs) attract many pelagic fish species and provide a focus for recreational and commercial fishing. These usually consist of a raft of rope, matting and other material, which is anchored to the seabed. FADs employed in the US Virgin Islands, for example, have been shown to increase catch rates of jacks (Carangidae) and mackerels (Scombridae) (Friedlander *et al.*, 1994). The reasons why most pelagic fishes aggregate in the vicinity of FADs are unclear but a few smaller species may use FADs as spawning sites. Thus flying fish (*Hirudichthys affinis*) in the

Caribbean are caught by setting nets around floating mats of sugar cane leaves that are offered as spawning sites.

Many species of crabs and lobsters are also limited to specific habitats, in the form of rocks and cobble containing crevices or burrows for safety from predators and for mate guarding during breeding (Bauer & Martin, 1991). This makes the animals susceptible to traps, which can target such areas. Molluscs such as mussels and periwinkles rely on shallow rocky habitats, and those that choose intertidal sites are particularly vulnerable to exploitation. On the other hand, some spawning sites are safe from certain forms of fishing gear, as in the case of rocky habitats, used by temperate rays (*Raja*) to anchor their egg cases. These are unsuitable for trawling, thereby safeguarding adults from a major form of fishing in temperate waters. Artificial reefs also attract fish and can make them more accessible to fishers. There has been debate as to whether they simply attract fish or actually increase production. A recent review by Pickering & Whitmarsh (1997) concluded that the effects of artificial reefs are largely the result of attraction: i.e. they work through taking advantage of behavioural traits. As such, they are a threat rather than a conservation benefit if fishing is not controlled adequately.

HABITAT CHOICE: NURSERY AREAS

Many marine species use confined nursery areas as juveniles. These are often inshore areas, such as estuaries, mangrove swamps, creeks and saltmarshes, offering safety from predators, better food supplies and seasonally higher temperatures for faster growth. However, the confinement of juveniles in these areas also makes them particularly susceptible to fishing, pollution, barrage construction, water abstraction and reclamation (Boehlert & Mundy, 1988). On the British coast, for example, European bass densities are 1000 times greater in estuarine nurseries than on adjacent open coasts (Jennings, 1992). While regulations protect bass from fishermen in many of these habitats there have been no attempts to ensure that the habitat itself is also protected (Jennings *et al.*, 1991).

Power stations and power-generating barrages can cause high levels of mortality in fishes which use estuarine nurseries and an understanding of the fishes' behaviour can provide a basis for reducing such mortality. Young salmon, for example, are killed on power station cooling-water intakes as they travel from rivers to the sea and a Welsh power station was even shut temporarily as a result of Atlantic salmon (*Salmo salar*) smolts accumulating on filter screens (Turnpenny *et al.*, 1985). Behavioural responses of fish can be used to reduce the probability of mortality within nursery areas. Thus sluices, fish by-passes, bubble screens and low-frequency sound emissions can deter some species of fish from passing into turbines or into power station cooling water intakes (Davies, 1988).

REPRODUCTIVE BEHAVIOUR

A recent review of fishes has suggested that species with parental care may be particularly sensitive to threats to populations (Bruton, 1995). In the case of marine species, it is not clear whether this is owing to parental behaviour *per se*, or the fact that species exhibiting care of the young typically inhabit coastal areas, which are more susceptible to pollution. Pollution has been shown to have numerous sub-lethal effects on reproductive behaviour, including disruption of courtship and care (Jones & Reynolds, 1997). Few exploited species exhibit parental care, though if the retention of embryos and birth of live young is considered a form of 'care', many species of sharks and tropical rays qualify (Dulvy & Reynolds, 1997). Species with parental care generally have low reproductive potential, owing to associations with low fecundity, long development times of embryos, and advanced age at maturity. Populations of such species may therefore be more likely to decline under exploitation, and take longer to recover. The most obvious examples of this are large species of whales, which have lactation periods ranging from 6 to 7 months in some species and as high as 2 years in the case of the sperm whale, *Physeter macrocephalus* (Tillman & Donovan, 1986). Indeed, there is anecdotal evidence that reproduction in some cetaceans is disrupted by removal of other members of social groups. For example, it has been suggested that older female sperm whales may share in suckling calves (Tillman & Donovan, 1986).

In the fishery for the Dungeness crab (*Cancer magister*) in British Columbia, Smith & Jamieson (1991) have reported that size limits effectively bar the smaller females from the catch, while adult males are caught. As a result, females dominate the adult population and males greater than the size limit are rare. An assessment of male mating activity and the sizes of mating pairs suggested that mature females would have difficulties finding a sexual partner in intensively exploited fisheries and that this would adversely affect total egg production.

Some commercially important species of shrimps and fishes change sex as they become larger. Sex change also occurs in many species caught for the aquarium trade. The impact of this on exploited populations depends on whether the larger (targeted) sex is male or female, and on whether the trigger for sex change is a social cue (e.g. the absence of larger males) or a strictly size-related cue (Sadovy, 1996). For example, in the case of shrimps, stronger fishing pressures on the larger females may reduce the population and slow the rate of recovery because females are more likely to limit the population than are males. This can be deduced from the potential rates of reproduction of each sex and hence the direction of sexual selection (Clutton-Brock & Parker, 1992; Reynolds, 1996). Sex change in some fishes is determined by local demographic conditions, including sex ratios (Warner, 1988; Ross, 1990). In species where females change to males according to social cues when the latter are removed, this could alleviate potential fertilization problems following the loss of the larger males as a result of fishing. Thus, knowledge of the behavioural mechanisms of sex change helps in understanding population responses to exploitation.

The same logic about which sex is likely to limit the population applies to sessile marine invertebrates, but here there is evidence that fertilization can indeed be reduced at low densities (Levitan *et al.*, 1992). In bay scallop (*Argopecten irradians*), for example, recruitment is positively related to adult abundance (Peterson & Summerson, 1992). Some invertebrate species exhibit behavioural responses that may increase reproductive success at low densities. Thus immature giant scallops (*Placopecten magellanicus*) appear to aggregate with adult scallops when their densities are low. These small-scale aggregation patterns were proposed as an adaptation which increases fertilization success (Stokesbury & Himmelman, 1993). Studies of such behaviours may increase our understanding of population consequences of fishing.

INTRODUCED SPECIES

There has been growing concern about the effects of introduced species in the marine environment (e.g. Carlton & Geller, 1993; Carlton, 1996; Beardmore *et al.*, 1997). For example, aquaculture programmes are responsible for the accidental introduction of large numbers of farmed Atlantic salmon into the wild each year (Gausen & Moen, 1991). These fish are adapted, both genetically and phenotypically, to artificial rearing facilities

(Fleming, 1995). Behavioural studies have shown that domesticated females are competitively inferior against wild fish, constructing fewer nests, covering nests less efficiently, and being more likely to have their nests destroyed by other females (Fleming *et al.*, 1996). Males performed even worse, attaining only 1–3% of the fertilization success of their wild counterparts, in part owing to inferiority during the intense sexual selection that occurs over competition for mates. Although this may seem good news for the genetic integrity of wild populations, even small amounts of reproductive success can lead to significant introgression of genes for domestic behaviour and morphology.

FUTURE BEHAVIOURAL-RESEARCH NEEDS

Historically, fish populations were rarely considered to present a conservation problem, because empirical evidence often suggested that there was no risk of extinction and that populations would recover as soon as fishing mortality was reduced (Beverton, 1990). However, in the last decade, political and social pressures have forced fishery managers to consider the wider impacts of fishing on marine ecosystems (Norse, 1993). By-catch mortality of dolphins, seals, turtles, sharks and other species have led to growing conservation concerns. This has been coupled with the recognition that fishing may cause ecosystems to shift to radically different species compositions, even if the total fish biomass remains constant (Figure 14.5). Thus, fisheries scientists are now forced to consider wider conservation issues and not simply the management of single fish populations. Much marine management and conservation is still based on population and trophic models, but there are areas in which an understanding of behaviour can lead to improved marine conservation.

Behavioural interactions lie at the heart of all predator–prey relationships, and the capture of marine species by humans is no exception. Behavioural interactions also underlie density-dependence, which in turn underlies the ability of populations to withstand and recover from exploitation. We have identified a number of key areas in which an understanding of behaviour could help to improve marine conservation. The behaviour of target and by-catch species determines their susceptibility to capture in fishing nets and initial studies have shown that remarkable reductions in by-catch mortality are possible if nets are designed to account for these behaviours (Wardle, 1993). Although turtle excluder devices are now a legal

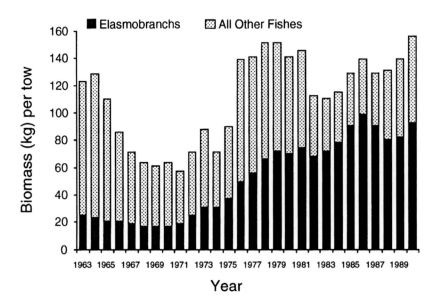

■ Elasmobranchs   ▨ All Other Fishes

Figure 14.5 Total biomass (kg) per tow (time-series smoothed) of fishes caught in autumn bottom-trawl surveys on Georges Bank, north-west Atlantic. Elasmobranchs include skates (*Raja*), spiny dogfish (*Squalus acanthius*) and other sharks, and 'other fishes' include primarily cod (*Gadus morhua*), haddock (*Melanogrammus aeglefinus*), winter flounder (*Pseudopleuronectes americanus*), and yellowtail flounder (*Limanda ferruginea*). Based on Murawski & Idoine (1992).

requirement in some shrimp fisheries, there is considerable scope to reduce the high levels of by-catch in many other fisheries (Alverson *et al.*, 1994). A knowledge of diurnal and seasonal migration patterns in fishes and their responses to habitat are needed if artificial reefs, marine reserves and protected nursery grounds are to be effective conservation measures. Further studies of reproductive competition on the spawning grounds of diadromous species are vital for making effective recommendations for conserving the genetic integrity of wild fishes in the face of expanding aquaculture programmes. Moreover, an understanding of the behavioural responses of fishes at low population levels helps to explain why catch rates can remain high at low abundance and why fisheries-based assessment data should often be treated with caution.

The role of behaviour in population recovery is less well understood. Although there is good evidence for behavioural competition for food, mates, and habitats, it is very difficult to carry out the experiments necessary to quantify density-dependence in any habitat, let alone one where it is often impossible even to see the study animal! One way forward may be to make comparisons of population dynamics of species that differ in their extent of habitat specificity, and hence limitation. Such comparative studies could take advantage of existing data from population surveys in conjunction with measurements of habitat quality. Habitat manipulations and direct observations are often possible in species that stay in one place, including intertidal molluscs, crustaceans that use shelters, and many reef inhabitants, such as echinoderms and fishes. Finally, recent advances in computer technology have led to satellite tracking of species that surface periodically, including cetaceans, turtles, and sharks, as well as computerized data storage tags that record temperature and pressure, enabling researchers to retrace the routes taken by migrating individuals when the tags are returned by commercial fishers (Metcalfe & Arnold, 1997). These are exciting advances because for the first time they allow us to compose a picture of populations and ecosystems based on the behaviour of individuals. A fuller understanding of these links should provide a sound basis for understanding how humans affect marine organisms, and how better to conserve them.

ACKNOWLEDGEMENTS

We thank Valerie Debuse and Nick Dulvy for helpful suggestions, Tom Webb for help with the figures, and Ian Fleming and Jeff Hutchings for comments on the manuscript.

PART V

Conservation applications of behaviour

15

Communication behaviour and conservation

PETER K. MCGREGOR, THOMAS M. PEAKE & GILLIAN GILBERT

INTRODUCTION

Communication is widely held to be one of the most important and ubiquitous of behaviours and one that underlies every major aspect of animals' lives (e.g. Hauser, 1996). Signals have evolved to transmit information and to modify the behaviour of receivers (e.g. Stamp Dawkins, 1995). It is not surprising, therefore, that these same signals can provide human observers with information of value to conservation, for example, many species of birds can be identified by their songs alone, a fact used by many surveys (e.g. Emlen & DeJong, 1992). Conservation can also make use of the behaviour-modifying properties of signals, for example, playing back song to elicit territorial responses from males in order to accurately map songbirds' territories (Dhondt, 1966; Falls, 1981) and count secretive species (e.g. Ratcliffe et al., 1998).

The aim of this chapter is to highlight particular aspects of the relationship between communication behaviour and conservation that are rarely dealt with in the literature (cf. Clemmons & Buchholz, 1997). This chapter is not an extensive review of such practical uses (e.g. McGregor & Peake, 1998), nor is it a manual for their implementation (for such a manual see Gilbert et al., 1998). We generally use studies on bird vocalizations with which we have been involved as examples to discuss a range of general issues, some of which are practical, some are theoretical and some are speculative. However, all of these issues show the close association between communication behaviour and conservation.

INDIVIDUAL IDENTIFICATION

In this section we discuss the approaches that have been used to identify individuals and their associated costs and benefits. When discussing individual distinctiveness it is important to distinguish two ways in which such features can be used, namely to discriminate between individuals and to identify individuals. Discrimination simply involves deciding whether two sets of observations (e.g. sound spectrograms) represent one or two individuals. Identification involves deciding which individual is represented. Discrimination is a much less demanding task but also much more limited in its potential application in conservation; essentially it is only useful for census purposes. Identification has much more to offer conservation but presents difficulties at several levels of consideration, therefore it is discussed at length below. We end the section with specific examples in which individually distinctive vocalizations have been used in a conservation context.

Why identify individuals?

It is generally true that more types of ecological data and of higher quality can be collected from study populations when individual animals can be identified (Hammond et al., 1990; Newton, 1995; McGregor & Peake, 1998). A large number of techniques have been devised in order to identify individuals, the majority of which involve catching animals and adding external marks or devices (e.g. Table 2.1 in McGregor & Peake, 1998). These sorts of technique can provide invaluable information, but may be inappropriate for a variety of reasons. For example, marks may alter the behaviour of animals (McGregor & Peake, 1998) or their prey (Norman et al., 1999). Also some species may be difficult to catch or difficult to track post-release and it may be desirable to avoid any disturbance associated with capture. For some of these species, identification based on individual characteristics (e.g. Table 2.2 in McGregor & Peake, 1998) may be a feasible alternative to marking and/or could provide important additional information.

Differences in species-specific acoustic signals would seem to have considerable potential in this respect; vocal individuality has been shown in a large number of bird species (Falls, 1982) as well as many mammals including cetaceans (Janik et al., 1994), primates (Butynski et al., 1992) and bats (Masters et al., 1995). Acoustically distinct signals have the potential to be used as an alternative to marking and have provided valuable conservation information on species that breed out of sight in dense vegetation such as bitterns Botaurus stellaris (McGregor & Byle, 1992; Gilbert et al.,1994) and corncrakes Crex crex (Peake, 1997; Peake et al., 1998).

Qualitative techniques

Researchers have been aware of the potential of individually distinct vocalizations to monitor individuals for more than 20 years (Beightol & Samuel, 1973), however it is only relatively recently that access to affordable recording and analysis equipment has meant that these sorts of techniques can be realistically considered in conservation efforts run on tight budgets. Levels of vocal individuality have been determined in a number of ways, with the early methods involving the visual comparison of spectrogram representations of signals by human observers (Hutchinson *et al.*, 1968).

The main attraction of this method is that the human eye can make use of many features of signal structure that may be very difficult to quantify. For example, in Figure 15.1 it is easy to see the within-individual similarities and between-individual differences in the shapes and relative positions of elements in spectrograms of yodel calls of the black-throated diver, *Gavia arctica*. It is not easy to imagine how such features of spectrograms could be described adequately by simple measures taken from the spectrograms. However, it has been shown that such complex and consistent patterns can be accurately matched by eye, even by inexperienced observers (Gilbert *et al.*, 1994). Accurate matching of vocalizations that produce simple spectrogram patterns is less likely and the likelihood decreases as the number of spectrograms involved increases (Gilbert *et al.*, 1994). There are also general difficulties in visual comparison of spectrograms in ensuring consistent assessment within and between observers and with the subjective assessment of the degree of similarity between two spectrograms (e.g. Gilbert, 1993).

Multivariate measures

Quantitative approaches avoid many of the problems associated with qualitative methods and multivariate methods of analysis are now particularly common. The multivariate technique most commonly reported in the literature is discriminant function analysis (DFA, e.g. Galeotti & Pavan, 1991; Bauer & Nagl, 1992), which creates linear combinations of measured signal parameters in such a way as to best separate signals recorded from different individuals.

It is worth sounding a cautionary note about techniques like DFA because of their prevalence in the literature. This analysis has proven useful when it has been unclear which spectrogram measurements provide the key to individuality, in particular when the most useful measurements may

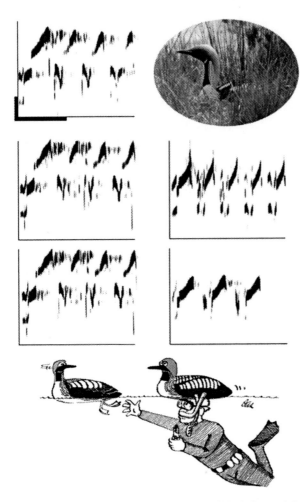

Figure 15.1 An illustration of the potential of black-throated diver, *Gavia arctica*, yodel vocalizations to identify individuals. Conventional marking with leg rings is problematic because black-throated divers rarely venture onto land except to nest (top right; from C.H. Gomersall, RSPB Images), therefore rings are difficult to see. Also individuals are difficult to catch; the cartoon by Ken Otter (above) shows one of the more successful catching techniques developed by Sjolander in Sweden (Gilbert, personal communication). By contrast, yodels can be recorded from distances that avoid disturbance and yodels are individually distinctive (Gilbert *et al.*, 1994). Spectrograms made from recordings (Canary 1.2.1; 24 kHz sampling rate, 350 Hz bandwidth, grid resolution 2.9 ms by 43 Hz; scale bars in top left indicate 500 Hz and 2 s) show the high level of similarity in three yodels recorded from the same individual (left column) and the considerable differences between yodels from two different individuals (right column).

change according to which individuals are included in the analysis (Gilbert, 1993). They also demonstrate whether individual variation is present at a level that will be useful for discriminatory or re-identification purposes. However, even when used with caution alongside visual comparison of spectrograms, there are problems with using DFA as a re-identification tool because DFA can only classify signals to individuals that are already in the 'library' of known individuals. The signals of new individuals will not be recognized as such and they will be attributed to one of the known individuals in the library. It is highly unlikely that, for any populations of conservation concern, all individuals have been recorded and that the identity of recruited individuals is known, therefore, DFA has an extremely limited use in the identification of individual animals for conservation purposes.

Similarity measures

Techniques that provide a measure of the degree of similarity between two signals (or representations of signals) do not need the complete knowledge of individuals required by DFA. A number of studies have looked at measures of similarity either using measurements taken from representations of signals (e.g. Peake *et al.*, 1998) or by direct comparison of representations (e.g. cross-correlation of spectrograms, Lessells *et al.*, 1995; McGregor *et al.*, 1994).

In such studies, identification becomes a question of setting the threshold level of similarity between two signals above which the signals are considered to come from the same individual (McGregor & Peake, 1998). Viewed in such a way, the problem of setting threshold criteria of similarity for identifying individuals can be seen to be very similar to that involved in signal detection (Wiley, 1994), namely that it is impossible to simultaneously maximize rejection of misidentifications and maximize acceptance of correct identifications. This problem is shown in Figure 15.2, where there is overlap in the frequency distribution of within-individual similarity values and that for between-individual values; regardless of the criterion chosen, mistakes will occur (for further discussion see Figure 2.2 in McGregor & Peake, 1998).

Realizing the conservation potential of individual distinctiveness

A large number of published papers report high levels of individual distinctiveness in signals and conclude with the authors suggesting that the

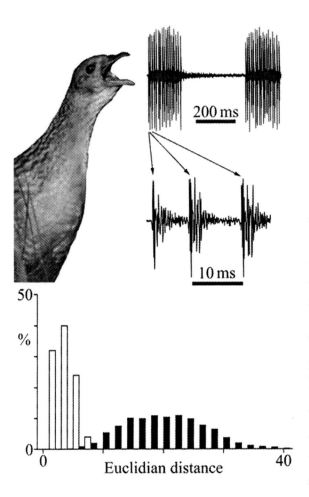

Figure 15.2 An illustration of how overlap in distributions of similarity within- and between individuals creates an optimization problem when setting threshold criteria for re-identification (further details in text). This example is based on the crake vocalizations of male corncrakes, *Crex crex* (left; from C.H. Gomersall, RSPB Images). The waveform representations show the two-syllable crake ('*crex crex*') and detail of the first three pulses from the first *crex* syllable (Canary 1.2.1 software, 24 kHz sampling rate, 350 Hz bandwidth, time resolution 0.7 ms). The frequency distributions plot similarity measures for comparisons of crakes produced by the same male (open histograms) and by different males (filled histograms). Similarity is represented by Euclidean distances where 0 = similar, 40 = dissimilar (see Peake *et al.*, 1998 for further details).

technique has potential to provide conservation information for the study species. However, as we saw with DFA techniques above, this conclusion is premature to the extent of being disingenuous. We are aware of only one census in which individually distinct vocalizations are routinely used; that of bitterns in the UK by the Royal Society for the Protection of Birds (Gilbert *et al.*, 1998). We suggest that it is time to move on from studies showing a level of individual distinctiveness to studies which specifically address the more important question: are these techniques useful in a conservation context?

There are several general reasons why the apparent potential of individually distinctive vocalizations is not realized. We briefly discuss some of these, before emphasizing some of the advantages of the technique that we consider ensure that it will have a place in conservation biology. We conclude this section with two specific examples that illustrate how these advantages and disadvantages interact to determine whether individually distinct vocalizations have practical conservation applications.

Some general disadvantages

Some limitations of the technique may be obvious from the outset. For example, in many bird species only males produce the long-range advertizing vocalizations that have generally proved individually distinctive (McGregor & Byle, 1992). Therefore only this subset of the population can be sampled by the technique. More subtle subsets may exist within populations (for example, unpaired territorial males, see Gibbs & Wenny, 1993) and it is important to be aware of these before conservation conclusions are drawn. When examining vocal individuality in any given species it is necessary to be assured that the original sound recordings came from different individuals; this may not always be easy. Where re-identification over time is desired it will be necessary to show that vocal distinctiveness remains constant over time. Some independent means of identifying individuals for repeat sound recording is therefore required. Some obvious general costs of the technique are the time taken, and expertise required, to analyse recorded material. When analysis is completed there is still the problem of setting re-identification thresholds (see Figure 15.2) and thus any identification has a level of ambiguity.

Ultimately the usefulness of the technique is determined by the balance of the costs of obtaining the information (in terms of the time and effort required to collect and analyse recordings; the levels of ambiguity involved in identification) and the benefits in comparison with other techniques (discussed below). We mention these points to stress the importance of feasibility studies considering all of such points from the outset and that the usefulness of these techniques is considered objectively.

Some general advantages

The benefits of the techniques can be considerable. In the majority of cases the only alternatives to vocal individuality involve capture on at least one occasion and subsequent marking. Both capture and marking can present substantial welfare risks to individual animals, which may be of particular concern in the case of endangered populations (McGregor & Peake, 1998). Although techniques such as radio-tracking can provide invaluable information, it may not be possible or desirable to use these necessarily disruptive capture and marking methods. Perhaps more importantly, a number of biases may be generated owing to several factors such as subsampling in the capture procedure, change in behaviour following capture and in response to marking. The range of biases reported in the literature is now so extensive that it has been suggested that all such techniques should be considered to be inherently biased (McGregor & Peake, 1998) and where known levels of accuracy are important the bias should be investigated before critical decisions can be based on the information obtained. If used instead of, or alongside, marking techniques, monitoring of individuals by individually distinctive signals should avoid or significantly reduce these kinds of problems and may have the additional benefit of allowing studies with larger sample sizes (this is particularly true for any technique involving adding transmitting devices).

First example: application to corncrake conservation

A study of corncrakes in which individuals were identified by their distinctive crake vocalizations (Peake *et al.*, 1998) was able to assess accuracy of the standard census method used with this species. The standard method was based upon craking rate and movement data collected from radio-tracked individuals (Hudson *et al.*, 1990). The comparison showed that the standard method underestimated numbers at the study site by 20–30%.

One reason for the difference was that radio-tagged males craked more commonly than the average for all males in the population. We suggest that the common technique of using playback to attract males for capture and subsequent fitting of radio-tags may have biased the catch towards particularly responsive (and hence more vocal) males. Whether this explanation is correct or not, the finding reinforces our point (above; McGregor & Peake, 1998) that any capture technique will result in bias. The ability to identify individuals without the need to catch them allowed the extent of bias to be estimated in this case.

A second reason for the difference was that movements of individuals within census periods were highly dependent upon the available habitat; Peake (1997) showed that males moved much greater distances in areas of 'poor' habitat than those in areas of 'good' habitat. This result suggests that habitat quality affects census accuracy in general. Longer distance movements by individuals increase the chances of those individuals being counted twice, hence censuses carried out in poor habitat areas could arrive at higher numbers than those carried out in good habitat, irrespective of the actual numbers of birds present. More importantly, changes in habitat over time could result in highly spurious conclusions being drawn; e.g. it is quite conceivable that efforts to improve habitat within a given area could (based on census figures) apparently result in a drop in numbers of birds. Similarly, habitat degradation could result in an increase in the perceived number of birds.

In summary, the corncrake example clearly illustrates the main advantage of identification based on vocal distinctiveness; the data could not have been obtained in any other way. It also emphasizes the usefulness of the method as a means of calibrating currently accepted monitoring methods, an area of research with direct conservation application which is currently mostly overlooked.

Second example: application to black-throated diver conservation

The relatively low productivity of the British population of black-throated divers, *Gavia arctica*, was an important conservation issue (Campbell & Mudge, 1990). It also highlighted a monitoring problem. In order to assess whether the apparent lack of recruitment was a cause for concern, year-to-year identification of individuals was necessary to monitor site fidelity and survival. However, black-throated divers are almost impossible to catch (Figure 15.1). Male North American common loons, *Gavia immer*, were reported to have individually distinctive yodel vocalizations (Miller, 1988; Miller & Dring, 1988), therefore it was decided to see whether the yodels of male black-throated divers could also be used as a monitoring tool (Gilbert *et al.*, 1994). Spectrograms of black-throated diver yodels showed obvious differences between individuals and strong similarities in yodels from the same individual (see Figure 15.1) and individuals' yodels remained constant between years (Gilbert *et al.*, 1994; Gilbert, 1993). However, it was unexpectedly difficult to sound-record yodels; they were infrequent and unpredictable. The use of playback and stimulation using models was relatively unsuccessful in eliciting yodels because of the difficulty in accurately recreating the behavioural situations in which yodelling was most likely to occur. In Scotland, yodels were produced as part of the 'triumph' display when an intruder had been chased from a loch (Gilbert, 1993). Ultimately, two factors prevented the continuation of monitoring based on yodels: first, the realization that it would take a long-term intensive sound recording effort to build up a useful picture of the turnover in the Scottish black-throated diver population (Gilbert, 1993), and second, the successful use of artificial nesting rafts by the Scottish birds increased their productivity (RSPB, unpublished; M. H. Hancock, unpublished data) and lessened the conservation priority of the monitoring project.

In summary, this example shows that although a study of a congeneric species strongly suggested that identification based on vocal distinctiveness could be applied to a conservation problem elsewhere, a feasibility study discovered the unexpected difficulty of recording large numbers of yodels. It also showed that many factors unrelated to the technique (in this case, the use of breeding rafts) may be relevant to practical decisions about conservation utility.

QUALITY AND CONDITION

Many studies have reported correlations between measures of breeding success and song variation (e.g. Hasselquist *et al.*, 1996; Galeotti *et al.*, 1997) and it is generally accepted that females may use variation between males in aspects of songs to choose mates (e.g. Searcy & Yasukawa, 1996). In this section we look at the potential for song variation to provide information on the quality and/or condition of the singer and how such variation might affect data-gathering and decision-making for conservation.

Breeding status and singing

Obtaining some measure of the success or condition of breeding individuals will refine the information on which conservation decisions are based.

For those species that are secretive, difficult to catch and inhabit dense vegetation, it can be almost impossible to measure any aspect of breeding success. Fortunately, some such species give very clear indications of breeding success. For example, male corncrakes cease calling for seven to ten nights after pairing with a female (Tyler & Green, 1996) and similar indications are given by spotted crakes, *Porzana porzana* (Cramp & Simmons, 1980). Provided that the population is followed throughout the season, this behaviour can be used as a sensitive, bird-based indicator of habitat quality, because high-quality areas are likely to be settled by high-quality males who will attract mates first.

A failure to appreciate that the study species behaves by reducing or halting singing on pairing could result in very misleading conservation interpretations. For example, the naïve interpretation of the observation that one hay meadow always has a craking corncrake throughout the breeding season is that this meadow must be high-quality habitat for corncrakes, whereas in fact the opposite is true; this is a meadow where males fail to attract mates and therefore is probably of lower quality. A second example relates to censuses based on counts of singing males. A detailed study of ovenbirds, *Seiurus aurocapillus*, and Kentucky warblers, *Oporornis formosus*, found that unpaired males were three to five times more likely to be detected than paired males (Gibbs & Wenny, 1993).

Male condition, habitat quality and signalling

Signalling can incur significant costs in addition to the actual costs of signal production (which may be negligible for birds, see Leonard & Horn, 1995), for example increased susceptibility to predators and the seasonal cost of maturing the signalling apparatus (e.g. bitterns produce low-frequency booms from a modified neck musculature, which significantly increases adult male body mass, K.W. Smith, personal communication). Signalling behaviour may be considered condition dependent (e.g. Galeotti *et al.*, 1997), and this indication of signaller quality may be directly relevant to two conservation aspects.

The first of these is an early indication of population decline. In many long-lived species, individuals in poor condition may still be present and signal but fail to achieve reproductive success. If their signals reflect this poor condition (e.g. 'poor' booms by bitterns, see Gilbert, 1993), then conservationists can be alerted to the fall in condition considerably sooner than this will become apparent through a fall in numbers of signalling birds.

Such an early indication may also increase the chances of identifying the factors responsible for the change, such as habitat change, because the factors will be sought at the time the change in condition occurs rather than considerably later.

The second conservation aspect of signals indicating male quality or condition is that they could act as a more sensitive indicator of habitat choice, because individuals in the best condition will be more successful in securing the best-quality habitat (see above for a similar argument when individuals cease calling after attracting a mate). Closely following radio-tagged bitterns (Tyler et al., 1998) and corncrakes (Tyler, 1996) has given indications of which habitat features are important to these species and provided the basis of invaluable management advice for reserves and else-where. Individual differences in quality reflected in signalling abilities of these species may provide a more subtle tool with which to find more de-tailed habitat variables at different times during the season.

Finally, if it was found that signal variation in an endangered species indicated genetic fitness and predicted relative post-fledging survival (e.g. McGregor et al., 1981; Hasselquist et al., 1996), this would have obvious implications for captive breeding and re-introduction programmes.

Male quality and playback

Signals have evolved to modify the behaviour of receivers and as such they can have a conservation role. For example, playback can be used to catch individuals, manipulate sexual receptivity, and attract settling birds to par-ticular areas. The success of playback in altering behaviour in the desired way can be increased if the meaning, variation and context of the signal are understood. For example, playing back a signal indicating an individual of large body size or good condition may repel, rather than attract, others. Similarly, if settling birds use signals as an indication of prior occupation, playback may deter birds from settling in the protected areas, which appear to be settled already. A similar playback solution has been suggested to prevent lions, *Panthera leo*, moving from protected areas into the surround-ing areas that do not have resident populations of lions because here they are shot. Lions are known to use aspects of roaring to assess the likely odds of winning an encounter (McComb, 1992), therefore playing lion roars from areas without lions may deter movement from protected areas by giving the appearance that the surrounding areas are also settled by lions (M. Garstang, personal communication). Judicious incorporation of the assessment cues lions are known to use (McComb, 1992) into playback (e.g. to indicate higher odds of losing encounters outside protected areas) is likely to increase the success of any such manipulation of lion ranging behaviour.

GEOGRAPHIC VARIATION AND GENETICS

Animal vocalizations commonly vary over the geographic range of the species. Geographic variation has been described in song birds (e.g. Martens, 1996), some groups of non-song birds (e.g. Bretagnolle, 1996), terrestrial mammals (e.g. pika, *Ochotona princeps*, Conner, 1982a) and marine mammals (e.g. killer whales, *Orcinus orca*, Ford, 1991). Such variation can provide information of value to conservation efforts because it may reflect underlying differences in population genetics and the extent of isolation between groups or populations (e.g. Conner, 1982b; McGregor *et al.*, 1997). Currently the relationship between geographical variation in vocal behaviour and conservation is best described in birds. We discuss here the conservation implications of geographic variation in bird vocalizations that vary over large (macrogeographic) and small (microgeographic) distances.

Macrogeographic variation

Geographic variation in vocalizations over relatively large distances (for example, an appreciable portion of the species range) is referred to as macrogeographic. The term regiolect is increasing used (e.g. Martens, 1996) to indicate large areas covered by a similar vocal variant and to emphasize the difference from microgeographic dialect patterns of variation (see below).

In the case of the chiffchaff, *Phylloscopus collybita*, genetic analysis of the cytochrome b region of the mitochondrial genome has shown that song regiolects are so genetically distinct that they might represent at least four different biological species (Helbig *et al.*, 1996). There are obvious conservation implications for the realization that one species is in fact at least four reproductively isolated units. It is likely that similar effects will be found in other widely distributed song birds (Martens, 1996).

Such effects in non-song birds were considered to be unlikely because non-song birds were thought to have vocalizations that were geographically invariant (e.g suboscines, Kroodsma, 1988). However, recent studies have shown clear geographic variation in vocalizations of petrels (Procellariiformes excluding Diomedeidae) (Bretagnolle, 1996). Although geographic variation in non-song birds is more often described as minor (e.g. shorebirds, Miller, 1996) it may have conservation interest. For example, a study of the details of corncrake vocalizations found significant differences between the main European populations (Peake & McGregor, 1999). The extent to which these differences are related to genetic differences is currently being investigated (R. Green, personal communication). If there is a degree of correspondence between genetic 'distance' and the similarity of crake vocalizations, sound recording would be a useful tool in rapidly assessing the degree of isolation between populations of conservation concern without the need for capturing birds and relatively costly genetic analysis. At worst, macrogeographic variation may be a useful indicator of where to focus catching and genetic sampling efforts.

Microgeographic variation

Geographic variation in vocalizations over relatively short distances (often taken to mean within the dispersal capabilities of the species) is referred to as microgeographic (Mundinger, 1982), with the term 'local dialect' being reserved for a mosaic pattern of such variation in which shared acoustic themes differ discretely from vocal characteristics of other adjacent groups (e.g. Marler & Tamura, 1962; see also Figure 15.3). Given the powerful role of song variation in reproductive isolation, it seems likely that local dialects have important consequences for gene flow, despite claims to the contrary (Martens, 1996).

The local dialects described for corn buntings *Miliaria calandra* (McGregor, 1980) are considered to be particularly clear (Slater, 1989). A study in North Uist, Western Isles, UK, related the clear local dialects in the area (Figure 15.3) to patterns of breeding and dispersal (McGregor *et al.*, 1997). As extra-pair paternity was negligible in this population (Hartley *et al.*, 1993), parentage could be assigned to young on the basis of their nest site. The relationship between local dialects, breeding and dispersal was clear; > 90% of males and ~ 80% of females bred in the local dialect in which they were hatched, that is, local dialects of males and of the mates of females fairly reliably identified the origins of an interbreeding population. Therefore, despite the appearance of a single interbreeding population in a study area with rather homogeneous habitat and no obvious barriers to dispersal, behaviour associated with song variation effectively fragmented the corn buntings into three populations: the central Reserve local

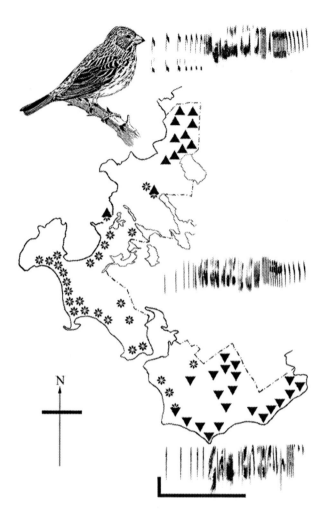

Figure 15.3 Local dialect (microgeographic) variation in the songs of corn
buntings, *Miliaria calandra*, recorded in the Balranald area of North Uist in 1990.
Spectrograms show exemplars of the three dialects (symbols: Tigharry, triangle;
Reserve, star; Paible, inverted triangle) (Canary 1.2.1; 24 kHz sampling rate,
350 Hz bandwidth, grid resolution 0.7 ms by 94 Hz; scale bars show 2 kHz and
1 s). The map shows the position of each territorial male as the symbol of the local
dialect it sang. Three males sang two dialects (combined symbols). The broad line
shows the coastline, narrow lines enclose lochs and the dotted/dashed line shows
the inland limit of suitable habitat. The cross-bar of the North–South arrow
indicates 1 km. For further detail see McGregor *et al.* (1997). The drawing of the
corn bunting is by Maggie Laing.

dialect of ~ 30 males, the Paible dialect of ~ 20 males to the south and the Tigharry dialect of ~ 10 males to the north (Figure 15.3).

Similar patterns of local dialects have been reported for corn buntings in several areas in Europe (McGregor, 1994) and a similar level of reproductive isolation by dialects has been shown in Sussex (McGregor et al., 1988). Corn buntings are declining rapidly in mainland Britain (Donald et al., 1994) and it is possible that this decline has been accelerated by local dialect-driven population fragmentation (McGregor et al., 1997).

Some populations of UK corn buntings do not have clear dialects and occur at relatively low density (Holland & McGregor, 1997). It has been argued that such a pattern could be used to infer recent changes in habitat suitability leading to recolonization by corn buntings (Holland et al., 1996). Such a use of patterns of microgeographic variation is rather similar to the use of similarities between local dialects in white-crowned sparrows, Zonotrichia leucophrys nuttalli, to infer the pattern of resettlement of fire-climax chaparral habitat in coastal California (Baker & Thompson, 1985).

HUMAN INTERFERENCE WITH COMMUNICATION

So far in this chapter we have considered the conservation value of information present in signal variation or in animals' responses to signals. In this section we discuss how human activities can have consequences for conservation through adverse effects on animal communication behaviour. Experimental studies of brief noise disturbance such as low-flying jet aircraft tend to show a short period of behavioural and physiological disturbance with long-term habituation to the noise (e.g. Weisenberger et al., 1996). While it would seem obvious that more continuous noise would limit the transmission of acoustic signals, establishing direct interference with communication behaviour has proved to be difficult. A study in progress on a human activity which drastically affects signalling (the removal of fiddler crab claws) may help to elucidate the range of effects of conservation importance when communication is disrupted.

Effects of traffic noise on populations

A reduction in the density of breeding birds adjacent to busy roads is considered to be a common phenomenon with appreciable effects (Reijnen et al., 1997). For example, in the Netherlands the effect was detected in 60%

of the 43 species studied, with density reductions of 20–98% occurring at 250 m from roads (Reijnen et al., 1995) and with detectable effects up to 3.5 km from motorways (Reijnen et al., 1996). Pollution, visual disturbance and direct mortality through collisions with cars are unlikely to have such an effect (Reijnen et al., 1995). The role of traffic noise in reducing habitat quality is supported by negative correlations between noise levels and breeding density (Reijnen et al., 1997) and by the findings that in areas close to busy roads willow warblers, *Phylloscopus trochilus*, breed less successfully, tend to be first year birds and move away in subsequent years (Reijnen & Foppen, 1991). However, it has proved difficult to show direct effects of noise on bird singing behaviour similar to those reported for calling anuran amphibians (Barrass, 1993). For example, F. T. Awbrey (personal communication) found no simple effect of levels of traffic noise on vocal behaviour of several species of song bird in southern California. Although some species showed significant increases in calling rate in noisier locations, other species showed the opposite trend and none of the relationships were particularly strong.

Effects of noise in the ocean on marine mammals

Noise could adversely affect communication behaviour by restricting the effective range of signals (Klump, 1996) and therefore the extent of signalling between groups of individuals making up communication networks (McGregor & Dabelsteen, 1996). An idea of the scope for such effects comes from a consideration of the large baleen whales. For example, blue whales, *Balaenoptera musculus*, have vocalizations that have the potential to traverse oceans (Payne & Webb, 1971) but the current levels of anthropogenic ocean noise (Gordon & Moscrop, 1996) almost certainly mean that such signalling distances are not realized and therefore the extent of their communication networks and the information available from them is restricted. Although the conservation implications of restricting such networks are likely to be more subtle and longer term than direct physiological effects of very loud sounds (Mayer & Simmonds, 1996), they remain to be investigated. Some information may come from environmental impact studies of introducing sources of very loud sounds to the ocean such as the Acoustic Thermometry of Ocean Climate (ATOC) project and very loud sonar.

Effects of removing claws of fiddler crabs

Male fiddler crabs (genus *Uca*) are characterized by their major chela (Christy, 1988), a greatly enlarged claw that can constitute up to 40% of male body mass and be up to 30 times larger than the other chela (Crane, 1975; Hyatt & Salmon, 1978; Greenspan, 1980; Rozenberg, 1997). This enlarged claw can no longer be used for feeding, rather it functions in visual signalling to attract females and in male–male fights over breeding burrows. In Algarve, Portugal, major chelae of the European fiddler crab, *Uca tangeri*, are a valuable resource. They are sold for more per kg than many species of fish caught locally (R. F. Oliveira, personal communication). Males are caught at low tide; the major chela is pulled off and then the crab is released. This human activity is probably the main selection pressure on the adult population as, even in a protected area, up to 40% of males had a missing or regenerating major chela (J. L. Machado & R. F. Oliveira, unpublished data). In addition to obvious adverse effects on survival (survival seems to be high, but data are scarce) and destruction of breeding burrows (males are often dug out of burrows where they have sought refuge), removing the major chela prevents males from signalling. Males regenerating the major chela continue to attempt to signal, to fight and to court, but other males and females respond to them as if they were females. In addition to such disruption of the usual patterns of behaviour there are likely to be more subtle effects through the operational sex ratio and selection against males with the largest chela. Population sizes in the area where claws are removed have suffered a tenfold decrease in recent years and studies are underway to determine the nature of the link between claw removal, subsequent behavioural disruption and changes in population size (Oliveira *et al.* 1999). The chances of unequivocally demonstrating the importance of such links are much higher in this case than in instances of anthropogenic noise interference with acoustic signals (see above) because of the drastic nature of human interference with communication in this fiddler crab.

CONCLUDING THOUGHTS

In this chapter we have tried to illustrate the potential for some aspects of communication behaviour to have a role in conservation. As with so many other relatively new fields of modern biology, from molecular techniques through remote sensing to mathematical models, we recognize that the emphasis should be very firmly on the word potential. Those in the front line of conservation often need tried and tested examples of a new method before they can justify its implementation. There is usually a gap between research which hints at the conservation implications of communication behaviour and examples of application. Anyone who seriously believes in the value of their research to conservation should convince those active in conservation of this value with a proper exploration of its potential, ideally through joint research. Many of the examples we have used (black-throated diver, corncrake and bittern) have resulted from successful collaboration between the RSPB and the Behavioural Ecology Research Group at the University of Nottingham. We know of other examples of this type of useful collaboration in the UK and, on the basis of our experience, we would strongly recommend this as an excellent way of combining behaviour with conservation, and one that deserves to be practised more widely. It is also worth remembering that there are many field workers, particularly abroad, faced with monitoring problems of difficult and endangered species, but without the time and resources to convince funding bodies of the usefulness of such a new approach to solving conservation problems. We hope we have given enough of an indication of the tangible benefits of communication behaviour to conservation to promote its more widespread application.

A recent explanation of the conservation utility of population consequences of behavioural decisions taken by individuals ended with the caveat that the approach required a thorough understanding of the adaptive repertoire, or the natural history, of the species of interest (Goss-Custard & Sutherland, 1997). The same caveat applies just as strongly to using aspects of communication behaviour. For example, a naïve interpretation of the observation that corncrakes call from one particular area throughout the season is that the area must be particularly good for corncrakes, but as we know that corncrakes cease calling on attracting a mate (Tyler & Green, 1996), the area in question is probably particularly bad for corncrakes because the males there do not attract mates. Although we have dealt mainly with the vocal communication behaviour of birds, our comments should apply to most signalling modalities in most taxa. The take-home message for using communication behaviour in conservation is the same as that for any subject of study – there is no substitute for the application of sound science.

ACKNOWLEDGEMENTS

Most of the detailed studies we discuss were funded by CASE awards between the RSPB and BBSRC (SERC) to G.G. and T.M.P. The analysis equipment used was funded by NERC, The Royal Society and the Nuffield Foundation. We thank Ken Otter, Ken Smith and Andy Terry for comments on the manuscript. We are very grateful to Maggie Laing and Ken Otter for the drawings.

16

Reducing predation through conditioned taste aversion

DAVID P. COWAN, JONATHAN C. REYNOLDS & ELAINE L. GILL

INTRODUCTION

Negative consequences arising from eating a particular food can result in subsequent avoidance of that food (e.g. Schaeffer, 1911). In the 1930s this process was identified as a constraint on the effectiveness of fast-acting rodenticides (Hamilton, 1936). Animals that become ill after consuming a sub-lethal dose of rodenticide bait learn to avoid that bait and become 'bait-shy' (Chitty, 1954). This phenomenon underlies the domination of rodent control by anticoagulant rodenticides, since the 1950s, because the 3–5 day delay between anticoagulant consumption and the onset of symptoms reduces the likelihood of an association being made (but see Smith *et al.*, 1994). The association of the taste of a food with illness leading to reduced food consumption was recognized as a distinctive form of conditioned aversion by Garcia *et al.* (1955). Subsequently this has most commonly been described as conditioned taste aversion (CTA) although a confusing variety of other terminology has been used, including taste conditioned aversion, conditioned food aversion, conditioned flavour aversion and learned food aversion. The common elements of these are the pairing of taste as the conditional stimulus with illness as the negative unconditional stimulus, leading to subsequent aversion to food with the same taste. Conditioned aversion of this general form is widespread across taxa from molluscs to humans (Gustavson, 1977), and is one of the most intensively studied learning processes, with the laboratory rat (*Rattus norvegicus*) as the predominant subject (Riley & Tuck, 1985). It is widely believed that the CTA response has evolved to reduce the risk of poisoning and plays a leading role in determining dietary preferences (Rozin & Kalat, 1971; Kalat, 1977). Aversive behaviour in general might be exploited to modify the feeding behaviour of species to meet wildlife management objectives. For instance, some feeding deterrents and repellents can create conditioned aversions (Rogers, 1978; Avery & Decker, 1994; Watkins *et al.*, 1994). It is important to recognize that, in these cases, the conditional stimulus is the taste of the repellent rather than that of the food to which it is applied. There is thus no expectation of reducing consumption of untreated food unless through association with food cues other than taste. A more ambitious aim is to generate an aversion to the taste of the food itself, thus leading to the protection of untreated food. The context in which this was first proposed was reducing coyote (*Canis latrans*) predation on live prey (Gustavson *et al.*, 1974). In this chapter we explore the possibilities of using this approach to reduce predation in conservation contexts.

WHAT DO WE WANT TO ACHIEVE?

Although the intention of CTA management techniques is always to reduce predation that is considered undesirable, the intervention may be on behalf of either the predator or the prey. For example, persecution of raptors, because of perceived threats to human interests, is widely believed to have contributed to their decline in the UK, particularly in the nineteenth century (Sharrock, 1976). Even today populations of some species, such as the hen harrier (*Circus cyaneus*), are still constrained by human persecution (Etheridge *et al.*, 1997). The illegality of such activities is difficult to enforce. The reasons for persecution may be predation on only a single species, e.g. red grouse (*Lagopus lagopus scoticus*), of great significance to game interests but of modest importance for the overall diet of the predator. If CTA could be used to remove this item from the predator's diet then persecution, if genuinely focused on the perceived threat to a game species, would be reduced.

In other contexts we might seek to intervene on behalf of the prey. Predation has been considered a potential threat to many endangered species. Vulnerable prey include reptiles (Nellis & Small, 1983), birds (O'Donnell, 1996) and mammals (Strachan *et al.*, 1998). Predators also include reptiles (Savidge, 1987), birds such as corvids (Slagsvold, 1980) and raptors (Paine *et al.*, 1990) and mammals such as mustelids (O'Donnell, 1996), canids (Patterson, 1977) and commensal rodents (Atkinson, 1985). The life-history stage at which predation occurs is also important. For example, amongst birds, eggs and nestlings are exposed to different and generally greater predation risks than adults. The variety of location, capture and consumatory behaviours associated with this diversity of predator–prey interactions means that the CTA approach has to be tailored to specific circumstances. Importantly, these circumstances may be very different from those relating to CTA in the laboratory rat, which has formed the basic model for our current understanding of the phenomenon.

A clear conservation objective for reducing predation must be established. The perceived link between predation and threats to bird populations has led to predator removal being instigated in conservation contexts (Reynolds & Tapper, 1996). This has met with some success when focused on non-native predators, particularly on islands (Newton, 1993). Here predator removal seems to be reasonable in terms of both conservation goals and ecological principles. However, reviews by Newton (1993), Reynolds & Tapper (1996) and Côté & Sutherland (1997) found that while removal of native predators frequently enhances the breeding success and post-breeding population sizes of prey bird species there are few clear demonstrations of increases in breeding densities. Meeting a conservation goal of increased breeding population size thus cannot reliably be expected to follow reduction of predation through either predator removal or the use of CTA. Some circumstances offer greater ecological prospects of success than others, for example, where it is known that the prey population is being held below carrying capacity by predation in a well-defined area such as a nature reserve (Suarez et al., 1993). The potential advantages of CTA rather than predator removal in these circumstances are two-fold. First, the induction of transitory illness in the predator necessary to achieve a CTA has some ethical implications, but these doubts might be considered more than offset, compared to lethal control, by the additional expectation of 'quality' life for the predator. The theoretical ecological advantage is more straightforward. The removal of predators often affords only a transitory reduction in predation unless pursued relentlessly, particularly for territorial species where culled residents are rapidly replaced by the expansion of adjacent territories or through immigration. By contrast, if a territory holder has a CTA to a particular prey, it will indirectly protect those prey in its territory from predation by other individuals through its continued presence there; in effect the poacher becomes the gamekeeper.

CHARACTERISTICS OF CTA

In order to move towards the development of practical solutions to real wildlife management problems we need to understand the distinct and essential properties of CTA. There are a number of key features of CTA that distinguish it from other forms of associative learning:

1. There can be a delay of up to about 12 hours between exposure to the conditional stimulus and the onset of illness as the unconditional stimulus (Smith *et al.*, 1994).

2. Only a single pairing of the conditional taste stimulus with subsequent illness is necessary to lead to aversion (Garcia *et al.*, 1955).

3. Physiological evidence indicates that the induction of CTA involves the *area postrema* in the brain stem, located next to the site of a 'chemoreceptor trigger zone' for nausea and emesis (Borison & Wang, 1953; Bernstein *et al.*, 1986; Mestel, 1997).

4. The aversion develops even when the unconditional stimulus of illness occurs under anaesthetic (Roll & Smith, 1972), and in humans the aversive response can be categorized as subconscious.

5. In some species, once an aversion has been established, internally induced nausea occurs when the conditional stimulus is subsequently encountered. This is clear through self-reporting in humans (Mestel, 1997). There are also descriptions of retching and vomiting and other behaviours, such as vigorous head-shaking, suggestive of self-induced nausea when the referent conditional stimulus is re-encountered, e.g. coyote (Gustavson, 1977), mongoose (*Herpestes auropunctatus*) (Nicolaus & Nellis, 1987), cougar (*Felis concolor*) (Gustavson *et al.*, 1976) and raccoon (*Procyon lotor*) (Semel & Nicolaus, 1992). These observations imply that the aversion is internally reinforced by re-exposure to the conditional stimulus.

In wildlife management, the aim is to generate a CTA with the above characteristics by the addition of an undetectable chemical to a particular food or prey item, which induces a mild and transitory illness after consumption. It is expected that animals consuming treated food will subsequently avoid eating prey that tastes like the original conditional stimulus. The key feature is that, unlike a repellent, the presence of the CTA agent does not change the taste of the referent food in a way that allows the animal to distinguish treated from untreated food. In its simplest form the conditional stimulus might be a bait consisting of the meat of the prey species to be protected, together with an effective dose of a CTA agent.

SUGGESTED CTA AGENTS

The nature of the illness-inducing chemical is critical to the prospects of using CTA to modify predatory behaviour and it must meet the following criteria:

1. *Efficacy.* The chemical must induce a robust, long-lasting CTA after an oral dose.
2. *Safety.* The effective dose should be well below the LD_{50} for the target species, and for any non-target species potentially at risk of exposure. It should not cause chronic (long-lasting) illness or detrimental effects in target or non-target species or their offspring, nor should it persist in the environment.
3. *Detectability.* The effective dose must be undetectable in terms of taste by the target species when mixed with bait for consumption; this may be a property of the compound itself, or its taste may be masked by formulation, e.g. microencapsulation.
4. *Environmental stability.* The chemical should remain unchanged in baits placed in the field.
5. *Delayed activity.* The onset of illness should be sufficiently delayed to allow the consumer to finish its meal but nonetheless sufficiently swift to induce a CTA.

A variety of CTA agents have appeared in the literature. The following is not an exhaustive synthesis but summarizes the characteristics of the main molecules that have been used to explore the characteristics of aversion learning processes and have been tested or suggested as candidates for wildlife management applications of CTA.

Apomorphine and lithium chloride

Apomorphine and lithium chloride have been extensively used to study CTA and are capable of generating long-lasting aversions in a number of species (Wittlin & Brookshire, 1968; Revusky & Gorry, 1973; Mason & Reidinger, 1983; Conover, 1989; Zahorik et al., 1990; du Toit et al., 1991). Apomorphine is unstable, rapidly oxidizing in air and light whereby it loses its emetic effect; we know of no studies that have generated CTA using oral administration rather than injection of this chemical. Lithium chloride (LiCl) has a salty taste and produces a reliable CTA through oral dosing only when 200–500 mg/kg body mass are ingested. It is possible to produce large meat baits with relatively strong tastes in which the presence of LiCl is undetectable (Gustavson et al., 1974; Nicolaus et al., 1982; Forthman-Quick et al., 1985). Even with large baits, however, aversions may develop to the taste of LiCl rather than to the taste of the untreated bait or prey (Burns,

1980). Furthermore, rendering LiCl undetectable in small, bland-tasting baits such as eggs is particularly problematic (Nicolaus et al., 1989a). There is also concern regarding the humaneness of the malaise caused by LiCl relative to the illness caused by other CTA agents (L. K. Nicolaus, personal communication).

Carbachol

Carbachol is a cholinergic agonist that generates CTA through oral dosing at lower concentrations than LiCl (Myers & De Castro, 1977; Nicolaus et al., 1989b; Avery & Decker, 1994). However, it is detectable in some baits (Nicolaus & Nellis, 1987). Furthermore, the onset of illness is quick and results in cessation of feeding, therefore if the rate of food consumption is low animals may not consume sufficient carbachol to generate a full CTA response (Nicolaus et al., 1989b). Most importantly there is only a limited margin of safety between the effective and lethal dose (Nicolaus et al., 1989b).

Thiabendazole

Thiabendazole is a mammalian anthelmintic and also a systemic fungicide. The anthelmintic therapeutic dose ranges from 45 mg/kg body mass for the horse (Equus caballus) to 160 mg/kg body mass in the rat. It has been evaluated as a CTA agent in a number of species. Conover (1989) reported no CTA effects in raccoons when fed eggs containing thiabendazole. Strong aversions were reported for two wolves (Canis lupus) towards two foods treated with thiabendazole but only weak responses were observed towards two other treated foods presented when the wolves were hungry (Zeigler et al., 1983). New Guinea wild dogs (Canis familiaris hallstomi) and dingos (Canis familiaris dingo) developed aversions to untreated lamb meat after two exposures to lamb meat containing thiabendazole (Gustavson et al., 1983). Polson (1983) showed that black bear (Ursus americanus) damage to beehives could be reduced using baits containing thiabendazole and beeswax. Aversions to a novel food have been generated in laboratory rats through oral intubation with thiabendazole (E. L. Gill, unpublished data). Those receiving 100 mg/kg body mass developed aversions to a novel food that only lasted for two no-choice post-treatment tests. Some animals that received a single dose of 200 mg/kg body mass showed a long-lived aversion that lasted for over 10 post-treatment tests but in others aversion attenuated after 4–5 post-tests. Although thiabendazole is considered undetectable by some authors this has not been formally established and Polson (1983) described some bears as being reluctant to eat baits, perhaps owing to the unpalatability of thiabendazole. There is a high margin of safety for thiabendazole, with the LD_{50} being more than 3000 mg/kg body mass for the Norway rat, house mouse (*Mus domesticus*) and European rabbit (*Oryctolagus cuniculus*). Furthermore, a great deal of the data required for pesticide registration is already available for thiabendazole given its use as a veterinary medicine. Therefore, despite reservations regarding its detectability and the variability of the reported CTA responses, thiabendazole warrants further investigation for wildlife management applications.

Cinnamamide

Cinnamamide is a synthetic derivative of cinnamic acid, a naturally occurring plant secondary compound. It is repellent to both birds and mammals, acting through its taste, odour and post-ingestional effects (Crocker *et al.*, 1993; Gill *et al.*, 1994; Gurney *et al.*, 1996). Two studies of cinnamamide as a CTA agent have been carried out using a single dose of 160 mg/kg body mass administered by oral intubation. A strong CTA to water containing saccharin was established in house mice, which lasted throughout a succession of two-choice post-treatment tests over 64 days, during which plain water was available (Watkins *et al.*, 1999). A CTA to a novel food was also generated in laboratory rats (E. L. Gill, unpublished data). In the first post-treatment test, using a no-choice protocol, consumption of the referent food by cinnamamide-treated rats was negligible. However, during the next two tests consumption increased such that by the fourth, 4 weeks after the conditioning event, there was no significant difference between treated and control groups. There is a good safety margin between the effective dose and the LD_{50} (1600 and 1850 mg/kg body mass for mice and rats respectively). Cinnamamide is detectable by birds and mammals; consequently its taste and odour would need to be masked by formulation, such as microencapsulation, before the compound could be considered further as a CTA agent for field studies.

17α-ethinyloestradiol

The most effective CTA agent identified to date is 17α-ethinyloestradiol, a synthetic oestrogenic hormone. Its taste appears to be undetectable for a range of mammalian species, and a single dose can generate long-lasting aversions (Nicolaus et al., 1989a; Semel & Nicolaus, 1992).

Recent studies of the laboratory rat have confirmed this potential (E. L. Gill, unpublished data). A strong CTA to a novel food was established in laboratory rats by oral intubation with 4 mg/kg body mass of oestradiol. Consumption of novel food, by oestradiol-treated rats, was negligible in all no-choice post-treatment tests while that of control rats was at least that of the treatment day and generally increased over five post-treatment tests. During each 30 minute post-treatment test period, all treated rats investigated the novel food in their bowl, some dug in the food bowl with their forepaws, some chewed their bedding (pica), and most eventually retired to their bedding and either slept or sat inactive. These behaviours were not seen in control rats, most of which ate almost continuously for 30 minutes. Further post-treatment tests carried out every 3–4 weeks with the treated group showed that their reduction in consumption of novel food persisted to a sixth post-test (10 weeks after treatment). After nine post-tests (19 weeks after treatment), five out of the 12 treated rats ate < 1 g of novel food, and 25 weeks after treatment three rats ate < 1 g. 17α-ethinyloestradiol thus induced a longer-lasting and more robust CTA than either cinnamamide or thiabendazole. One year after the initial treatment, a similarly robust CTA was re-generated with oestradiol, using the same individual rats and the same, now familiar, food.

We have also carried out studies of oestradiol-induced CTA in the red fox (*Vulpes vulpes*) (E. L. Gill, unpublished data). Four captive foxes (three males, one female), which had not been fed for 16 hours, were individually presented with 85 g pheasant meat (a novel food) treated with 17α-ethinyloestradiol dissolved in polyethylene glycol to deliver 4 mg/kg body mass. Consumption was measured 30 minutes later and any remaining meat removed. All foxes ate treated pheasant meat within 30 minutes, although one male ate only half the ration. The other two males vomited about 40 minutes after ingestion. No other signs of malaise were recorded, and all foxes ate their familiar food when it was provided 4–5 hours later. A week later, the foxes were given unadulterated pheasant meat for 30 minutes. All individuals approached the meat and then retreated, shaking their heads, without tasting it. The foxes were given unadulterated pheasant meat on eight subsequent occasions over a 52-week period and all reacted in the same way. Some individuals also covered the food in wood shavings, one urinated on it and another turned the bowl over to cover the meat.

Although oestradiol is generally a highly effective CTA agent it may not induce aversion in all mammals; Franklin's ground squirrel (*Spermophilus franklinii*) and striped skunk (*Mephitis mephitis*), for example, failed to develop CTA to eggs (Penner, 1995). The effective dose of oestradiol is well below the compound's acute oral LD_{50} in the rat (1200 mg/kg body mass) but, as a synthetic reproductive hormone, the compound has the potential to affect reproductive processes and systems (Yanagimachi & Sato, 1968; Yasuda *et al.*, 1981). Although no ill-effects have been noted in captive animals, concerns have been expressed that oestradiol may not be suitable for extensive field use, especially where baiting coincides with reproductive stages, or the target species is protected or endangered. This highlights the need to identify alternative compounds, which meet all of the above criteria, in order to use CTA to manipulate predatory behaviour in the field. Identification of the properties of oestradiol that make it such an effective CTA agent may assist in the identification of alternative, environmentally safer compounds.

Other possibilities

The modes of action of compounds that induce CTA are not known, and may be various. It is generally accepted that nausea and/or vomiting play a major role in the generation of CTA, although vomiting is not necessarily a desirable effect because it may reduce the dose of compound absorbed by the consumer, thus affecting its CTA properties (L. K. Nicolaus, personal communication). The mechanism of nausea caused by a toxin can take a number of routes, which are not fully understood: probably the most rapid involves the stimulation of toxin-sensitive cells in the gut wall (the enterochromaffin cells), which release serotonin, a nerve-signalling chemical, which stimulates the vagus nerve connecting the stomach to the brain stem. The vagus nerve stimulates the chemoreceptor trigger zone part of the brain stem, (the *nucleus tractus solitarius* (NTS)), which leads to nausea and emesis (Mestel, 1997). Toxins may also reach this trigger zone *via* the bloodstream, entering the brain stem close to the NTS at the *area postrema*. Understanding the physiology of nausea and emesis may help in the identification of new CTA compounds. Oestrogenic compounds can raise levels of histamine in reproductive tissue. Treatment with an antihistamine drug (chlorpheniramine) prior to administration of a CTA dose of oestradiol cypionate eliminated the CTA effect in female rats (Rice *et al.*, 1987). Histamine is also released as a result of exposure to X-rays (Levy *et al.*, 1974), which itself causes CTA (Garcia *et al.*, 1955) but not always nausea (Smith, 1971). Histamine is produced in the body by the decarboxylation of the amino acid histidine, increasing gastric secretions, thus possibly affecting appetite or inducing a mild nausea. The involvement of histamine in CTA generation by oestrogenic compounds may explain some of the activity of oestradiol and possibly the apparent variation between individuals and species in nausea or malaise noted after treatment. The role of histamine in the generation of CTA needs to be investigated further with a view to utilizing this physiological route for the identification of potential novel CTA agents. Serotonin is also involved in the regulation of feeding and emesis (Mestel, 1997). Hence, the elevation of serotonin levels in the body by preventing its re-uptake is likely to induce nausea. There are a number of drugs that inhibit serotonin re-uptake, e.g. paroxetine (Thomas *et al.*, 1987), fluvoxamine (Wild *et al.*, 1993), and fluoxetine (Prendergast *et al.*, 1996), thus studies of these and similar molecules as possible CTA agents might be rewarding.

WHAT FACTORS DETERMINE THE STRENGTH AND PERSISTENCE OF AVERSIONS?

Practical application of a suitable CTA agent will need to take into account the factors that determine the strength and persistence of CTA responses. Not all tastes are equal in their associability with subsequent illness. In general, tastes that are preferred appear to generate more robust CTA responses (Green & Churchill, 1970). This is somewhat paradoxical given that many naturally occurring toxins taste bitter and are inherently less preferred (Galef & Osborne, 1978). Nevertheless, that preferred tastes are more effective in generating CTA is a practical advantage.

The strength of the unconditional stimulus, i.e. the illness, is related to the strength of the consequent CTA (Dragon, 1971). However, illness intensity is not easy to quantify and Nachman & Hartley (1975) found no clear link between the gross symptoms of illness induced by a range of rodenticides and their ability to induce CTA. Little is known about the specific qualities of illness that lead to CTA. However, it is generally acepted that those associated with nausea and emesis are more potent although emesis is not a prerequisite component (Gustavson, 1977). Nevertheless, not all drug-induced illnesses can act as the unconditional stimulus for CTA (Bures & Buresova, 1977).

The optimal time between the exposure to the conditional stimulus and the onset of illness appears to be between 1 and 6 hours; thereafter the strength of the CTA wanes (Garcia et al., 1974). In practice this would be achieved through appropriate choice of CTA agent and dose.

The strength or intensity of the conditional taste stimulus is also considered to be positively related to the robustness of the consequent CTA (Nowlis, 1974). Thus it may be easier to generate aversion to the strong taste of, for instance, an adult bird rather than the relatively bland taste of an egg. The effect of stimulus strength is also mediated by its novelty. Once rats become familiar with a particular taste they are less prone to develop a CTA in response to illness than if the taste is novel (Nachman et al., 1977). This may arise through 'learned safety' (Kalat & Rozin, 1973) where learning that a taste is 'safe' becomes increasingly incompatible with associating that same taste with negative consequences. The stronger responses to novel tastes may also reflect greater attention being paid to novel cues (Nachman et al., 1977). In some practical applications it may be possible to minimize the effect of familiarity by, for instance, inducing an aversion to a particular type of egg prior to the breeding season of the species concerned. In others the behaviour we might seek to modify will often be towards a prey type that is already familiar to the predator. However, it is possible to induce a robust CTA to a familiar food (Rusiniak et al., 1976). What appears to be important is unambiguity of association between a particular taste and subsequent illness along with the inherent distinctness of the taste: strong familiar tastes produce stronger aversions than weak familiar tastes (Nachman et al., 1977).

CUE SALIENCY – IS TASTE IMPORTANT?

Taste has long been considered the primary cue for conditioned aversion in the rat (Garcia & Koelling, 1967). This is not, however, the case for all species and Wilcoxon et al. (1971) found that colour functioned equally well as taste in bobwhite quail (Colinus virginianus). Taste is therefore not a necessary component of food aversion learning. What might be unique is the form of the unconditional negative stimulus in terms of internal perception of illness. Perhaps certain illness 'types' initiate a retrieval process for memories of events that have occurred over a longer period of time than when the unconditional stimulus is external (Nachman et al., 1977). For many species gustatory cues, perceived during food ingestion, are likely to be of special relevance and have the greatest saliency for association with subsequent illness through such a retrieval process. It is now recognized that aversions can be triggered by other food-related cues especially if those cues are presented in the same context as the critical taste cue. Domjan (1973) found that rats made ill in the presence of an odour subsequently avoided a compartment containing that odour; however, this aversion was much stronger if the animals had been made ill after drinking water in the presence of the odour. Rusiniak *et al.* (1976b) also showed that aversions to taste potentiate odour aversions, while Galef & Osbourne (1978) demonstrated that taste facilitates the association of visual cues with illness. In a practical setting, we thus need to ensure not only that the taste cue we use to elicit the aversion is sufficiently similar to the dominant taste of the prey species, but also that other cues represented by the prey are incorporated into the learning process we initiate in the predator.

CAN CTA INHIBIT PREY-SEEKING BEHAVIOUR?

Feeding can be regarded as a two-stage process. First, there is food-seeking and appetitive behaviour, which will be largely directed by visual, auditory and olfactory cues. Second, there is consumptive behaviour, where acceptance or rejection of a food item will be determined by gustatory cues, notably taste together with some olfactory component, which collectively represent flavour (Gustavson, 1977). Clearly, intervening solely at the consumatory stage would be likely to do more harm than good in a practical context. CTA thus needs to inhibit seeking and capture of prey to be of use in manipulating predatory behaviour.

It is known that appetitive as well as consumatory behaviour can be inhibited as a consequence of conditioned aversions. Burghardt *et al.* (1973) showed that garter snakes (*Thamnophis sirtalis*), which received a single LiCl injection contingent with earthworm (*Lumbricus terrestris*) consumption, continued to seize earthworms but took longer to swallow them or dropped the earthworms after striking. After additional post-treatment exposures the snakes rejected the earthworms without attack. E.L. Gill (unpublished data) has investigated whether aversions towards dead mealworms (*Tenebrio mollitor*) in laboratory rats inhibit killing of live mealworms. Rats that were known to eat both dead and live mealworms were each given five dead mealworms. All rats ate their mealworms within 3 minutes. Thirty minutes after finishing the meal, rats in the treated group

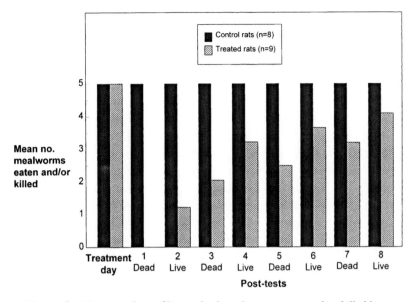

Figure 16.1 Mean numbers of live or dead mealworms eaten and/or killed by rats during eight post-tests at weekly intervals following gavage on the treatment day with 4 mg/kg body mass oestradiol (treated rats) or carrier solution (control group) 30 minutes after feeding on five dead mealworms.

($n = 9$) were given a single gavage of 17α-ethinyloestradiol (4 mg/kg body mass) in a carrier of polyethylene glycol and water, control group rats ($n = 8$) were given a sham gavage of carrier only. For each post-treatment test, rats were individually presented with five dead or live mealworms for 15 minutes. In the first test, rats received five dead mealworms. Individuals in the control group ate all their mealworms in < 4 minutes; the treated group did not eat any mealworms within 15 minutes, although five individuals picked up at least one worm. A week later, five live mealworms were presented to each rat. Individuals in the control group ate all their mealworms within 4 minutes. Eight of the treated rats picked up ≥ 1 live mealworm, four of which either killed or ate ≤ 2 mealworms each. During subsequent tests consumption of both dead and live mealworms by treated rats gradually increased (Figure 16.1) and all individuals picked up mealworms. However, by the end of the experiment, one of the treated rats still did not kill or eat any live or dead mealworms.

Brower (1969), in considering how generalization from taste cues to visual cues led blue jays (*Cyanocitta cristata*) to avoid consuming not only poisonous but non-poisonous monarch butterflies, recognized that at first the birds associated the taste of a poisonous insect with illness, and then subsequent encounters allowed the birds to associate distinctive visual cues, leading to rejection on sight of both poisonous and non-poisonous insects. A similar two-stage conditioning hypothesis was proposed by Gustavson et al. (1974). Here the initial CTA was followed by exposure to the taste alone, which became 'punishing'; at this stage other key cues, notably visual and olfactory, become conditioned. An alternative view is that the weaker non-gustatory cues are simply potentiated by the taste aversion at the initial pairing of taste with illness (Rusiniak et al., 1976b). In either case the consequence is 'avoidance at a distance' (Nicolaus, 1987). This is illustrated by the vigorous head-shaking of our foxes when they smelt, without tasting, pheasant meat after their initial conditioning experience.

The sequence of events whereby taste aversions generalize to other cues, leading to avoidance at a distance, may be part of the natural processes that produce forms of aposematism, notably Batesian mimicry (Nicolaus et al., 1983). In the practical application of CTA we are trying to establish a similar process whereby non-toxic mimics, in the form of live prey, are subject to less predation as a consequence of their similarity to a toxic model in the form of a treated bait. This has been achieved in certain settings. Gustavson et al. (1974, 1976) inhibited killing of lambs and rabbits by coyotes, and sheep by wolves, through conditioning with either bait packages or dead carcasses of the referent prey containing LiCl or by LiCl injection paired with prey carcasses. Most predators still killed after an initial exposure to LiCl conditioning but this was subsequently suppressed by one additional conditioning treatment. Dead baits do not present the full range of cues used during appetitive prey-seeking behaviour. For animals to associate these additional cues with the taste and perhaps olfactory cues that have been conditioned by baits, they may have to perform subsequently a full sequence of appetitive behaviours. However, previous memory associations of the referent taste with these appetitive behaviours may lead to the inhibition of killing by a single initial experience. That killing can occasionally be inhibited by the initial conditioning experience alone (e.g. Gustavson et al., 1974, 1976) suggests that this sometimes occurs. When visual cues are associated with the conditioning stimulus then inhibition of killing by a single conditioning event might be achieved more reliably. Brett et al. (1976) showed that attacks on mice by buzzards (primarily *Buteo jamaicensis*) could be inhibited if distinct taste was paired with distinct colour in the conditioning stimulus associated with LiCl-induced illness. Clearly, the closer the model bait resembles the mimic prey we wish to protect, in terms of the cues that are used to direct the predator's appetitive behaviour, the more likely it is that attack will be inhibited. In this sense, cues might be considered additive in determining the closeness of the mimic to the model (Martin & Lawrence, 1979). Ideally the model conditioning stimulus should be live prey: this is an unlikely practical prospect.

The initial extension of these encouraging findings to the field produced equivocal and controversial results. Two particular problems with such studies are apparent. First, in some cases the doses and distribution of LiCl in baits led to its detection and thus live prey failed to mimic the taste of model baits (e.g. Conover et al., 1977; Burns, 1980). Second, claims of reduced predation in large-scale field trials have been criticized for the lack of adequate experimental controls (e.g. Conover et al., 1979). However, in a series of carefully designed experiments, Lowell Nicolaus and his colleagues have demonstrated that CTA can be used to manipulate the behaviour of free-living predators and established two important principles. First, they showed unequivocally that killing of live prey can be inhibited under field conditions (Nicolaus et al., 1982). In this study free-living raccoons, known to kill and eat chickens regularly, were exposed to dead chicken baits laced with LiCl. No attacks on live chickens by known individuals were subsequently recorded during the month after the exposure to the baits. Second, they demonstrated that conditioned aversion using model baits can cause extensive reductions in predation on mimics in the field. Nicolaus et al. (1983), in the first of a series of studies, showed that free-ranging crows (Corvus brachyrhynchos) that ate green-painted chickens' eggs containing a CTA agent subsequently avoided both toxic and non-toxic green-painted eggs. The subsequent studies used toxic egg models to reduce attacks on non-toxic eggs by a wide range of free-living avian and mammalian predators including common ravens (Corvus corax) (Nicolaus et al., 1989b), raccoons (Semel & Nicolaus, 1992), mongooses (Nicolaus & Nellis, 1987), crows (Dimmick & Nicolaus, 1990) and multi-species guilds of predators (Nicolaus, 1987; Nicolaus et al., 1989b). The approach has recently been examined by others in a conservation context (G. Bogliani & F. Bellinato, unpublished data). There were differences apparent between these studies. For instance, the persistence of the aversions varied from 14 days (Nicolaus & Nellis, 1987) to a year (Dimmick & Nicolaus, 1990). Furthermore, in some cases the aversion generalized to eggs that did not look like treated eggs (Nicolaus et al., 1989b) while in others this did not occur (Nicolaus et al., 1983). Such differences perhaps reflect the complexities of working with a diversity of free-living species but understanding their causes will be important in developing an effective management technique. These studies, with the exception of Nicolaus et al. (1982), related to egg predation and, perhaps most significantly, used illness-inducing agents other than LiCl, i.e. carbamates, lindane or oestradiol. Eggs may appear less challenging targets than active, live prey. Nevertheless, these studies demonstrate that, through CTA, appetitive behaviour, in terms of shell breaking, can be consistently inhibited in the field.

CONSTRAINTS ON EFFECTIVENESS IN THE REAL WORLD

We are still some way away from having practical field techniques that exploit CTA. Our ability to manipulate predatory behaviour in natural settings will not only depend on the host of parameters that are known to influence the establishment or otherwise of CTA under controlled laboratory conditions but also the myriad of interactions that these factors will have with ecological variables of the real world. Such factors include the population densities of predators and prey, the nature and abundance of alternative prey, the mobility of predators and their prey, and their social and territorial behaviours.

The Game Conservancy Trust is researching predation by red foxes on adult female grey partridge as a model system for the exploitation of CTA. Predation by foxes occurs mainly during spring and early summer when the partridges are incubating. This is before fox cubs leave their natal den, so only territory-holding adult foxes are involved. During this season each fox territory is occupied by a breeding pair of adults, sometimes with additional 'helpers', which may occasionally also breed successfully (Reynolds & Tapper, 1995). The model system seems suited to the CTA approach because foxes are territorial, and because predation by foxes has a critical impact on partridge population dynamics even when partridges form only a small fraction of fox diet (Reynolds & Nicolaus, 1994; Reynolds & Tapper, 1995, 1996; Tapper et al., 1996).

Efficient predator targeting

The first practical aim is to deliver an effective dose of aversive chemical with the correct associated cues to all or the majority of foxes with access to the area in question. But economy is also necessary. Since we still know little about the generality of conditioned aversions, we must assume that CTA baits will be freshly killed examples of the referent species, 'laced' with the CTA agent. Because the aim is actually to protect the referent species, such baits are valuable and economy is obviously desirable. Pre-bating seems an obvious way to meet all these requirements, and we have been exploring a pre-bating system for foxes (M. J. Short & J. C. Reynolds, unpublished data). One aim is to establish habits of bait-site visitation in territory-holding foxes, allowing a higher success rate with scarce CTA baits than would be achieved by random placement. Fox territory size varies from 0.2 to 40 km in different landscapes (Corbet & Harris, 1991), and the intensity of territory use also varies as a result (compare Schofield, 1960; Mulder, 1985; Artois *et al.*, 1990; White *et al.*, 1996). So for any one area it is important to estimate a bait spacing that will give a reasonable probability of treating the majority of adult foxes. In our study site in rural Dorset, radio-tracking demonstrated that individual foxes did not cover the whole of their territory each night. When we made track beds on field margins (likely fox route-ways) at an average density of about 12 per km, only 20% (range 11% to 23% in six trials) of plots were visited by a fox or foxes each night. The addition of a fox-specific odour cue ('Fox Gland Lure 100', S. Hawbaker & Sons, Pennsylvania) made no significant difference to this probability. Following the first visit by a fox or foxes, the probability of revisitation on succeeding nights was also un-changed (range zero to 60%; mean 20%). By contrast, for buried food baits (canned dog food buried under about 1 cm soil) with no other odour cue, the probability of revisitation was significantly higher (range 10% to 90%, mean 50%). Longer-range olfactory attractants, or combinations of these with short-range attractants or food rewards, have yet to be tested. Several epi-sodes recorded on video at bait sites already suggest that complex social behaviour may interfere with our goal of accessing all individual foxes pres-ent. For instance, we have recorded the scent-marking of food bait by one individual and later clear avoidance of the scent-mark itself (and thus the bait site) by another individual. In the context of rabies vaccination baits, Trew-hella *et al.* (1991) also found evidence of monopolization of baits by dominant individual foxes. Furthermore, our data suggest that the addition of fox-specific odour may, in fact, decrease visitation rate by resident foxes; this requires further study.

Non-target species

Another aim of a pre-baiting system is to exclude non-target predators. Arguably any animal that eats an aversive bait should be considered as a target species. However, some species involved in this way may have a vulnerable conservation status and/or statutory protection. Since aversive baits are necessarily dead, other involved species may be only scavengers. Furthermore, because baits are designed to deliver a safe but aversive dose of chemical with regard to the target predator's body size, the involvement of non-target species may result in the delivery of either an ineffective or a lethal dose.

The use of a pre-bating system creates a risk of attracting a wider range of species by prolonging baiting, and the opportunity for precise targeting by using species-specific cues. In the model fox/partridge system referred to above, most non-target species (mice, rats, corvids and buzzards) can be excluded by burying bait. However, badgers will also dig out buried baits; they are more common than foxes in most parts of Britain, are arguably a partridge predator but also have statutory protection. Despite video surveillance and the use of plastic marker beads we have no field evidence of badgers digging out meat baits of the kind we use. Nevertheless, in view of their statutory status, and for the sake of efficiency, we have begun to explore the use of fox-specific olfactory cues to deter badgers from using pre-bait sites. Using the fox-specific odour cue alone (i.e. interest but no food reward), we found that on 52% of occasions where a fox passed within 0.5 m ($n = 74$, in six independent study areas) the lightly buried cue was detected and investigated further. This often (19% of visits) meant they scratched out the plaster disc on which the lure was presented; the disc was also sometimes chewed or carried, or the plot scent-marked. By contrast, no passing badgers ($n = 90$, in six independent study areas) showed any sign even of orientation towards the cue. In work currently underway, we are finding that if the odour cue is food-based (anchovy) rather than fox-specific, badgers showed a very clear interest.

THE WAY FORWARD

This area of research needs to progress on two fronts. Fundamental studies of the nature of CTA are important on several counts: they may lead to the formulation of powerful and safe products to generate aversions; they will demonstrate the generality or specificity of taste-conditioned aversions; they will also refine expectations of how CTA might best be exploited in the field. Equally, key questions about predator behaviour and field logistics, which may ultimately limit or prevent the efficient exploitation of CTA in wildlife management, can only be answered in the field. In both aspects of research, continued funding is unlikely unless unambiguous, encouraging results accumulate from well-designed experiments.

ACKNOWLEDGEMENTS

The Central Science Laboratory studies referred to here were funded by the Land Use and Rural Economy Division of the UK Ministry of Agriculture, Fisheries and Food. We are indebted to Lowell Nicolaus for personally educating us regarding both the power and subtlety of conditioned taste aversion processes. We should also like to thank Anne Whiterow and Mike Short for their assistance with our laboratory and field studies.

17

Retaining natural behaviour in captivity for re-introduction programmes

MICHAEL P. WALLACE

BACKGROUND

As profound threats to ecosystems increase worldwide, the populations and species they contain are increasingly threatened. Along with the rise in numbers of endangered species, there is a parallel need to develop efficient methods for captive propagation and for the preparation of captive-born progeny for release to the wild. The process is slow, however, and, as will be discussed in this chapter, techniques are tested and refined on a species-by-species basis owing to the myriad of reasons for endangerment and vast array of difference in biology, behaviour and life-history.

Griffith *et al.* (1989) reported generally poor results in re-introduction programmes they investigated when the main criterion for determining success was a wild, self-sustaining population. An objective analysis of re-introduction programmes is difficult for several reasons, as pointed out by Beck *et al.* (1994). Even for species with high reproductive rates, building a viable, self-sustaining population takes time and most projects have not been in progress long enough to fully assess their success or failure. Except for the well-tested hacking technique for raptors, with a several thousand year history of falconry tradition, most programmes are still in the process of refining rearing methods, release strategies and post-release monitoring and training techniques. Also, many of the early attempts at reintroducing animals were not as well organized and documented as today's efforts. Without a more rigorous scientific approach to record-keeping and thorough monitoring after release, analysis of the results is difficult. Basing future efforts on the outcomes of the earlier, incomplete programmes, is risky and the potential for wasting precious time, funding and political support for such endeavours is high.

There are many aspects to the rearing and release process that need consideration: genetic and demographic management of the breeding pairs, proper diet, hygiene and standardized husbandry techniques are important to produce healthy individuals. Husbandry techniques that result in healthy young are generally all that are needed for captive life, but if we intend progeny of captive animals to be released to the wild, behavioural considerations become a priority. The rearing conditions must be well thought out so that normal husbandry does not conflict with the behavioural needs of a potential release animal. It is more difficult for veterinarians or animal keepers to cope with the health needs of a new-born mammal or freshly hatched chick if, as a release candidate, it needs to be kept visually isolated from people.

When rearing animals for captivity, selective pressures in the wild are generally not considered. The lack of natural selective pressures may erode, over a few generations, the genetics of important morphological, behavioural or physiological traits necessary for survival in the wild if those traits are not relevant to captive life. An r-selected species, one that has a short generation time, is more prone to the possibility of 'selection for captivity' (Derrickson & Snyder, 1992; Jimenez et al., 1994). Beside the long-term potential for selective pressures to make some captive animals more suitable for the captive environment, the behavioural ontogeny of an individual may lack critical elements during development that could give the reintroduced animal the edge it needs to survive. Once released into natural habitat, it must cope with hunger, predation, disease, parasites and inclement weather, let alone competition with other members of its' own species to survive and reproduce.

To be successful, wild animals have evolved an array of finely homed behavioural and physiological responses that adapt them to environmental stresses. The development of complex behaviour patterns results from the interaction between an animal's genetic make-up and the experience it gains as it interacts with the rearing environment (Polsky, 1975). Some behaviours cannot be experienced in adult life if the animal is not exposed to repeated cues throughout its juvenile development (Gossow, 1970). It follows that the behavioural development in a poorly planned captive environment will probably result in individuals unprepared for life in the wild (Derrickson & Snyder, 1992).

BEHAVIOURAL CONSIDERATIONS

In order to provide rearing conditions that maximize the survivorship of released animals, there must be a thorough understanding of many aspects

Table 17.1
Natural history strategy: r – and k – selection

r – selection	k – selection
Short life span	Long life span
Produce many young	One or few eggs per season
Give little training to young	Longer parental dependence of young
Survivorship is low	High survivorship
Populations fluctuate greatly	Population stabilizes near caring capacity
Reach sexual maturity quickly	Reach sexual maturity slowly
Intrinsic rate of population increase – high	Intrinsic rate of population increase – low
Examples:	Examples:
masked bobwhite quail	condors
Bali mynah	eagles
Guam rail	cranes
black-footed ferret	swans
golden lion tamarin	rhino
	oryx

of the species' behaviour and the environment it is adapted to. The most successful releases occur if the animal is reared in a manner consistent with its natural inclinations. Two features, the concept of a life-history strategy and the degree of sociality, can be helpful in understanding how a species has adapted behaviourally to a particular niche.

Life-history strategy

The way a species evolved in and interacts with its niche, it's life-history strategy, gives some perspective on a species' means of survival. Most vertebrate animals can be characterized or having a life-history pattern that lies along a continuum from those with a high intrinsic rate of population increase with regular or stochastic fluctuations in population size (r-selected) to those with populations that fluctuate little around a relatively stable environmental carrying capacity (k-selected) (Pianka, 1970). Species can be categorized (Table 17.1) as to their tendency to show one series of traits or the other. This comparison is useful to develop an understanding of how a species has adapted to a particular niche and possibly to acquire clues on how to rear and release a particular species.

Small species with a rapid physical development and high propagation growth rates such as golden lion tamarin (*Leontopithecus rosalia*), black-footed ferrets (*Mustela nigripes*), masked bobwhite quail (*Colinus virginianus ridgway*) and Bali mynah (*Leucopsar rothschildi*) are likely to require release of large numbers because of a corresponding, naturally high mortality rate. Parental attention in these species after fledging may be limited, making the parental dependency period relatively short. Once released, there is often little for the researcher to do other than to monitor the outcome. For k-selected species, the opposite is true. Cranes, condors, elephants or rhinos would require greater efforts to optimize the survivorship of each individual through more intensive preparation for release and post-release management. The longer dependency period these offspring normally receive from their natural parents may indicate extensive pre-release and post-release training such as is necessary to increase survivorship in swans, condors and cranes (Lewis, 1990; Abel, 1993; Wallace, 1994).

Sociality

The degree and type of sociality a species exhibits is important to consider while developing a release technique. The release process for the non-social raptor is quite different than for flocking species such as parrots that communicate auditorily and visually in the wild (Temple, 1978; Synder *et al.*, 1987) or cathartid vultures that forage and roost communally (Wallace & Temple, 1985). Depending on how sociality in any species might increase survivorship, social expression may be very different and its effect on any release programme will vary. For example, two cathartid vultures, the American black vulture (*Coragyps atratus*) and the turkey vulture (*Cathartes aura*) are both considered highly social, often roosting and foraging communally. The black vulture must hunt for food as a group, although the flock may sometimes be widely dispersed, while turkey vultures are not obligated to forage communally and often forage alone. Its abilities to find food through the use of olfaction, not only makes this scavenger unique among birds, but also allows the release of turkey vultures to be a relatively simple process. While black vultures were only successful if released in pre-socialized groups, young turkey vultures released as single birds survived quite well (Wallace & Temple, 1983, 1987a).

Imprinting

The concept of imprinting was first described in birds as a sensitive period in behavioural ontogeny when critical information is permanently learned

(Lorenz, 1952; Hess, 1959). Information as to species identity, habitat type and even orientation in the landscape (philopatry), may be learned during sensitive periods that vary in length and intensity depending on the species (Hess, 1972; Immelmenn, 1975). Animals that have a very short nestling phase such as pheasants and waterfowl tend to 'imprint' very strongly but over a short period of days or even hours. Species that spend a long period of time in the nest, such as condors, which fledge at 6 months of age and have a long parental dependancy period of several more months, seem not to have a strongly defined critical period for determining species identity. They must be kept isolated from human contact over a longer period of time to avoid malimprinting to the keepers raising them.

MANAGEMENT TECHNIQUES

Enriched environment

What Miller, Reading & Forest (1996) called an 'enriched environment' for enhancing the captive rearing situation can mean a range of changes, from structural complexity in the physical environment that may enhance a particular behaviour, to actual training to modify or teach a behaviour that might increase survivorship in the wild. The degree of enhancement to the rearing environment that might produce the best survivorship in released animals again varies with each species, from none to quite elaborate schemes. Some have attempted to raise release candidates in a simulated portion of the species' natural habitat, such as with black-footed ferrets by using a closed arena that includes prairie dogs and their tunnels (Biggins *et al.*, 1990, 1993). Others have allowed release candidates access to alternative environments that contain the same basic features to the species' own habitat into which it is later released. Golden lion tamarins, for example, are released over several months, first in the confines of participating zoos across the United States prior to their permanent release in their Brazilian native habitat (Kleiman *et al.*, 1986).

The environmental enhancement of some species can include more specific training to improve foraging or predatory skills and/or predatory avoidance. Release candidates can also be taught or conditioned to avoid dangers in the environment, such as the use of power poles as perches, in the case of California condors (*Gymnogyps californianus*) (Figure 17.1).

Three main rearing techniques are basically used in captivity to produce

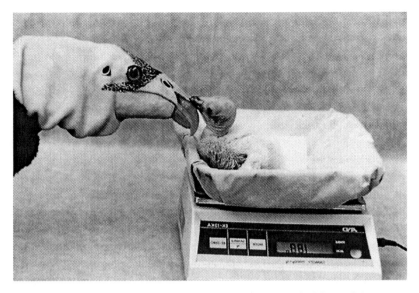

Figure 17.1 Californian condor, *Gymnogyps californianus*, chick being fed using a hand puppet to prevent habituation to people and encourage imprinting on their own species (photograph by Clair Mirande).

animals behaviourally acceptable for release: parent-rearing, cross-fostering and isolation-rearing.

Rearing techniques

In parent-rearing, the natural parents of the offspring rear the young. In a slightly different technique, foster-parent rearing by conspecifics, the young are raised by other adults of the same species, which produces essentially the same behavioural atmosphere. Both methods insure the best possible species imprinting and socialization and are used whenever possible.

Cross-fostering is a technique to rear and release young when it is not practical, or efficient, to allow the actual parents to participate in the care and fledging process of their offspring. Cross-fostering has been used when captive parents are required to produce more young than they can rear on their own. Captive Andean condors (*Vultur gryphus*) have been subscribed to 'parent-rear' California condors in captivity and wild prairie falcons (*Falco mexicanus*) have been used to rear peregrine falcons (*Falco peregrinus*) produced in captive breeding programmes (Sherrod *et al.*, 1982). Subsequent pair formation of falcons fledged by this method appears to be normal and directed appropriately toward their own species. More social species, such as cranes, have shown mixed results in cross-fostering experiments (Lewis, 1990).

When parents or foster parents are not available or practical, isolation-rearing may be a viable alternative for producing behaviourally adequate young for release. Especially in captive breeding programmes of endangered species, an important goal is to produce as many offspring as possible, because there is always some risk that genetically important founders in the population may die prematurely. In birds, particularly, multiple clutching – the removal of eggs from the parents to induce the production of more eggs than would normally be laid – is an important tool for dramatically increasing the reproductive rate. It also usually necessitates hand-rearing many of the offspring. Through the use of one-way windows and video monitors, animals can be reared in visual and auditory isolation from human contact. Raising social species in the presence of conspecifics and in visual contact with adults in an outer cage can help enhance the social environment as well as allowing the animal to integrate as soon as possible into a group with more experienced individuals.

Hard versus soft releases

The terms hard and soft release refer to the process of actually liberating the animal. Hard implies a release with virtually no waiting period at the site of release. It is commonly felt that soft releases should be conducted whenever possible. In a soft release, the animals are held in enclosures at the site for acclimatization to local environmental conditions, familiarization with landmarks, post-transport recuperation and establishment of social and reproductive bonds (Brambell, 1977; Caldecott & Kavanagh, 1983; Kleiman, 1989; Chivers, 1991; Wallace, 1994).

While reintroductions are essentially management programmes, with the goal of re-establishing populations, they must be scientifically conducted with studbook-type databases on losses, rescues, re-releases, births and genetic lineage. Re-introductions must have long-term post-release monitoring if they are to have the best chances of success. Without a commitment to effective monitoring, it may be impossible to access the behavioural reaction to the new environment and how problems of low survivorship may be remedied.

CASE HISTORIES

Case histories of several programmes can help illuminate these concepts.

Bali mynah

Bali mynahs are reared to fledging age by their parents in holding pens at release areas within the Bali Barat National Park on the island of Bali. The family groups are managed so the offspring are isolated from visual and auditory contact with humans. The young fledge from natural-looking cavities in the pen and are allowed to acclimatize to their surroundings for a period before release. Food and water are kept available at the site until their foraging reaches a level of efficiency where food subsidy is no longer necessary. Bali mynahs released in this way in 1992 quickly assimilated into a wild flock. Pairs formed, and young were successfully reared by released birds in their first year. Unfortunately poaching for the pet bird trade continues to threaten both the wild and released birds (Siebels, 1993).

Guam rail

The loss of the Guam rail (*Rallus owstoni*) on its home island of Guam is primarily owing to an extremely effective predator, the brown tree snake (*Boiga irregularis*) that was introduced from other regions of Indonesia. Aggressive, but relatively easy to breed in captivity, these rails are reared at North American zoos by their parents in off-exhibit and exhibit facilities in view of the public. Sometime after fledging, offspring chosen for release are gathered and shipped to Guam and held in staging pens for a period to adjust to the climate (Derrickson, 1986). Insects, skinks, geckos, and other natural foods move freely in and out of the holding pens giving the birds an appropriate search image before their release on the remote and snake-free island of Rota (Shepherdson, 1994).

Since breeding is relatively rapid and the logistics of parent rearing are easy, no concentrated effort has been put into working out isolation rearing methods for this species. The island of Rota is sparsely populated by people and, if given a choice, rails tend to be secretive, even after rearing in view of the public. As long as the release candidates can be exposed to wild food, survivorship does not seem to depend on supplemental feeding after release or any specific training of this species to natural foods. Of the 94 birds released, only two seemed to have problems foraging and died of starvation.

Some were seen to begin foraging immediately upon release. The biggest limitations to long-term population growth in the wild has been predation by feral cats and too rapid a dispersal from the release site. The cats have been removed from the island and a softer release method, by allowing more time in the pre-release pen, tends to manage dispersal (K. Brock, personal communication).

Golden lion tamarin

The first efforts to release golden lion tamarins were largely unsuccessful. Despite some efforts at pre-release training, the reintroduced animals showed striking deficits in locomotion and orientation and only five of the 14 released tamarins were still free-ranging 8 months after release (Kleiman et al., 1986). Eventually only three survived and were recaptured. In response to the failure of the initial release attempts, the zoo rearing environment was enhanced with moving limbs and vine-like ropes to challenge their locomotor development. Because the tamarin pairs are naturally reluctant to leave the area around the roost box, a post-release free training exercise could be developed. Instead of being released in Brazil, release candidates were first freed in a remote area of the National Zoo. The area was supplemented with feeding devices to stimulate foraging, and ropes and vines to mimic natural rain forest. As they expanded their range over a 2 month period, foraging, locomotor and orientation competence increased. Less time and labour-intensive then previous training methods, this post-release training produced animals that were far more successful once released in their natural habitat. Several zoos around the United States are now used as post-release training centres for this primate (Bronikowski et al., 1989).

Red wolf

Where tamarins require an elabourate, free-ranging training programme, captive-reared red wolves (Canis rufus) released into North Carolina were first acclimated in 225 square meter pens for periods lasting from 2 weeks to over 2 years. Human contact was minimized and their diet changed from dog food to an all-meat diet. The schedule varied to simulate a feast or famine feeding regime and wolves were provided the opportunity to hone predatory skills by having access to live prey. Wolves were provided supplemental food in the area after release. Of the 36 animals released, 21 died and seven were removed from the wild for management reasons. The six that survived long-term were among the wolves that experienced longer acclimatization periods but many of the deaths were a result of intraspecific strife between recently released adults and established wolves (44%). At least 20 pups were eventually produced by the six survivors. The wild-born pups showed higher survival rates than wolves re-introduced into their first year (Phillips *et al.*, 1995).

Houbara bustard

For Galliformes, anti-predator training can significantly increase survivorship. Post-release survival of Houbara bustards (*Chlamydotis macquecniri*) was improved through exposure to a live red fox (*Vulpes vulpes*) during a pre-release training episode. High mortality (38%) to wild red fox predation occurred when birds were hand-reared in groups and released without any training. Suspecting that hand-reared bustards might be showing less appropriate responses to predators, Van Heezik *et al.* (unpublished data), tested the effects of isolation rearing, anti-predator training with a model fox and anti-predator training with a live fox. In a large 15 m × 40 m pen, a hand-reared fox, on a long lead, was allowed to stalk, pursue and, occasionally, catch bustards. While birds reared with minimal contact showed an increase in stress levels when shown a model fox in captivity, compared to bustards reared in daily contact with people, the nervousness did not translate to better survivorship from fox predators after release. Nor did the training with the fox model where the authors believed habituation may have been involved. Survivorship was increased significantly when the live fox was used in the training bouts.

Masked bobwhite quail

As an r-selected species, the endangered masked bobwhite quail reproduces at a high rate. The recovery plan for this species calls for 2000 birds to be released per year. Pre-release training was used to lower the expected high mortality rate resulting from predation. Response time to predators was decreased through the use of hunting dogs to simulate wild canids and trained hawks to simulate danger from wild raptors (Ellis *et al.*, 1978). These techniques proved effective, but labour intensive and not as efficient as a cross-fostering technique used through most of the project. Groups of incubator-hatched masked bobwhite quail were housed with vasectomized Texas bobwhite quail (*Colinus virginianus*) prior to release. Once released the adult male quail tended to shepherd the dependent young to food and cover and teach appropriate responses to predators (Dobrott, 1993).

Whooping crane

Cross-fostering worked well with this species in this situation. The birds had no chance to breed with their sterile foster parent and spent much of their time in contact with their own species, which encouraged later pair formation. An attempt to cross-foster whooping cranes (*Grus americana*) to sandhill cranes (*Grus canadensis tabida*) nests, as eggs, was not as successful. The experiment was an attempt to develop a distinct population of whoopers that could follow migratory patterns of their sandhill foster parents, which nest in Idaho and winter along the Rio Grande in west central New Mexico (Drewien & Bizeau, 1978). Low nesting survivorship owing to coyote predation and bad weather dropped the population from 209 chicks to 84. Most of those that fledged died from collisions with power lines and wire fences. The cranes that survived to maturity appeared to have difficulty forming pair bonds with their own species (Lewis, 1990).

While cross-fostering experiments with cranes were not entirely successful, other methods to build wild populations are being developed. Extra young whooping cranes, produced by multiple clutched captive parents at the International Crane Foundation in Baraboo, Wisconsin, have been isolation-reared using hand puppets. To enhance the young cranes' environment, whole costumes and working models of parents were used in the crane habitat, enabling the attendant to achieve more than proper imprinting socializations and avoidance of tameness to humans. A full body costume or white sheeting, loose fringes where the wings would appear, and a puppet head on one hand, enabled the attendant to exercise fledgling cranes and escort them into appropriate feeding sites in a marsh habitat. The fledglings were conditioned to contact calls from the costumed attendant through the use of a tape recorder, as well as alarm calls played in response to the surprise appearance of non-costumed people alone or with dogs (C. Mirande, personal communication). While the contact calls effectively encouraged social bonding to the costumed attendant, the alarm calls, when played under the appropriate circumstances, elicited an escape response from the young birds. This artificial stimulus, in the appropriate

Figure 17.2 Whooping cranes, *Grus americana*, can be raised by people disguised as cranes who direct them to good feeding locations and mimic alarm behaviour in the presence of danger (photograph by Mike Wallace).

environmental context, from the surrogate parent apparently helped the chicks generalize their social attachment to wild conspecifics after release (Figure 17.2). Once with wild birds, they successfully learned migration patterns and fine-tuned several skills.

Although this technique is labour intensive, less vigorous methods of hand-rearing and releasing cranes produced birds showing too much tameness to people and inappropriate feeding behaviour leading to the failure of the effort (Nesbitt, 1979).

Trumpeter swan

High survival rates were achieved in trumpeter swan (*Cygnus buccinator*) releases conducted by the University of Wisconsin Wildlife Ecology Department at Madison by extensive use of models and decoys. Like the crane experiments, the 'enhanced' environment, for these isolation-reared chicks, was a portion of the natural habitat. After keeping chicks in pens and imprinting them to mounted specimens of adult swans, a floating swan decoy was effectively used to guide the cygnets to potential feeding areas in the marsh. To accomplish this and remain hidden from the birds, the attendants used a blind covering, a fisherman's float, and hip waders to walk on the bottom or float through the marsh. The cygnets followed the decoy tethered to the end of a stick and towed by the attendant. The mere presence of the decoy near likely feeding areas tended to intensify the feeding response of the cygnets at that location. They were also led, by the decoy, to daytime roosting islands in the marsh, where maintenance behaviours and sleeping could be done safely, reducing the danger from most predators. When potential predators, such as raccoons (*Procyon lotor*) or coyotes inadvertently appeared on the shore lines, or when a person with a dog staged an appearance, the decoy parent reacted with vocalizations and avoidance manoeuvres that led the cygnets to safety. Until they could fly on their own, this decoy parent would lead the fledglings back to the safety of the cage each night (Abel, 1993). Unless such labour-intensive methods as described above can significantly increase survivorship, they are not cost effective and cannot be justified. Using this fledgling technique, a 75% first year survivorship was reported, whereas swans raised by their parents, or other adults that fledged from urban ponds, showed only 25% survivorship in their first year. Strong philopatry is shown in this species. The decoy method has the potential to establish or increase populations rapidly in specific, appropriate habitats and achieve, more efficiently, an expansion of its range (Abel, 1993).

California condor

While cranes and swans can be taught much about their environment, concerning food acquisition and predator avoidance, prior to their abilities to fly, somewhat different strategies have been employed to release condors, which normally learn much information through a lengthy parental dependancy period after they have fledged. Between 1992 and 1994, 13 California condors were released to the wild. Over the 6 month nestling phase, condor chicks were raised by the use of an adult puppet and integrated into same-age social groups. At 6 to 10 months of age they were tagged and transported to release sites in the Southern California mountains. While isolation-rearing with conspecifics seems to produce a socially adjusted release candidate, the species is normally guided by parents and other older birds after fledging to the wild. Unless there is an existing population to provide guidance to released condors, as was the case with Andean condors (Wallace & Temple, 1987b), these social scavengers have no role models to guide their natural, intense curiosity. Of the 13 birds released, four died from power-line collisions and one from ethylene glycol ingestion.

The power-line collisions problem seemed associated with an unnatural propensity of the young released condors to seek out power pole cross-arms as day perches and overnight roosts. Both behaviours raised the probability of a collision as they flew to and away from the perch. An experiment to use negative conditioning was conducted at the Los Angeles Zoo, beginning in 1994 (Wallace, 1997). Instead of being transferred to the field release pen at fledging age and released to the wild after an acclimation period, groups of condors were fledged into a large (15 m × 15 m × 33 m) flight pen in a remote, off-exhibit portion of the zoo. Several natural perches were arranged throughout the pen in the form of live and dead tree branches. A 4-metre pole with a cross-arm and ceramic insulators, resembling a power pole without actual wires, was placed in a prominent position within the pen. An electrified wire on the perching surface of the mock power pole gave a mild, but uncomfortable, shock to any condor that landed on it. With a time-lapse camera and identification tags on their wings, we recorded that it took zero to six encounters with the mock power pole, over a 3 week period, for the condors to permanently learn to avoid the pole. Some condors learned by apparently watching other birds' reactions to the pole.

The remaining eight condors from the previous release were captured, because they continued to use power poles in the wild and would have a strong influence on subsequently released birds. Since 1995 over 40 California condors have been given 6 months to over 1 year exposure to the mock power pole before release to the wild. Although previously released condors used the poles on a daily basis, birds conditioned by the mock power pole have not been seen landing on real power poles in the wild. Two additional condors have died from striking power lines since the use of aversion training began. While power poles aversion training appears to reduce the condor's exposure to the wires, and the number of strikes, it does not eliminate the problem because they still encounter a number of power lines that traverse hilly terrain used by the birds while slope soaring.

CONCLUDING THOUGHTS

It becomes apparent when reviewing re-introduction programmes that the most successful tend to be correlated with extensive history and experimentation. For species with long generation times, pertinent data are much slower to accumulate than with species that breed rapidly and mature early.

Knowing the target species, ecology, behaviour and life-history strategy are essential to a basic understanding of the animal, without which success with reintroduction efforts are unlikely. The most cost-effective methods will be species specific. No one technique will necessarily be efficient and successful in every situation, but one will emerge as the most effective for a specific set of conditions.

18

Consequences of social perturbation for wildlife management and conservation

FRANK A. M. TUYTTENS & DAVID W. MACDONALD

INTRODUCTION

Human involvement, direct or indirect, in almost every ecosystem is so profound that laissez-faire conservation is more likely to lead to neglect than to pristine Utopia. Doing nothing is rarely an option. Whether the goal is to preserve the imperilled or to control the pestilential, management is, to a greater or lesser extent, necessary. Responsible management requires its practitioners to predict effectively the consequences of their intervention. In attempting to manage vertebrate populations these consequences will often take a demographic form – indeed, the objective may be to increase or decrease the number of a population. However, those seeking to manipulate vertebrate populations must be alert to the complexity of individual behaviour. Behavioural response to intervention may radically affect the numerical outcome of attempts to manage a population. In particular, attempts to reduce populations may stimulate changes in the behaviour of those remaining – a perturbation effect – which diminish the efficiency of the attempted management, perhaps even making the intervention counter-productive. Our purpose in this chapter is to review evidence for this general hypothesis.

This topic is important because, if the hypothesis is correct, it has bearing on a huge diversity of interactions between people and wildlife. Wild animals are killed by humans worldwide for subsistence or profit, for sport and disease control. Although controversial, the sustainable harvest of wildlife has been advocated as a pragmatic alternative to either over-consumption of populations or 'non-interference' preservation of habitats (Robinson & Redford, 1991; Beissinger & Bucher, 1992; Kock, 1996). Most studies of harvesting examine only the effects of the removal of individuals upon the predicted subsequent size of the targeted population (Olsson *et al.*, 1996, and references therein). Here, we argue that size is not everything, and that social perturbation caused by control operations could have far-reaching implications for wildlife management and conservation, and might even contribute to the failure of control objectives. Despite recent interest in the topic (Macdonald, 1995; Olsson *et al.*, 1996; Tuyttens & Macdonald, 1998a), relatively little is known about the effect of management on social behaviour, within the larger topic of how social behaviour contributes to population demographics and dynamics (Miller & Ozoga, 1997). To explore this topic we have identified six population parameters that might be affected by human intervention. These may be demographic (density, age/sex structure, and the pattern and rate of dispersal), or socio-spatial (space use, breeding system, and sociality) and others, not included in this review, such as community structure, are inter-connected in complex ways such that changing one aspect often affects many others. Subsequently, we illustrate the implications of such perturbations for the control of pest species, the conservation of desirable and/or harvested species, and the management of wildlife diseases.

POPULATION DYNAMICS

Population density and recovery

The degree to which control affects population density depends on the severity of the control applied, the ability of the survivors to compensate by increased fecundity or survival, and the immigration:emigration ratio. These processes have been the topic of an immense literature spanning the work of Lack (1954) to, for instance, Caughley & Sinclair (1994). Rapid compensatory density-dependent responses are often the cause for the failure to maintain pest species at low levels (Green, 1984; Gao & Short, 1993; Williams & Twigg, 1996). Insofar as fecundity is linked to longevity (and, broadly, to body size), the potential for a compensatory response to population reductions may not be as great in long-lived (or K-selected) species as in short-lived (or r-selected) ones (King & Moors, 1979; Clark, 1990). If it were not for recruitment from other areas ('sources' – sensu Pulliam, 1988) many populations of animals ('sinks') could not support the heavy hunting pressure they receive (e.g. Cary *et al.*, 1992; Hodgman *et al.*, 1994; Novaro, 1995; Slough & Mowat, 1996). Over- exploitation of species with economic or aesthetic value is a major cause of local extirpation and even extinction of large vertebrates (Martin, 1984; Redford, 1992; Bodmer *et al.*, 1997). The recovery of persecuted populations may be dampened through any disturbing effects that attempted control may have on the behaviour of survivors. Batcheler (1968) suggested that the deaths of brushtail possums, *Trichosurus vulpecula*, some months after a poisoning campaign, resulted from the deterioration of nests, trail networks, territorial marks and scent posts, which fell into disarray after the death of a large proportion of the population. This anarchic loss of social structures is a phenomenon that could apply in some measure to many controlled populations, and has not been adequately tested in any.

Age/sex structure

Population control often distorts the age/sex structure of the population, sometimes deliberately. Fertility control can be expected to result in an older population (Tuyttens & Macdonald, 1998a), while unselective lethal control *per se* should have no effect on age/sex structure (Caughley, 1977). However, depending on the characteristics of the targeted population and the type of control employed, the survival rates of all age/sex classes are rarely reduced *pro rata*, and unselective control may not be common. For example, game sportsmen in England formerly snared large numbers of red foxes, *Vulpes vulpes*, over winter – the time when they saw most foxes and therefore were most stimulated to act against them. The snaring was seemingly unselective, and there was certainly no intention to select for a particular sex. Yet these winter fox snaring culls were often heavily skewed towards killing males, probably because they were predominantly the dispersing sex, and were roaming, perhaps less vigilantly, in search of females. Many other mammal populations that are hunted or trapped are female-biased because disproportionately more males are harvested. This could be a result of males having larger home ranges or greater mobility (marten, *Martes americana*: Hodgman *et al.*, 1994; bobcat, *Lynx rufus*: Rolley, 1985; lynx, *Felis lynx*: Slough & Mowat, 1996), being dominant (stoat, *Mustela erminea*, and long-tailed weasel, *Mustela frenata*: Powell, 1979), being less wary (roe deer, *Capreolus capreolus*: Andersen, 1953), or being specifically targeted. The latter is especially true for sexually dimorphic species, where males may be considered desirable trophies. An obvious example is that of male Asian elephants, *Elephas maximus*, that are selectively targeted by ivory poachers as females do not possess tusks

(Sukumar, 1989). It is also easier to kill a lone adult male than a female in a herd. In addition, the 'high risk–high gain' strategy for promoting reproductive success employed by such polygynous animals may bring males into greater conflict with people (Sukumar, 1991) and make them more prone to human predation (Berger & Cunningham, 1995).

Changes in fecundity or in age-specific survival with density could explain the frequently observed young age distribution of harvested populations (Clout & Barlow, 1982; Rolley, 1985; Gese *et al.*, 1989). Such an age bias could also be the result of older individuals being selectively targeted, for example, because they have larger tusks (African elephants, *Loxodonta africana*: Poole & Thomsen, 1989) or provide more meat (capybara, *Hydrochaeris hydrochaeris*: Moreira & Macdonald, 1996). The specialized demands of the laboratory animal trade for the larger wild-caught primates illustrate that sometimes females and young may be selectively targeted as well (Bowles, 1996).

Dispersal

Age- and sex-specific dispersal rates may also lead to unstable age/sex structures in a population following attempted control. Rapid recolonization of a recently depopulated area, predominantly by dispersing immature male brushtail possums, explained the initial strong male-bias, which contrasts with the stable female-biased sex ratios characteristic of undisturbed areas (Clout & Efford, 1984). Partly because of the brushtails' dispersal capability, the species is a remarkable survivor, although its adaptability to perturbations varies throughout its range (Kerle, 1984). Male-biased dispersal, as in the possum, is very common in mammals (Greenwood, 1980). Two main hypotheses seek to explain this sex-bias: intra-sexual competition for mates, and inbreeding avoidance (Johnson & Gaines, 1990; Wolff, 1994). It is noteworthy that the largest dispersive age/sex class is not necessarily the class of animals that first recolonizes depopulated areas. For example, although in Eurasian badgers, *Meles meles*, dispersal is generally more common among adult males than adult females (Woodroffe *et al.*, 1993), the first recolonizers of cleared areas are almost exclusively female (Tuyttens *et al.*, 1997). This may suggest that for adult male badgers the number of mates is the primary resource to be maximized, while female space use is determined to a greater extent by the availability of food and dens – a pattern that is widespread in mammalian social systems.

Despite complications such as males searching for females and other examples of conspecific attraction (Stamps, 1988), dispersal tends to be a density-dependent phenomenon. Gaines and McClenaghan (1980) explained how the relationship between dispersal and population density depends on the proximal mechanism underlying dispersal. The Social Subordination Hypothesis (Christian, 1970), for example, predicts that intraspecific competition and aggression forces subordinate animals to disperse most during phases of peak density. The Genetic–Behavioural Polymorphism Hypothesis also predicts that dispersers are social subordinates. It assumes, however, that dispersal behaviour has a large genetic component that is selectively favoured during the early increase phase of fluctuating populations, while aggressive spacing behaviour, favoured when density continues to increase, sets the stage for a declining phase. A change in social behaviour, induced for example by a continued removal operation encouraging the influx of animals that are less aggressive and that have a higher reproductive rate, could lead to an excessively high population density and perhaps even trigger a plague (Krebs et al., 1978; Krebs & Chitty, 1995). Bekoff's (1977) Social Cohesion Hypothesis proposes that, particularly in species that live in highly social groups, the major stimulus for dispersal is the failure to develop strong social ties to the natal group prior to emigration rather than agonistic interactions at the time of emigration. Removal of individuals may prevent, or perhaps even encourage, the formation of social ties and hence affect dispersal. Otherwise the latter hypothesis predicts no association between population density and dispersal (Gaines & McClenaghan, 1980).

There is a vast literature on vertebrate dispersal and it is not our purpose to review it here. Rather, we point out merely that the process is complicated, that several hypotheses plausibly explain various features of dispersal in different taxa and, importantly, all these processes are likely to be affected by attempted population management. Clearly, a good understanding of the proximate as well as the ultimate explanations of dispersal and spatial organization is of crucial importance for predicting the speed and pattern of recovery of persecuted populations. Most vertebrate species show dispersal patterns that are density-dependent or that are non-random for other reasons (Johnson & Gaines, 1990; McCullough, 1996). Models that assume random dispersal and colonization patterns often underestimate the harvest rate that can be sustained by local sink populations (Cary et al., 1992), but overestimate the persistence of harvested metapopulations (Smith & Peacock, 1990; McCullough, 1996).

SPATIAL AND SOCIAL STRUCTURE

Space use

Increased movement in an enlarged home range seems to be a general behavioural adaptation to human disturbance in cervids (Jeppesen, 1987). Red deer (*Cervus elaphus*) react to hunting disturbance by seeking dense cover within their home ranges or by taking flight, often over more than 5 km to refuge areas where they may stay for more than a week (Jeppesen, 1987). When hunting becomes intense deer become nocturnal and wary, leave preferred areas, and expend more energy as a result of increased stress and movements. Black rhinoceros, *Diceros bicornis*, have also been observed to abandon local sites in response to simulated poaching, with females covering up to 40 km in a day (Berger & Cunningham, 1995).

Control operations may also alter space use indirectly, for example, through effects on population demography or resource availability. Home range sizes are generally inversely correlated to population density (e.g. raccoon, *Procyon lotor*: Ellis, 1964; California vole, *Microtus californicus*: Ostfeld, 1986), and to resource availability (e.g. brushtail possums: Warburton, 1977; Davies, 1978). However, most removal studies fail to disentangle the relationship of home range size to population density, age/sex structure, resource availability and human disturbance. The experiments by Mares and Lacher (1987) are perhaps an exception and allowed the authors to conclude that resource abundance rather than competitor density determined chipmunk, *Tamias striatus*, home range size. Seasonal, sexual, age, and habitat-specific strategies result in additional complexity (Carpenter, 1987; Erlinge *et al.*, 1990; Salvioni & Lidicker, 1995).

Understanding of the ultimate determinants of space use is a prerequisite to predicting the effect of control operations on the spatial organization of a population. Katnik *et al.* (1994) reason that the social systems of solitary carnivores are more sensitive to harvesting if the primary determinant of spatial relationships among males is mate access (as for the fisher, *Martes pennanti*, and stoat), rather than energetic requirements related to food availability (as for the marten). Territoriality based on the defence of economically defendable resources may be abandoned when competition for these resources is reduced as a consequence of population reductions. Such a change in spatial organization from defended exclusive territories to undefended overlapping ranges has been associated with attempted control in, for example, the Eurasian badger (Cheeseman *et al.*, 1993) and the bobcat (Zezulak & Schwab, 1979). Packs of gray wolves, *Canis lupus*, in exploited populations have larger and less stable territories than in saturated populations where prey abundance is more or less constant (Ballard *et al.*, 1987). The pattern of recolonization of exploited areas depends on prey density and age/sex composition of the surviving wolves (single males and pups dispersed, while other sex–age combinations remained in the areas where they had been persecuted and rapidly accepted immigrants). Adjacent animals may or may not expand their ranges to include vacated niches depending on whether they are 'expansionists' (e.g. coyotes, *Canis latrans*) or 'contractionists' (e.g. Eurasian badgers) (Kruuk & Macdonald, 1985). Even among expansionists, it is conceivable that the territorial system will eventually break down when a minimum population size is reached (R. List & D. W. Macdonald, personal communication).

Breeding system

Reduction in group size by human persecution has been associated with changes in the breeding system in some populations of monk seal, *Monachus schauinslandi*, and Weddell seal, *Leptonychotes weddellii*, from polygyny to monogamy (Jouventin & Cornet, 1980). Culling may also increase the probability of polygamy, for example by changing the operational sex ratio (e.g. hen harriers, *Circus cyaneus*: Balfour & Cadbury, 1979; Picozzi, 1984; coyotes: Kleiman, 1977, cited in Lott, 1991), by altering the age-structure (Komdeur & Deerenberg, 1997) or by increasing the *per capita* food availability (e.g. red foxes: Zabel & Taggeart, 1989). Komdeur and Deerenberg (1997) report how higher food availability may be associated with sequential polygyny or polyandry in bird species where uniparental care has little effect on reproductive success (e.g. Northern jacanas, *Jacana spinosa*, and tree swallows, *Tachycineta bicolor*), and with the appearance of helpers at the nest in species that otherwise show biparental care (e.g. Northwestern crow, *Corvus kaurinus*, Seychelles magpie robin, *Copsichus sechellarum*).

The traditional breeding territories of male pronghorn antelopes, *Antilocapra americana*, break down under heavy hunting pressure, which forces the females and territorial males to move to more protected areas and to form large unstable 'disturbance groups'. If hunting disturbance is short-lived no courtship activity takes place in these disturbance groups and pronghorns re-establish their previous social system (Deblinger & Alldredge, 1989). If disturbance persists throughout the rut, however, breeding takes place in the disturbed groups, which facilitates mating by subordinate males that previously were excluded by dominant territorial males (Copeland, 1980). Reductions of the average age and size of breeding males similarly affected the breeding system of harvested populations of elephant seals, *Mirounga leonina* (Bonner, 1989). The reproductive consequences of simulated heavy exploitation of female and male white-tailed deer, *Odocoileus virginianus*, have been investigated by Ozoga & Verme (1984, 1985). Lack of maternal domination and access to uncontested fawning territory were important factors for improving the reproductive success of 2.5-year-old does following complete removal of all other members of their matriarchal group. As in the pronghorns, if too few mature males remain to suppress younger males, the breeding system changes from a strict dominance competition to a 'scramble' competition among the young males to mate oestrous does (Miller & Ozoga, 1997). Especially in herds in southern latitudes, young male age structures and skewed sex ratios can delay and prolong the breeding season. Extended rutting activity is likely to reduce the physical condition and survival of males, females and fawns. The same herd structure in northern areas may not be affected because of overriding environmental constraints on reproductive timing. These perturbation effects are not confined to prey species. One might speculate, for example, that sport hunting of prime male lions would lead to adolescent males precociously, but unstably, seeking to hold female prides.

Sociality

Coyote groups in high-density populations are more likely to retain yearlings as helpers than groups in populations brought to low density by human persecution; in the latter, the basic social unit is a mated pair and solitary nomads are common (Andelt, 1985). Hunting disturbance can disrupt pair-bonds and family structures in a variety of waterbirds, and this may affect their survival and reproductive potential (Madsen & Fox, 1995) and eventually alter their social structure (Lott, 1991). Lott (1991) cautioned that some forms of human persecution are experienced as increased predator pressure. This could encourage the formation of larger groups from smaller ones that have been reduced by hunting, trapping or fishing and could decrease group stability. Such coalescence of smaller social units into larger and more visible aggregations has been reported for red deer populations subjected to hunting-with-hounds (Langbein & Putman, 1996). Elevated levels of inter-specific aggression by adult female Asian elephants where poaching and human disturbance is common has been interpreted as an extension of their natural anti-predator behaviour (Sukumar, 1989).

Heavy exploitation might alter the internal characteristics of a population by removing territory-holding individuals, creating behavioural instability and keeping the social organization in a perpetual state of flux. This has been suggested to occur in the populations of many mustelids (Powell, 1979; Hornocker *et al.*, 1983; Arthur *et al.*, 1989; Cheeseman *et al.*, 1993), and some felids (Hornocker & Bailey, 1986; but see Litvaitis *et al.*, 1987). Social disruption following hunting has been evoked to explain increased agonistic interactions among male mountain lions, *Felis concolor* (Hornocker & Bailey, 1986). In general, the rate of social interactions declines rapidly once stable social and spatial relationships have been established (Stamps, 1994). Interactions between familiar animals are apt to be different quantitatively and qualitatively from those between unfamiliar animals.

The killing of old resident males encourages immigration of young males which, in some social systems, kill infants to bring females into oestrus. The effect of infanticide on the rate of recruitment is an important issue for the management of, for example, lion, *Panthera leo*, and brown bear, *Ursus arctos*, populations (Wielgus & Bunnell, 1994; Arcese *et al.*, 1997). Models that ignore the impact of increased infanticide in such hunted populations are likely to underestimate the minimal viable population size (MVP) (Arcese *et al.*, 1997) and overestimate the number of animals that can be hunted annually.

IMPLICATIONS FOR WILDLIFE MANAGEMENT AND CONSERVATION

Pest control and management of harvested species

Some animals adjust better to anthropogenic disturbances than others. Clearly there are large populations of opportunistic mammals thriving alongside people, and these stem from diverse taxa: urban foxes, even hyaenas, dockland rats and temple-dwelling Hanuman langurs. Aside from such commensalism, populations subjected to continual and long-term persecution may adapt to overcome human pressures through learning or natural selection, or both (Hornocker *et al.*, 1983). As is obvious to the naturalist who sees a fox in full flight from a person at 500 m on farmland, whereas the outrageously tame foxes of the Alaska islands sniff his boots, populations subjected to shooting, trapping or poisoning may develop appropriate caution, for example trap-shyness (coyotes: Andelt *et al.*, 1985; bobcats: Knick, 1990; lynx: Slough & Mowat, 1996) or bait-shyness (rodents: Gao & Short, 1993) respectively. The evolution of neophobic behaviour is a significant impairment to the efficacy of programmes to eradicate wild rats, *Rattus norvegicus* (Brunton *et al.*, 1993). Coyotes have been reported to be more active during the daytime in areas where hunting pressure is less (Andelt, 1985), and this phenomenon of activity shift can certainly result from natural predation (Fenn & Macdonald, 1995). Hunting disturbance can alter diurnal rhythms, increase escape flight distances, and modify the distribution and abundance of waterbirds in space and time (Madsen & Fox, 1995).

All these adaptations, combined with density-dependent compensatory mechanisms, may affect the resilience of pest and quarry species to control operations. The control of self-regulating populations can be unproductive because culling removes surplus animals that would die or disappear regardless. The removal of dominant or territory-holding individuals may even be counterproductive and result in a temporary increase in population density owing to increased survival of, and immigration by, transients and/ or subadults (Young & Ruff, 1982; Smal, 1991). Smuts (1978) also concluded that gradual removal of lions in small parts of their range is of questionable value in terms of increasing the producitvity of prey animals because it increases the proportion of subadults or cubs which, in relation to body mass, eat more than older individuals.

It may become a moot point as to whether to interpret a given change as behavioural plasticity or social distortion. In disturbed populations full compensatory growth predicted from the removal of density-dependent inhibitions (food, space, pathogens) may not straightforwardly be realized if population reduction has also triggered destabilizing perturbations. Intensive culling of lions over a large area produces many vacant areas that attract foreign lions, which then interact aggressively, possibly suppressing both birth rate and cub survival (Smuts, 1978). It is also possible, therefore, that perturbation enhances the chances of success of attempts to eradicate undesirable animals, as is illustrated by the case of the coypu (*Myocastor coypus*), which was introduced to Great Britain, where it caused environmental damage. Gosling & Baker (1989a) reported that because of differences in ranging behaviour male coypus were captured at a higher rate than females. Males thus became the limiting sex as the population was reduced to very low densities. Female reproductive success declined and as a result the coypu population in Britain declined faster towards extinction than expected.

It is easy to envisage costs of perturbation expressed in stress, lost forag-ing time, or increased expenditure of energy on aggression or flight. For animals that benefit from living in groups, the removal of one individual is unlikely to translate into a simple *pro rata* loss to the survivors: the loss of a helper will have different consequences to that of a breeder, a subordinate or a dominant, and the social disruption and selective costs may also affect the survival chances of other group members very differently. The salient point is simple: management actions that treat all individuals as equal may go awry when applied to species with complicated social structure. The consequences of killing a territory-holder will differ from those of killing a transient, as will those of killing a dominant rather than a subordinate. These consequences may not matter in some contexts, but it would be poor management, and perilous, to assume that they never do. In African eleph-ants, calves raised in small families that have been fragmented by poaching have a lower chance of survival than calves raised in large, intact families (Poole & Thomsen, 1989). Indeed, the social disruption of many elephant populations is so severe that it has been estimated that if all poaching was stopped now, it would take 20–30 years to restore their social structure (Poole & Thomsen, 1989). Behavioural diversity is a major determinant of population viability and should therefore also be included in conservation practice and politics (Buchholz & Clemmons, 1997).

The more a population is reduced in size the more vulnerable it is to extinction as a result of demographic, environmental and genetic stochas-ticity (Gilpin & Soulé, 1986). Inbreeding and the loss of genetic variation have been associated with lower fitness and increased risk of extinction (Frankham, 1995b). Social behaviour and, more specifically, the response of animals to human activities, need to be considered in models of viability and genetic variability of persecuted populations. Historically, these subtle-ties have not been included in 'off-the-shelf' packages for Population Viabil-ity Analysis, several of which may give different results from the same data (Macdonald *et al.*, 1998). These simplifications have prompted criticism of PVA models unless they are written specifically to accommodate the intri-cacies of species-specific behaviour (Groom & Pascual, 1998). The import-ance of decreased heterozygosity increases nonlinearly as the effective population size (N_e) decreases (Crow & Kimura, 1970; Lacy, 1987). N_e is the size of a 'genetically idealized' population that would give rise to the same variance of gene frequency, or rate of inbreeding, as the real population (N). Usually N_e is much less than N because it depends on the number of breed-ing males and females in the population and their social behaviour. We suggest that control operations often cause larger reductions in the ratio of N_e/N than is anticipated. This is first of all because N_e is extremely sensitive to fluctuations in population size (Frankham, 1995c). Even if the population size averaged over the years is equal, populations that have been over-exploited and then been allowed to recover will have a lower N_e than populations that have been sustainably harvested. The younger age structure and skewed sex ratios that characterize heavily exploited populations also negatively affect N_e. The breeding system will often determine the genetic diversity transmitted to the next generation, and as discussed above, this may also be altered directly or indirectly by control operations. Sometimes hunting directly influences N_e/N, as is illustrated by the selective removal of the most heterozygous bighorn rams, *Ovis canadensis*, which tend to have the largest horns (FitzSimmons *et al.*, 1995). Estimates of population fitness, minimal viable population size and sustainable harvest may be inaccurate if these factors are not taken into account.

For those populations of otherwise desirable animals for which control is unavoidable, it is important that management strategies are chosen with extreme care. However, the consequences of social perturbation are still little known. The removal of whole herds at once may be preferable from an animal welfare viewpoint. However, all traditional knowledge specific to that group will be lost from the population. Translocations are apparently also not without problems. For example, young translocated elephants may turn into rogues when they are no longer disciplined by their mothers. In the Kruger National Park trials are under way to apply fertility control to elephants. As breeding of any one individual would only be prevented on a temporary basis, this strategy might best safeguard genetic variability within the population. However, the behavioural and social effects of such a fertility control campaign are still unknown. Since reproductive success is the cornerstone of natural selection, one might expect an individual's status to be affected by persistent failure to breed.

Control of wildlife diseases

Many populations of wild animals are controlled because of the diseases they harbour. For directly transmitted diseases it is often assumed that reducing host density will reduce or eliminate the disease (Bailey, 1975; Anderson & May, 1979). However, the relationship between host density and disease dynamics can be very complex. In addition, disease dynamics may be affected in numerous and unexpected ways by the control operations (Macdonald, 1995; Tuyttens & Macdonald, 1998a). Although the epidemiological consequences of distorted age/sex structures and altered socio-spatial behaviour have been ignored in most disease–host models, they could considerably influence the success achieved by some control methods as compared to others (Swinton et al., 1997; Hitchcock et al., 1999; Tuyttens & Macdonald, 1998a).

Several authors have argued that culling for rabies control could be less effective or even counter-productive if it increased contact among surviving foxes; the often cited, but rarely demonstrated creation of a vacuum effect, which draws in animals from other areas, is only one of several mechanisms whereby this could occur (Bacon & Macdonald, 1980; Mollison, 1985; Macdonald, 1995). Of course, just as perturbation effects have the potential to be counter-productive, they may also be productive: socially mediated turmoil in a rabid fox population might increase contact rate, leading to enhanced prevalence, and a more complete die-off, which more effectively eradicates the epizootic. The point, then, is that perturbation effects are not necessarily disadvantageous, but they do merit consideration.

We are currently investigating whether such disturbances may contribute to the failure of government-organized Badger Removal Operations to control bovine tuberculosis (Tb) in badgers, and hence cattle, in SW England (Tuyttens, 1999). The low rate of dispersal (Woodroffe et al., 1993) and the stable territorial organization characteristic of undisturbed high-density populations (e.g. Kruuk, 1989) may not be conducive to the spread of Tb from one social group to another. We seek to test the hypothesis that perturbation caused by the removal of social groups may alter demographic and socio-spatial aspects in such a way that the spread of Tb may be facilitated, or at least altered. For directly transmitted infections, this could happen if social perturbation increased either the contact rate between animals or the likelihood of disease transmission when such a contact occurs, or both. The contact rate between badgers of different social groups could be increased, for example, if the territorial system breaks down, if home ranges are enlarged and overlap more, if badgers make more and further forays outside the group's range, or if badgers interact more in order to re-establish spatial and social relationships in the aftermath of a removal operation. The chance of disease transmission on encounter could be increased if removal operations lead to more aggression (e.g. owing to the disruption of owner–intruder asymmetries), to a suppression of the immune system owing to stress related directly to the removal operation, or to an increased investment in reproduction, to a younger population consisting of more immunologically immature cubs, to altered behaviour at latrines, and so on. A simple deterministic model suggests that if such perturbations have a strong effect on the probability of disease transmission they could neutralize the potential benefits of attempted lethal control and under certain scenarios render control counter-effective (Swinton *et al.*, 1997). These results have been supported by at least two other models incorporating spatial and stochastic aspects (White & Harris, 1995; Hitchcock *et al.*, 1999). Hence, it seems warranted to investigate the feasibility of alternative strategies such as vaccination or sterilization and whether they induce less perturbation (Tuyttens & Macdonald, 1998a). Non-lethal methods that are equally efficient at solving wildlife problems but cause less perturbation and suffering are likely to have greater public acceptability (Tuyttens & Macdonald, 1998b).

CONCLUDING REMARKS

First principles in ecology and behaviour make it plausible that population control may unintentionally, and perhaps unexpectedly and undesirably, alter many demographic and socio-spatial parameters of the targeted population. These disturbances could have far-reaching implications for the management and conservation of wildlife. That these possibilities are real is beyond question. Our purpose in this review has been to explore whether data exist to evaluate this possibility. The elephant and badger examples show clearly that census data on population size may provide only limited information regarding the long-term viability of exploited populations, or the degree to which wildlife diseases are controlled by reducing the density of the host population, respectively. Too often control of wildlife populations is undertaken with little knowledge or consideration of the potential consequences these actions may have on the social system of survivors. However, the success of many control operations, as well as the viability of the populations concerned, may crucially depend on these effects. Attempts should therefore be made to evaluate and predict these effects before control is commenced. A good understanding of the proximate and ultimate mechanism underlying the socio-spatial behaviour of the targeted population will be helpful, if not necessary, in this respect. Our point is a simple one, and can be expressed simply as good management: the populational, social and behavioural characteristics of many mammal species are intricate and complicated. There may well be scales of operation, or circumstances, under which these complexities have minimal effect on management. On the other hand, there are circumstances under which management of a complicated system will be most efficient when attention is paid to perturbation effects and their potentially counter-productive consequences. This caution can be phrased straightforwardly in terms of management efficiency, but that does not detract from the fact that it has an ethical dimension: where better understanding of social systems can lead to a smaller cull, a lesser intrusion or a shift to non-lethal management systems, then, all else being equal, there will be ethical as well as efficiency gains.

ACKNOWLEDGEMENTS

We thank Luigi Boitani, Gus Mills, Mark Pagel, Marsha Sovada, Jeremy Wilson and Jerry Wolff for comments upon a draft of this paper. Support for our work on perturbation in badgers, in collaboration with CSL, was kindly provided by the Ministry of Agriculture, Fisheries and Food and other financial support provided by the People's Trust for Endangered Species.

19

Animal welfare and wildlife conservation

ELIZABETH L. BRADSHAW & PATRICK BATESON

INTRODUCTION

The importance of taking animal welfare into consideration when dealing with the problems of conservation, particularly those involving captive wild animals, is increasingly recognized (see Gipps, 1991). Nonetheless, concerns about animal welfare are often seen as being opposed to the drive for conservation; this is because concerns about welfare are focused on the individual, whereas those to do with conservation are focused on populations. In this chapter we examine why this opposition may have arisen, lumping under 'conservation' concern not only for local populations but also for whole species and other 'ecological collectives' (Rawles, 1996). We suggest that it is helpful to separate animal rights issues from those to do with animal well-being and briefly review how well-being may be assessed. Next we consider where improvements in animal well-being overlap with conservation and where the tensions arise. We illustrate the complexities that arise from these tensions by referring to our own work, namely, the welfare of red deer (*Cervus elaphus*) hunted with hounds. Finally, we present a method of relating the qualities of welfare practice with those of conservation practice.

Acrimonious disputes have arisen between organizations which, in their different ways, have sought to protect animals and the natural environment from human activities (Rawles, 1996; O'Regan, 1997; Rodriguez, 1997). In part, the disputes have arisen from confusion about what comes under the umbrella of 'animal welfare'. Campaigns for improved 'welfare' often run together ethical concerns about rights to life with those about the state of living animals. Since these are different issues, a major gulf has arisen between the 'rightists' and the 'welfarists' (Garner, 1996). The rightists take an absolute view of the moral status of animals. They refuse to sanction the exploitation of animals for the benefit of humans or other animals (e.g. Regan, 1983). Animals should just be 'left alone' (see Jasper, 1996, p.129). The welfarists argue, in contrast to the rightists, that utilization of animals is not in itself problematic, providing the benefits to humans are great and the animals are cared for humanely (e.g. Singer, 1990). Many who are involved in improving the standards for keeping and treating animals derive their motivation from a sense of responsibility for animals rather than a preoccupation with animal rights (e.g. Dawkins, 1990).

Ethical considerations largely underpin concerns about both animal well-being and conservation. In both cases, however, the sciences of animal welfare and conservation biology are concerned with the orderly application of method to questions about the states of individuals or of populations. The development of animal welfare as a science has occurred later than that of conservation biology (Soulé & Wilcox, 1980). Nevertheless, considerable progress has been made in clarifying the methods of welfare assessment, a subject that we review briefly below. Improving the well-being of animals is now recognized as an achievable objective, which may be set alongside the objectives of conservation. Without hiding residual tension, we draw attention to the ways in which the goals of animal welfare and conservation are complementary.

ASSESSMENT OF WELL-BEING

Suffering is at the forefront of public concern about animal welfare and an assumption that it can and will be assessed is embodied, for instance, in legislation in the United Kingdom about the use of animals in research, care of domestic pets and farm animals, and much else. The dictionary definition of suffering is perfectly plain, namely: 'To have (something painful, distressing, or injurious) inflicted or imposed on one' (*Shorter Oxford English Dictionary*). Pain and distress are subjective sensations. How may they be measured?

Marian Dawkins (1980, 1983) pioneered the approach of measuring animals' choices for situations that might be subjectively unpleasant to them. More recently she has borrowed from economics the idea of inelastic and elastic demand (Dawkins, 1990). She reasoned that people continue to buy items such as bread in an inelastic way whether the price is high or low. On the other hand, items regarded as luxuries are no longer bought if the price rises too much; the demand is elastic. This idea is helpful, for example, in assessing whether an animal will escape from what may be a painful experience irrespective of how much work it has to do in order to effect escape. If sufficiently high priority is given to getting away from the noxious event, the animal will 'pay' as much as is needed to get away. On the other hand, if escaping the experience is not as important to the animal as might have been supposed intuitively, the requirement will be more elastic.

A somewhat different approach to the assessment of welfare was pioneered by Broom (1986). An individual's welfare may vary on a continuum from poor through to good, so that reference is made to 'improving welfare' or 'ensuring that welfare is good' rather than simply to preserving or ensuring welfare. The quality is determined by the animal's 'state as regards its attempts to cope with its environment'. Using this definition allows welfare to be measured in a orderly way that is independent of moral considerations or projections into the animals of human feelings, emotions and intentions. However, such assessments of animal welfare would apply just as much to snails (Cooper & Knowler, 1991) as they would to birds or mammals. Given the public concern about suffering, Duncan (1993) argued that feelings ought to be brought into the definition of welfare. Broom & Johnson (1993) agreed that sentience enters into welfare assessments and, in a subsequent review, Broom (1998) has considered the issue of feelings from the standpoint of their utility to the animal.

It is possible to ask about the extent to which physiological states associated with the subjective sense of suffering experienced by humans are found in animals that might be suffering. Such assessments are based on the plausible assumption that many physiological mechanisms concerned with maintenance of internal state are less likely to change in the course of evolution than those involved in the control of behaviour. This is because the internal environment is much the same in animals that are adapted to a wide variety of different habitats. The big evolutionary changes have largely come in the behavioural mechanisms that relate to the external environment. A powerful line of evidence for suffering comes, therefore, from using correlations between the application of stressors and the physiological states of humans as a guide to what happens to animals in comparable circumstances.

The general point is that the basis for judgements about suffering in animals can be made relatively transparent. Humans are prepared to generalize from their own feelings, emotions and intentions to other human beings. If it is rational to do that, it is no less rational to extend the generalization to other species (Bateson, 1991). Adopting a human-centred approach in order to investigate an animal's capacity for experiencing suffering leads to asking whether the animal has anatomical, physiological and biochemical mechanisms similar to those that in a human are known to be correlated with such experiences. The approach also raises the issue of whether the animal performs in similar ways to humans who are believed to be suffering. It is possible to provide criteria that are based on modern methods of measuring behaviour and analysing the functional character of the nervous system. If the animal stops activities that it habitually performs in conditions that might be supposed to produce pain, if it learns how to avoid such conditions, and if it has parts of its nervous system dedicated to avoidance of damage or disturbance of its internal state, then there are grounds for worrying that it might feel something. These concerns are made much more acute if it has a large brain relative to its body and shows some of the cognitive capacity seen in humans.

The subjective experiences of an animal, if it has any, may be totally different from humans, reflecting its different way of life and the different ways in which its body works. The inherent plausibility of this thought raises an additional approach to suffering that considers the animal's functional requirements in relation to its survival and reproductive success (see Bateson, 1991; Barnard & Hurst, 1996; Timberlake, 1997). When an animal does not behave as humans would in the same circumstances, scientists should be sensitive to its requirements, its evolutionary history and the details of its social life. Therefore, assessments of suffering will also depend on good observational data about the natural behaviour of the species in question, its normal requirements, its vulnerability to damage and the ecological conditions in which it lives.

Is it possible to find a magic bullet for the assessment of welfare? It seems not. What emerges from many debates about what should and should not be measured in welfare studies is that a variety of approaches are more likely to benefit understanding than a single approach (Mason & Mendl, 1993; Appleby & Hughes, 1997). A multi-pronged approach is also more likely to reflect the various classes of problem that lie behind a concern for the quality of an animal's life (Fraser et al., 1997). All of the following approaches contribute to an assessment of adverse welfare: (a) measurements of physical damage to the animal; (b) measurement of the extent to which it has been required to operate chronically homeostatic mechanisms that would normally operate acutely; (c) measurements of physiological states that would be found in suffering humans; (d) considerations of the animal's behavioural ecology and normal social structure. When all these things are done, judgements about the quality of an animal's welfare are much more likely to win widespread agreement than if only one approach had been used.

CARE OF CAPTIVE ANIMALS

Zoos have become much more than places of entertainment or mere collections. Many play important roles in research and conservation (Seal, 1991). Conway (1980) suggested that captive-bred populations were variously to act as: (a) substitutes for wild populations in basic biological research; (b) substitutes for wild populations in the development of care and management techniques; (c) demographic and genetic reservoirs for wild populations; (d) 'ark' populations for species under immediate threat of extinction in the wild. Despite the responses of zoos to changes in public attitudes in the developed world, many still believe that zoos no longer have a place in modern society (see Mallinson, 1995). It is sometimes argued that conserving genetic stocks in captivity does not solve the problems that led to species becoming endangered in the wild through habitat destruction and unregulated exploitation and, therefore, distracts attention from what are regarded as the real problems. The critics also argue that zoos raise significant ethical issues that outweigh any conservation benefits. In part, these are rightist concerns relating to the morality of keeping wild animals in captivity. They may also be motivated, however, by the concern for the quality of life of captive animals – in other words, for their welfare.

Considerable efforts have been made to heed the needs of captive animals, which obviously vary from species to species (Kleiman, 1980). Initially, the focus was on improvements in nutrition and disease control, but over the years attention has been paid to all aspects of biology, especially behaviour. Knowledge of social behaviour may be used in cage groupings (Segal, 1989). Understanding of exploration and play leads to improvements of cage design and provisioning of toys (Thompson, 1996). Knowledge of foraging strategies leads to re-scheduling of feeding times (Bloomsmith & Lambeth, 1995) and the development of simple ways by which caged animals may work for their food (Reinhardt & Roberts, 1997). Generally, attention is being focused on environmental enrichment that enhances the functioning and survival of the species in question (Newberry, 1995). All of these advances in caring for captive animals may improve the quality of their lives. This in turn, combined with better under-standing of reproductive behaviour, may lead to improvements in the success of breeding programmes (Estep & Dewsbury, 1996). Here, then, good conservation of a species results from good welfare.

Tensions between welfarists and conservationists may be raised, first, by the capturing of wild, free-living animals to start or supplement captive populations and, second, the reintroduction into natural habitats of animals that have been bred or reared in captivity. The capture, restraint and transport of wild animals exposes them to a number of physical and psychological stressors such as extreme exertion, handling by humans, immobilizing drugs, novel noises or smells and temperature extremes. Such stressors bring with them the attendant risks of injury, shock, 'capture myopathy' and immuno-suppression (Jarret et al., 1964; Griffin et al., 1992; Beringer et al., 1996). If the species is new to captivity, basic aspects of its nutritional, housing and social requirements may not be fully understood, leading to high mortality rates (e.g. McLeod et al., 1997). Even when animals are fed well and kept free of disease, inability on their part to adapt to captivity may to lead to persistent poor welfare. Conversely, adaptation to a captive environment may reduce the animal's capacity to adjust to the wild, raising welfare concerns for those animals that are destined to be returned to their natural environment.

The problems of reintroducing animals to the wild are well-recognized. Many factors, such as the lack of specific foraging skills, reduced immunity to disease, and the animal's unfamiliarity with their natural predators and the terrain into which they are released, may all lead to high mortality (Scheepers & Venze, 1995; Woodroffe & Ginsberg, 1997). Unwitting artificial selection for behavioural characteristics suited to captive conditions such as reduced aggressiveness or timidity might be at the expense of being able to cope when the animals are reintroduced to the wild (Campbell, 1980). Even if the conservation goal is achieved and some reintroduced animals survive, the cost to their welfare may have been very great.

When conservation and welfare goals concerning captive wildlife come into genuine conflict, some trade-off has to be achieved. The precise character of the trade-off may depend on the rarity of the species in question and the severity of the welfare problems for the individual. Sometimes the conflicting objectives may be reconciled. In the case of captive-bred animals destined for release programmes, good welfare after release may well mean minimum intervention and minimum contact with humans. The solution is to adapt captive conditions to the needs of the animal, rather than the animal to the needs of captivity. For example, the maintenance of animals in natural social groups that allow them to express a wide repertoire of behaviour may not only make them a more interesting exhibit, but in addition improves the welfare of the animals without detracting from the conservation value to a zoo (Price *et al.*, 1994).

In summary, seeking to improve the welfare of wild-caught animals does not necessarily conflict with conservation aims. The measures taken to enhance welfare may, by improving survival and reproductive prospects, enhance the conservation value of the whole exercise.

PROTECTING WILD SPECIES

Questions about the nature and extent of human responsibility for the well-being of free-living wild animals are often, but not invariably, raised when human activities encroach on wildlife in some way. Conservation and animal welfare issues will overlap and may conflict when: (a) species are managed because they impinge on other species or on habitats perceived to be of greater value for conservation; (b) economic or recreational benefits are derived from wildlife; (c) individual animals are the victims of accidents, especially those resulting from human activity; (d) welfare is poor but not as a result of human activity.

Conservationists may seek to manage animals that affect adversely an endangered species or are destructive to valued habitats. They may do so directly by culling or the control of reproduction. They may also do so indirectly by the use of fencing or the use of deterrents. All methods of management have the potential to affect animal welfare. Awareness of the need to take into account animal welfare when assessing such operations has been increasing. For instance, eradication from Britain of muskrats (*Ondatra zibethicus*) in the 1930s was achieved with leg-hold traps and animals sometimes escaped with severed legs. To minimize suffering, cage traps were used in the subsequent eradication of coypu (*Myocastor coypus*) (see Gosling & Baker, 1989b). Such is the change in attitude that Caughley & Sinclair (1994) have argued that 'the wildlife manager's paramount responsibility is to ethical conduct rather than operational efficiency'. Nonetheless, reconciling welfare and conservation interests may be least tractable when conservation policy requires the culling of 'pests'. The goal of some conservationists may be to get rid of the unwanted species as quickly and as efficiently as possible (see Rawles, 1996). Even where the importance of

animal welfare considerations are noted, incorporating welfare considerations into management regimes may be hampered by shortage of both
human and financial resources (Veitch, 1985).

nIt is not always easy to think clearly about the issues. Part of the problem
is that we all tend to place animals into categories (pest or resource, sentient
or non-sentient, endangered or common, domestic or wild) and to treat them
accordingly (Caughley & Sinclair, 1994; Taylor & Dunstone, 1996). Proposals to eliminate feral cats and wild rats from an isolated, inhabited island to
protect endemic wildlife did raise local anxieties about animal welfare. However, these concerns related only to the pet cats, whose freedom to range
outdoors would be curtailed for some months if the elimination of feral
animals involved the use of poison (M. Brooke, personal communication).
Categorizing species as pests can, of course, have ramifications for conservation. Protecting black rats (*Rattus rattus*) in Britain, where they are rare, has
been hampered by their status as pests (Morris, 1993) and the possible
conflict between the conservation interests of this naturalized species and
those of native seabirds (see McDonald *et al.*, 1997). However, having the
status of a pest has greater implications for animal welfare than conservation
because few query the methods used to eliminate rats.

Non-lethal methods of control have potential for conservation use in
situations where culling has failed, is impractical or is publicly unacceptable (e.g. Cowan, 1996; Kirkpatrick *et al.*, 1996; Muller *et al.*, 1997). Manipulation of food aversion with behavioural techniques may have practical
importance in controlling browsing damage by ungulates (Arnould & Signoret, 1993) or predation of endangered species (Cowan *et al.*, Chapter 16
in this book), as may the use of fencing or other physical barriers (Yerli *et
al.*, 1997). Population control may be achieved by reducing fertility (Chambers *et al.*, 1997; Moore *et al.*, 1997). Immuno-contraception, whereby an
animal's own immune response is used to disrupt reproductive function, is
a recent development that may have potential in controlling populations of
some ungulate species (Kirkpatrick *et al.*, 1997), although long-term effects
on welfare must be monitored (Muller *et al.*, 1997). Fox breeding can be
stopped by dosing buried meat with cabergoline, a dopamine agonist which
inhibits the release of prolactin and causes the vixen to abort her foetuses
(Marks *et al.*, 1996); having aborted, she does not attempt to breed again for
a year. While such techniques may be preferred by animal welfare organizations, some methods may have a serious impact on both conservation and
animal welfare. For example, wildlife contraceptives have sometimes had
severe physiological, behavioural and pathological effects on target popula-

tions, and immuno-contraceptives may reduce resistance to disease (Nettles, 1997). In this case the attempt to reconcile conflicting interests would ultimately damage both welfare and conservation objectives.

Conflicting interests between conservation and welfare may also arise when humans attempt to derive economical or recreational benefits from wildlife. The sustainable harvesting of wild species for recreational or commercial reasons may be accepted on the grounds that this is the only realistic way in which sporting or industrial interests can be reconciled with conservation. However, it is still uncommon for the welfare of wild animals to be included in a formal consideration of the sustainability of harvesting practices (Papastavrou, 1996; Taylor & Dunstone, 1996). When conservationists are pragmatic about using sustainable harvesting as a means to an end, they can, thereby, come into direct conflict with welfare organizations. A protocol for assessing the impact on welfare of different forms of sustainable use has been proposed (Lindley, 1994). Sometimes, the conflict may be reduced because it is possible to minimize the impact of harvesting on welfare, as appears to be the case when night-shooting impala (*Aepyceros melampus*) for commercial ends (Lewis *et al.*, 1997). Unfortunately, this is by no means always feasible. For example, it seems impossible at present to kill whales humanely (Kestin, 1995; Blackmore *et al.*, 1997).

Other ways have been suggested of reconciling the concerns of animal welfare with those of conservation, recreation and business. Ecotourism, in particular, shows great potential. It has played an important role in mountain gorilla conservation (McNeilage, 1996), for example, and whale-watching operations can offer a benign economic alternative to whaling (Corkeron, 1995; Papastavrou, 1996). Nonetheless, careful monitoring of such schemes is required to prevent disturbance to wildlife, habitat destruction or disease transmission adversely affecting animal welfare and conservation goals (Corkeron, 1995; Dunstone & O'Sullivan, 1996; McNeilage, 1996).

Sometimes animals are inadvertently harmed by human activity. Contamination by fuel oil, injuries in road traffic accidents or poisoning by anglers' lead weights (Sainsbury *et al.*, 1995) have affected large numbers of individuals. Taking these animals into captivity for treatment is beset with problems. Careful thought has to be given to the length of time that the animal must spend in captivity, whether it can be returned to the wild, and its survival prospects if it is returned (Putman, 1990). Without such considerations, well intentioned actions can lead overall to poor welfare. When the actions have no direct or indirect conservation benefits, attempts to prolong the lives of injured or incapacitated animals may be especially questionable.

To take a specific and famous case, common sea birds contaminated by oil spills are the subject of great public concern. Much time, effort and expense is spent on cleaning and releasing affected individuals. Yet, the subsequent survival of these birds is poor (Sharp, 1996). Furthermore, focusing on 'feel good' welfare issues may divert attention from the root cause of the problem. Measures to prevent oil spills happening in the first place would have far greater implications for good welfare and good conservation than merely cleaning up oiled seabirds after the event. Nonetheless, some benefits may accrue to conservation if veterinary knowledge gained in rehabilitating common species is useful when planning the reintroduction or translocation of endangered species.

Wild animals may experience welfare problems that do not obviously arise from human activities, raising complex ethical issues that may encompass conservation. For example, should humans seek to eliminate disease from wild populations, given that good health is one component of good welfare? Some concerned with welfare may find it tempting to intervene if, by doing so, the welfare of individuals is apparently increased over the short term. Investigating causes of sickness within wild populations may have important implications for conservation (Kirkwood, 1993). However, it may not be advisable from the point of view of the long-term survival of a species to interfere actively with natural mortality patterns for purposes of short-term improvements to individual welfare or, indeed, species conservation – unless the species is already very rare (Wandeler, 1991; Kirkwood, 1993).

WELFARE AND CONSERVATION OF RED DEER

Culling of red deer in the United Kingdom is carried out using high-powered rifles as part (sometimes the sole part) of local management strategies to protect forestry, agricultural land or the natural environment. When Rawles (1996) used the example of culling red deer in the Scottish Highlands to illustrate the conflict between those concerned with individual animals and those concerned with managing the population, she described a division between the animal rightists and the conservationists. The rightists feel that deer should not be shot at all, even to prevent habitat damage by overgrazing. In contrast, the animal welfarists are much more ready to accept the need, in some circumstances, to reduce deer numbers and their primary concern centres on the methods used to achieve the cull. Providing deer are shot in such a way that death (or at least, unconsciousness) comes quickly, culling is an issue that may be addressed purely in terms of population management.

Red deer are tolerated on farmers' land for a variety of reasons. What happens to the conservation–welfare debate when the method of killing the animal, which may not be the most humane method, is given as the reason for conserving the animal in the first place? In North Devon and West Somerset, some 130 red deer are killed each year by hunting with hounds (Langbein & Putman, 1996). The hounds, guided by humans on horseback, are used to follow the scent trail of a deer, pursuing the animal until it comes to a halt and can be shot with a smooth-bore gun (rather than a rifle). The number killed by hunting with hounds is small in comparison with the annual number that are killed by stalkers with rifles, estimated to be in excess of 1000 for Exmoor and the Quantocks (Savage, 1993).

Even so, a justification for hunting with hounds is that this institution conserves deer (Lloyd, 1990). The social pressure generated by the recreational interest in hunting is, it is argued, the driving force in maintaining the number of red deer at their present levels. Although hunting with hounds has a welfare cost for individual red deer (Bateson & Bradshaw, 1997), the conservation benefits for the local population of deer may be great. If the argument is correct, any decision to ban hunting by legislation – or by the actions of landowners such as the National Trust – must involve a trade-off between the welfare of individual animals and the conservation of the local population. Before considering such a trade-off, some other questions must be resolved first.

What is the welfare cost of hunting with hounds? In our own research on red deer hunted with hounds, we asked four questions (Bateson, 1997; Bateson & Bradshaw, 1997; Bradshaw & Bateson, 2000).

- To what extent are the deer damaged by hunting?
- To what extent does the hunted deer differ from one that is severely injured?
- To what extent has the deer's evolutionary and individual history prepared it for being chased by hounds?
- To what extent are physiological states associated with the subjective sense of suffering, as experienced by humans, found in hunted deer?

Our study provided strong evidence for both muscle and blood damage that would compromise the deer's welfare at the end of a hunt if it escaped. The extensively hunted deer did not differ from severely injured deer on measures that allowed them to be compared; this provided an independent method of validation because both sides in the hunting debate agreed that the injured animals should be put down. On the basis of published evidence and our own observations of the behaviour and physiology of hunted red deer, we concluded that red deer are poorly adapted for prolonged chases by hounds. Finally, our own findings indicated that physiological states associated with suffering in humans are found in hunted deer. We concluded that the welfare costs of hunting are high.

The alternative to hunting with hounds, namely stalking with high-powered rifles, also has a welfare cost because not all deer are killed outright and some may escape wounded (Bateson, 1997; Bradshaw & Bateson, 2000). In assessing which culling method is more humane, it is necessary to take into account the additional numbers of deer that would be shot if hunting were to stop altogether. If hunting with hounds were to cease and the numbers of deer previously killed by the hunts were instead killed with rifles (in order to maintain cull rates), under what conditions would the welfare benefits of reduced hunting be outweighed by the welfare costs of increased shooting? When the overall welfare costs of culling before the end of hunting are the same as those afterwards, it is possible to derive a simple analytical equation (see Box 19.1). If all hunts resulting in a kill exceeded an acceptable welfare threshold and the welfare costs of hunting by hounds were the same as those of every gunshot wound, banning hunting by hounds would have a welfare benefit in all cases except those in which every stalker's shot led to a wounding. If wounding carries a higher welfare cost than hunting and the probability of wounding is estimated at 5% (Bateson, 1997), the average welfare costs of wounding would have to be 20 times as bad as those of hunting by hounds for a ban on hunting with hounds to have no welfare benefit. If, as seems likely, some deer escaping from the hunts suffered a welfare cost, then the welfare cost of wounding would have to be correspondingly higher because the stalkers only need to shoot as many as the hunts kill in order to maintain previous cull rates.

The advantage of the simple formulation used to derive these conclusions is that the numbers required to obtain an answer are already available and it is not necessary to calculate the units of welfare cost. A wounded deer may have to suffer its state much longer than a deer that is being hunted with hounds. It is difficult to ascertain with any confidence just how much worse than hunting is a lingering death after the lower jaw of a deer has been shattered with a rifle bullet. We have offered some estimates

Box 19.1 A simple welfare equation

The annual welfare costs of culling red deer before a cessation on hunting with hounds may be formalized as follows:

$$T_b = W_h \cdot P_h \cdot N_h + W_s \cdot P_b \cdot N_s$$

where:

T_b = total annual welfare cost before cessation

W_h = average welfare cost to those deer that suffer unacceptable welfare cost when hunted with hounds

P_h = probability of hunting leading to unacceptable welfare cost

N_h = total number of deer killed per year with hounds

W_s = average welfare cost to each stalked animal wounded by shooting

P_b = probability of being wounded by shooting

N_s = total number of deer shot per year

$$T_a = W_s \cdot P_a(N_h + N_s)$$

where:

T_a = total annual welfare cost after cessation

P_a = probability of being wounded by shooting

In this simple case, it is assumed that no deer escaping from a hunt suffers a welfare cost and every deer that is wounded by shooting eventually dies from its wounds. When the total culling rate is constant, cessation of hunting by hounds has no effect on the total welfare cost at the point of indifference (i.e. $T_b = T_a$):

$$W_h \cdot P_h \cdot N_h + W_s \cdot P_b \cdot N_s = W_s \cdot P_s(N_h + N_s)$$

Dividing through by $W_s \cdot P_a$

$$W_h \cdot P_h \cdot N_h / W_s \cdot P_a + P_b \cdot N_s / P_a = N_h + N_s$$

If the proportion wounded by stalking is unchanged by cessation of hunting, then let:

$$P_b = P_a = P_s$$

This then simplifies the equation:

$$W_h \cdot P_h \cdot N_h / W_s \cdot P_s = N_h$$

Dividing through by N_h and rearranging the equation:

$$P_s / P_h = W_h / W_s$$

This is at the point of indifference where no change in deer welfare occurs after a hunting with hounds ban.

If every hunt that leads to a kill is deemed to involve a welfare cost then a cessation of hunting with hounds would be beneficial when:

$$P_s < W_h / W_s$$

of the relative welfare costs obtained from the physiological measures of hunted deer and seriously wounded deer (Bateson, 1997; Bradshaw & Bateson, 2000). This comparison suggested that the welfare costs of extended hunting and wounding are the same at the moment of assessment. However, the time dimension is left out in such comparisons and must be considered.

If time spent suffering by each individual deer is treated as a separate data point, comparisons may then be made between the median number of suffering hours for hunted deer and stalked deer. Clearly, the wounding rate of the stalkers would have to exceed 50% of the total shot for the median number of hours suffered to exceed zero, if we assume that a cleanly shot deer dies without any suffering and all those that are wounded eventually die from their wounds. If all deer that are hunted suffer somewhat, whether or not they escape, then the difference in suffering caused by stalking and hunting is obvious. The wounding rate by the stalkers would have to be far higher than any estimate that has been suggested so far for cessation of hunting to have no welfare benefit.

The approach we used may be criticized on the intuitive grounds that many animals suffering for a short time could be regarded as equivalent to a few animals suffering for a long time. The quantification is difficult, however, because it does not follow that an animal that takes a long time to die from its wounds has suffered in the same way throughout that period. Moreover, one animal suffering for three weeks from a shattered jaw before it died from starvation would distort the picture for the population as a whole, particularly if the great majority experienced no suffering at all when they were culled. Unfortunately, we do not know with any degree of certainty the frequency of such occurrences, although such evidence as is available to us suggests that severe woundings leading to protracted deaths are very infrequent; nor do we know the duration of suffering of any wounded deer. The position is symmetrical for the hunting side of the equation. One hunted animal that incurred severe muscle damage before escaping might experience a sharp reduction in its welfare for days and, in bad weather conditions, might eventually die. So might an animal that was totally depleted of carbohydrate after being hunted all day in winter. Any severely stressed animal might also have a suppression of its immune system, increasing its vulnerability to disease (Martin, 1997). Faced with the difficulty of obtaining reliable empirical evidence, judgements about the welfare balance arrived at on the basis of the total numbers of suffering hours are liable to be made on the basis of a few anecdotes and are bound to be extremely unreliable.

Precision may never be given to calculations of the frequency and duration of welfare costs arising from the various methods of culling. If so, a decision to ban hunting hinges either on our own qualitative assessment that the welfare costs of hunting are greater than stalking or on a straight moral judgement. It is relevant in this context that the central test of cruelty in much of the legislation concerning the welfare of domestic and wild animals rests on the notion of humans causing unnecessary suffering, and further, that the regulations governing the treatment of domestic animals stipulate that animals should not be subjected to 'any avoidable suffering' (Radford, 1996). If it is widely accepted that hunting with hounds inevitably causes suffering, as now seems to be the case, then anybody who hunts does so in the knowledge that they cause suffering. By contrast, a stalker shoots an animal in the expectation that death will be instantaneous. If the animal is wounded by a shot, that is a mistake; it was never the stalker's intention to wound. This is perhaps the most marked moral distinction that may be made between the two methods of culling. The difference is accentuated because hunting with hounds is regarded as better sport when the chase is long, whereas stalkers will strive to improve their technique in order to minimize the risk of wounding. The suffering of hunted deer is, in the context of culling, both unnecessary and avoidable, given that more efficient culling methods exist.

This last point may be more compelling to the general public than any further attempt to formalize the welfare equation. However, the central issues in the stag-hunting debate that relate to animal welfare and conservation remain to be addressed. The preceding discussion was based on the assumption that the majority of culling will be carried out by competent marksmen using high-velocity rifles. What happens if an increase in shooting red deer, resulting from a cessation of hunting with hounds, entailed an increase in the use of shotguns, or greater activity on the part of incompetent marksmen? In general, these are problems that need to be addressed both by changes in the legislation governing the type of weapon used to shoot red deer – The Deer Act 1991 – and the requirements placed upon those who carry out the shooting – The Firearms (Amendment) Act 1997. A system of issuing tags for carcasses, such as already used in France and, at the time of writing, proposed for Scotland, would inhibit the activities of illegal poachers and serve to regulate the numbers of animals culled each year. If crop damage is one of the main concerns, then compensation schemes are required. Such changes in legislation have long-term implications for improving welfare, but will not prevent short-term conservation problems relating to local populations.

Lest the issue of conservation gets out of proportion, it must be emphasized that even the complete elimination of red deer from former hunting areas in North Devon and West Somerset would not affect the overall conservation status of this species in Britain. Significant concentrations of red deer are found in Scotland, NW England and East Anglia and in many parts of the South West where stag-hunting is not practised (Corbet & Harris, 1991; Langbein & Putman, 1996). For all that, red deer are a valued component of the local wildlife in the South West. Of equal importance to changes in legislation, therefore, is the establishment of effective local deer management schemes. Hunting has undoubtedly served an important role in generating an interest in the maintenance of the local population of red deer. This has been important in an area where much of the land is in private ownership, and the deer may range over the property of several landowners. Various management schemes involving local people have worked well in other parts of the country, although running such schemes is bound to be easier when the number of landowners involved is smaller than in the South West.

Despite all the possibilities for saving local deer populations in stag-hunting areas, farmers who cannot hunt red deer with hounds may immediately shoot the deer to minimize perceived damage to their crops or secure a quick financial gain from the sale of the carcasses. In doing so they would, perhaps deliberately, fulfil the prophecy of what would happen to the local red deer population if hunting were banned. Some evidence already suggests that this has happened after the National Trust banned hunting on its land in 1997 (The London Times, 14 February 1998). It is necessary to confront the short-term effects of intervening in what had been a delicate set of local arrangements. Given the strong moral case for culling red deer with rifles rather than by hunting them with hounds, how is the tension to be resolved? This dilemma comes to the heart of the debate between the welfarists and the conservationists. We consider next a way of coming to terms with the general problem.

INTEGRATING WELFARE AND CONSERVATION

We have argued that, in many cases, the well-being of animals and their conservation are not in opposition. The relationship between them is repre-

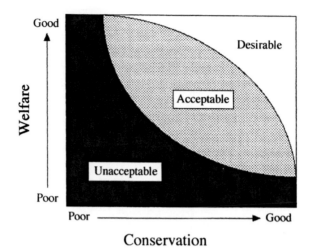

Figure 19.1 A decision space for considering the separate dimensions of
conservation and welfare. Practices that combine good conservation practices with
good welfare practices are judged to be desirable. Other practices may be
acceptable, but clearly the combination of poor welfare and poor conservation is
unacceptable.

sented in Figure 19.1. This provides a decision space for considering the
separate dimensions of conservation and welfare. Practices that combine
good conservation with good welfare are judged to be desirable. Other practices are less desirable and, clearly, the combination of poor welfare and
poor conservation is unacceptable when all other considerations are removed from the decision space.

It will sometimes be the case that judgements come down in favour of
animal welfare over conservation, or the other way around. For example,
eliminating a local herd of red deer may be regarded as poor conservation
practice, but in principle, at least, the welfare of individuals will not be
compromised, providing the deer are killed cleanly. Political and moral
pressures may drive decisions towards the top left hand corner of the space
shown in Figure 19.1. We have already argued in the case of the red deer
debate that the long-term consequences need not be as dire as that. Furthermore, Macdonald & Johnson (1996) have shown how these issues may be
addressed in fox-hunting and, by implication, other field sports.

Cases in which conservation interests may override welfare concerns
are not difficult to find. Eliminating introduced predators from islands to
protect endemic wildlife is often good conservation practice, but the welfare

of the predators may be very poor if gin traps or slow-acting poisons are used. The pressing time and financial constraints under which conservationists often operate inevitably leads to remedial or 'crisis' management (Wood, 1983; Given, 1993). In such cases the pressure to disregard individual welfare in the interests of population or even species conservation is high. The advantage of using the decision space shown in Figure 19.1 is that it does focus attention on the need to keep welfare and conservation concerns in mind at the same time.

CONCLUSIONS

In this chapter we have re-examined the relations between concerns about animal welfare and conservation. When the focus is on the well-being rather than the rights of the individual animal, an additional concern about conservation is not necessarily in opposition with good welfare. Indeed, we have suggested that the two issues should be seen as separate dimensions of an overall problem. In the case of red deer, providing welfarists accept the need for some deer to be killed, and wildlife managers accept a moral obligation to use the culling method that causes the least suffering, it seems that animal welfare concerns and conservation concerns are compatible. Good conservation practices can and should be implemented with close attention to the humane treatment of individual animals.

Difficult ethical problems of animal rights still remain to be resolved (Rawles, 1996) and, when these are crystallized as clear goals, they may continue to be seen as opposed to those of conservation. Moreover, the imposition of moral positions on those who need to be involved in the implementation of animal management may be counter-productive. Even so, we hope that our approach will help to highlight the areas where it is possible for conservation and animal welfare to be brought together with some degree of coherence.

Generally, poor individual welfare should only be tolerated if the conservation benefits are very great. Rather than treating welfare concerns as secondary or as an afterthought, conservationists increasingly acknowledge the ethical and practical reasons for taking an interest in the well-being of animals. In many instances, doing so benefits conservation as well as the welfare of individual animals.

ACKNOWLEDGEMENTS

We thank Harry Bradshaw, Mike Brooke and Murray Corke for commenting on a draft of this chapter. We are grateful to the National Trust for their support while we undertook the study of culling of the red deer.

References

van Aarde, R. J. & van Dyk, A. (1986). Inheritance of the king coat colour pattern in cheetahs *Acinonyx jubatus*. *Journal of Zoology* **209**, 573–578.

Abbot, J. & Mace, R. (1998). Managing protected woodlands: women, fuelwood and law enforcement in Lake Malawi National Park. *Conservation Biology* **13**, 418–421. Cambridge University Press, Cambridge.

Abel, R. (1993). Rearing trumpeter swans for reintroduction: methods using imprinting and a surrogate parent. Masters Thesis, University of Wisconsin, Madison, WI, USA.

Addison, J. T. (1995). Influence of behavioural interactions on lobster distribution and abundance as inferred from pot-caught samples. *ICES Marine Science Symposium* **199**, 294–300.

Akçakaya, H. R., McCarthy, M. A. & Pearce, J. L. (1995). Linking landscape data with population viability analysis – management options for the helmeted honeyeater *Lichenostomus melanops cassidix*. *Biological Conservation* **73**, 169–176.

Alam, M. (1995). Birth-spacing and infant and early childhood mortality in a high fertility area of Bangladesh – age-dependent and interactive effects. *Journal of Biosocial Sciences* **27**, 393–404.

Alcala, A. C. & Russ, G. R. (1990). A direct test of the effects of protective management on abundance and yield of tropical marine resources. *Journal du Conseil, Conseil International pour l'Exploration de la Mer* **46**, 40–47.

Alcock, J. (1987). Leks and hilltopping in insects. *Journal of Natural History* **21**, 319–328.

Alcorn, J. B. (1989). The traditional agricultural ideology of Bora and Huastec resource management and its implications for research. In *Resource management in Amazonia: indigenous and folk strategies* (Eds D. A. Posey & W. Balee), pp. 31–63. The New York Botanical Garden, New York.

Alcorn, J. B. (1991). Ethics, economies and conservation. In *Biodiversity: culture, conservation, and ecodevelopment* (Eds M. L. Oldefield & J. B. Alcorn), pp. 317–349. Westview Press, Boulder, CO.

Alcorn, J. B. (1995). Commentary. *Current Anthropology* **36**, 789–818.

Alerstam, T. (1990). *Bird migration*. Cambridge University Press, Cambridge.

Alerstam, T. & Högstedt, G. (1982). Bird migration and reproduction in relation to habitat for survival and breeding. *Ornis Scandinavica* 13, 25–37.

Alvard, M. (1993). Testing the 'ecologically noble savage' hypothesis: inter-specific prey choice by Piro hunters of Amazonian Peru. *Human Ecology* 21, 355–387.

Alvard, M. (1994). Prey choice in a depleted habitat. *Human Nature* 5, 127–154.

Alvard, M. (1995). Intraspecific prey choice by Amazonian hunters. *Current Anthropology* 36, 789–818.

Alvard, M. (1998). Indigenous hunting in the neotropics: conservation or optimal foraging? In *Behavioral ecology and conservation biology* (Ed. T. Caro), pp. 474–500. Oxford University Press, New York.

Alvard, M. & Kaplan, H. (1991). Procurement technology and prey mortality among indigenous neotropical hunters. In *Human predators and prey mortality* (Ed. M. C. Stiner). Westview Press, Boulder, CO.

Alverson, D. L., Freeberg, M. H., Pope, J. G. & Murawski, S. A. (1994). A global assessment of fisheries bycatch and discards. *FAO Fisheries Technical Paper* 339, 1–233.

Andelt, W. F. (1985). Behavioral ecology of coyotes in South Texas. *Wildlife Monographs* 94, 1–45.

Andelt, W. F., Harris, C. E. & Knowlton, F. F. (1985). Prior trap experience might bias coyote responses to scent stations. *Southwestern Naturalist* 30, 317–318.

Anderies, J. M. (1996). An adaptive model for predicting !Kung reproductive performance: a stochastic dynamic programming approach. *Ethology & Sociobiology* 17, 221–246.

Andersen, J. (1953). Analysis of a Danish roe deer population *(Capreolus capreolus (L.))* based upon the extermination of the total stock. *Danish Review of Game Biology* 2, 127–155.

Anderson, D. R. (1975). Optimal exploitation strategies for an animal population in a Markovian environment; a theory and an example. *Ecology* 56, 1281–1297.

Anderson, R. M. & May, R. M. (1979). Population biology of infectious diseases: part I. *Nature* 280, 361–367.

Andersson, M. (1994). *Sexual selection.* Princeton University Press, Princeton.

Andrewartha, H. G. & Birch, L. C. (1954). *The distribution and abundance of animals.* University of Chicago Press, Chicago.

Anon. (1994). Review of the state of world marine fishery resources. *FAO Fisheries Technical Paper* 335, 1–136.

Anon. (1997a). Fishery statistics 1995. *FAO Fisheries Yearbook* 81, 1–183.

Anon. (1997b). *Towards sustainable fisheries: economic aspects of the management of living marine resources.* OECD, Paris.

Appleby, M. C. & Hughes, B. O. (Eds) (1997). *Animal welfare.* CAB International, Wallingford, Oxon.

Arcese, P., Keller, L. F. & Cary, J. R. (1997). Why hire a behaviorist into a conservation or management team? In *Behavioural approaches to conservation in the wild* (Eds J. R. Clemmons & R. Buchholz), pp. 48–71. Cambridge University Press, Cambridge.

Arnold, R. A. (1983). Ecological studies of six endangered butterflies (Lepidoptera: Lycaenidae): island biogeography, patch dynamics, and the design of habitat preserves. *University of California Publications in Entomology* 99, 1–161.

Arnould, C. & Signoret, J. P. (1993). Sheep food repellents – efficacy of various products, habituation and social facilitation. *Journal of Chemical Ecology* **19**, 225–236.

Arthur, S. M., Krohn, W. B. & Gilbert, J. R. (1989). Home range characteristics of adult fishers. *Journal of Wildlife Management* **53**, 674–679.

Artois, M., Aubert, M. & Stahl, P. (1990). Organisation spatiale du renard roux (*Vulpes vulpes* L. 1758) en zone d'enzootie de rage en Lorraine. *Revue d'Ecologie la Terre et la Vie* **45**, 113–134.

Atkinson, I. A. E. (1985). The spread of commensal species of *Rattus* to oceanic islands and their effects of island avifaunas. *International Council for Bird Preservation Technical Bulletin* **3**, 35–81.

Aveling, C. & Aveling, R. (1989). Gorilla conservation in Zaire. *Oryx* **23**, 64–70.

Avery, M. L. & Decker, D. G. (1994). Responses of captive fish crows to eggs treated with chemical repellents. *Journal of Wildlife Management* **58**, 261–266.

Avise, J. C. (1994). *Molecular markers, natural history and evolution*. Chapman & Hall, New York.

Bacon, P. J. & Macdonald, D. W. (1980). To control rabies: vaccinate foxes. *New Scientist* **87**, 640–645.

Baguette, M., Convié, I. & Nève, G. (1996). Male density affects female spatial behaviour in the butterfly *Proclossiana eunomia*. *Acta Oecologia* **17**, 225–232.

Baguette, M., Vansteenwegen, C., Convié, I. & Nève, G. (1998). Sex-biased density-dependent emigration in a metapopulation of the butterfly *Proclossiana eunomia*. *Acta Oecologia* **19**, 17–24.

Bailey, N. T. J. (1975). *The mathematical theory of infectious diseases*. Griffin, London.

Bailey, R. S. (1991). Changes in the North Sea herring population over a cycle of collapse and recovery. In *Long-term variability of pelagic fish populations and their environment* (Eds T. Kawasaki, S. Tanaka, Y. Toba & A. Taniguchi), pp. 191–198. Pergamon Press, Oxford.

Baker, A. J. & Marshall, H. D. (1997). Mitochondrial control region sequences as tools for understanding migration. In *Avian molecular evolution and systematics* (Ed. D. P. Mindell), pp. 51–82. Academic Press, San Diego.

Baker, A. J. & Strauch, J. G. (1988). Genetic variation and differentiation in shorebirds. *Proceedings XIX International Ornithological Congress (Ottawa)*, pp 1639–1645. University of Ottawa Press, Ottawa.

Baker, A. J., Piersma, T. & Rosenmeier, L. (1994). Unraveling the intraspecific phylogeography of knots *Calidris canutus*: a progress report on the search for genetic markers. *Journal für Ornithologie* **135**, 599–606.

Baker, M. C. & Thompson, D. B. (1985). Song dialects of white-crowned sparrows: historical processes inferred from patterns of geographical variation. *Condor* **87**, 127–141.

Balfour, E. & Cadbury, C. J. (1979). Polygyny, spacing and sex ratio among hen harriers *Circus cyaneus* in Orkney, Scotland. *Ornis Scandinavica* **10**, 133–141.

Ballard, W. B., Whitman, J. S. & Gardner, C. L. (1987). Ecology of an exploited wolf population in south-central Alaska. *Wildlife Monographs* **98**, 1–54.

Barnard, C. J. & Hurst, J. L. (1996). Welfare by design: the natural selection of welfare criteria. *Animal Welfare* **5**, 405–433.

Barraclough, T. G., Harvey, P. H. & Nee, S. (1995). Sexual selection and taxonomic diversity in passerine birds. *Proceedings of the Royal Society of London B* **259**,

211–215.

Barrass, A. N. (1993). Micro-geographic variation in the reproductive behavior of male *Bufo woodhousei* and *Hyla cinerea* near highway traffic noise. In *Hubbs-Sea World Research Institute Workshop Proceedings* (Eds A. E. Bowles, F. T. Awlorey & D. Hunsaker), pp. 1–14. Center for Coastal Studies, San Diego, CA.

Batcheler, C. L. (1968). Compensatory response of artificially controlled mammal populations. *Proceedings of the New Zealand Ecological Society* **15**, 25–30.

Bateson, P. (1991). Assessment of pain in animals. *Animal Behaviour* **42**, 827–839.

Bateson, P. (1997). *The behavioural and physiological effects of culling red deer.* The National Trust, London.

Bateson, P. & Bradshaw, E. L. (1997). Physiological effects of hunting red deer, *Cervus elaphus. Proceedings of the Royal Society of London B* **264**, 1–8.

Batten, L. A. (1977). Sailing on reservoirs and its effect on waterbirds. *Biological Conservation* **11**, 49–58.

Bauer, H.-G. & Nagl, W. (1992). Individual distinctiveness and possible function of song parts of short-toed treecreepers (*Certhia brachydactyla*). Evidence from multivariate song analysis. *Ethology* **91**, 108–121.

Bauer, R. T. & Martin, J. W. (1991). *Crustacean sexual biology.* Columbia University Press, New York.

Bautista, L. M., Alonso, J. C. & Alonso, J. A. (1995). A field test of the ideal free distribution in flock-feeding common cranes. *Journal of Animal Ecology* **64**, 747–757.

Bayliss, P. (1989). Population dynamics of magpie geese in relation to rainfall and density: implications for harvest models in a fluctuating environment. *Journal of Applied Ecology* **26**, 913–924.

Beardmore, J. A., Mair, G. C. & Lewis, R. I. (1997). Biodiversity in aquatic systems in relation to aquaculture. *Aquaculture Research* **28**, 829–839.

Beauchamp, G. (1994). The functional analysis of human fertility decisions. *Ethology & Sociobiology* **15**, 31–53.

Beck, B. B., Rapaport, L. G., Stanley Price, M. R. & Wilson, A. C. (1994). Reintroduction of captive-born animals. In *Creative conservation: interactive management of wild and captive animals* (Eds P. J. S. Olney, G. M. Mace & A. T. C. Feistner), pp. 265–286. Chapman & Hall, London.

Beckerman, S. & Valentine, P. (1996). On native American conservation and the tragedy of the commons. *Current Anthropology* **37**, 659–661.

Beddington, J. R., Beverton, R. J. H. & Lavigne, D. M. (Eds) (1985). *Marine mammals and fisheries.* George Allen & Unwin, London.

Beightol, D. R. & Samuel, D. E. (1973). Sonagraphic analysis of the American woodcock's peent call. *Journal of Wildlife Management* **37**, 470–475.

Beissinger, S. R. & Bucher, E. H. (1992). Can parrots be conserved through sustainable harvesting? *BioScience* **42**, 164–173.

Bekoff, M. (1977). Mammalian dispersal and the ontogeny of individual behavioural phenotypes. *The American Naturalist* **111**, 715–732.

Bélanger, L. & Bédard, J. (1989). Responses of staging greater snow geese to human disturbance. *Journal of Wildlife Management* **53**, 713–719.

Bell, D. V. & Austin, L. W. (1985). The game-fishing season and its effects on over-wintering waterfowl. *Biological Conservation* **33**, 65–80.

Benirschke, K. (1977). Genetic management. *International Zoo Yearbook* 17, 50–60.

Bennett, A. T. D., Cuthill, I.C., Partridge, J. C. & Maier, E. J. (1996). Ultraviolet vision and mate choice in zebra finches. *Nature* 380, 433–435.

Bennett, P. M. & Owens, I. P. F. (1997). Variation in extinction risk among birds: Chance or evolutionary predispositions? *Proceedings of the Royal Society of London B* 264, 401–408.

Benson, K. & Stephens, D. (1996). Interruptions, tradeoffs and temporal discounting. *American Zoologist* 36, 506–517.

Berger, J. (1990). Persistence of different-sized populations: an empirical assessment of rapid extinctions in bighorn sheep. *Conservation Biology* 4, 91–98.

Berger, J. & Cunningham, C. (1995). Predation, sensitivity, and sex: why female black rhinoceroses outlive males. *Behavioural Ecology* 6, 57–64.

Beringer, J., Hansen, L. P., Wilding, W., Fischer, J. & Sheriff, S. L. (1996). Factors affecting capture myopathy in white-tailed deer. *Journal of Wildlife Management* 60, 373–380.

Berkes, F. (1987). Common-property resource management and Cree Indian fisheries in subarctic Canada. In *The question of the commons. The culture and ecology of communal resources* (Eds B. J. McCay & J. M. Acheson), pp. 66–91. University of Arizona Press, Tucson.

Bernstein, C., Kacelnik, A. & Krebs, J. R. (1988). Individual decisions and the distribution of predators in a patchy environment. *Journal of Animal Ecology* 57, 1007–1026.

Bernstein, C., Kacelnik, A. & Krebs, J. R. (1991). Distribution of birds amongst habitats: theory and relevance to conservation. In *Bird population studies* (Eds C. M. Perrins, J. D. Lebreton & G. J. M. Hirons), pp. 317–345. Oxford University Press, Oxford.

Bernstein, I. L., Courtney, L. & Braget, D. J. (1986). Estrogens and the Leydig LTW(m) tumour syndrome: anorexia and diet aversions attenuated by area postrema lesions. *Physiology & Behaviour* 38, 159–163.

Berthold, P. (1996). *Control of bird migration.* Chapman & Hall, London.

Beudels, R., Durant, S. M. & Harwood, J. (1992). Assessing the risks of extinction for local populations of roan antelope (*Hippotragus equinus*). *Biological Conservation* 61, 107–116.

Beukema, J. J. (1993). Increased mortality in alternative bivalve prey during a period when the tidal flats of the Dutch Wadden Sea were devoid of mussels. *Netherlands Journal of Sea Research* 31, 395–406.

Beverton, R. J. H. (1990). Small marine pelagic fish and the threat of fishing: are they endangered? *Journal of Fish Biology* 37A, 5–16.

Beverton, R. J. H. & Holt, S. J. (1957). *On the dynamics of exploited fish populations.* H.M. Stationary Office, London.

Biggins, D. E., Hanebury, L. H., Miller, B. J. & Powell, R. A. (1990). Release of Siberian polecats *(M. eversmanni)* on a prairie dog colony (Abstract). Seventieth Annual Meeting of the American Society of Mammalogists.

Biggins, D. E., Godbey, J. & Vargas, A. (1993). Influence of pre-release experience on reintroduced black-footed ferrets *(M. nigripes)*. Internal Report, 27 May 1993. US Fish and Wildlife Service, Fort Collins, CO.

Bininda-Emonds, O. R. P., Gittleman, J. L. & Purvis, A. (1999). Building large phylogenies by combining phylogenetic information: a complete phylogeny of the extant Carnivora (Mammalia). *Biological Reviews*, in press.

Birdsell, J. (1958). On population structure in generalized hunting and collecting populations. *Evolution* 12, 189–205.

Birkeland, C. (Ed.) (1997). *Life and death on coral reefs*. Chapman & Hall, New York.

Bishop, C. A., Murphy, E. F., Davis, M. B., Baird, J. W. & Rose, G. A. (1993). An assessment of the cod stock in NAFO Divisions 2J + 3KL. *NAFO Scientific Council Research Document 93/86*, Series No. N2271, 1–51.

Bjorge, R. R. & Gunson, J. R. (1989). Wolf, *Canis lupus*, population characteristics and prey relationships near Simonette River, Alberta. *Canadian Field Naturalist* 103, 327–334.

Black, J. M. & Owen, M. (1989). Agonistic behaviour in goose flocks: assessment, investment and reproductive success. *Animal Behaviour* 37, 199–209.

Black, J. M., Carbone, C., Owen, M. & Wells, R. (1992). Foraging dynamics in goose flocks: the cost of living on the edge. *Animal Behaviour* 44, 41–50.

Blackmore, D. K., Madie, P. & Barnes, G. R. G. (1997). Observations on the electric lance and the welfare of whales: a critical appraisal. *Animal Welfare* 6, 43–51.

Blanchard, B. M. & Knight, R. R. (1991). Movements of Yellowstone grizzly bears. *Biological Conservation* 58, 41–67.

Blindheim, J. & Skjoldal, H. R. (1993). Effects of climatic changes on the biomass yield of the Barents Sea, Norwegian Sea and West Greenland large marine ecosystems. In *Large marine ecosystems: stress, mitigation and sustainability* (Eds K. Sherman, L. M. Alexander & B. D. Gold), pp. 185–189. American Association for the Advancement of Science, Washington.

Bloomsmith, M. A. & Lambeth, S. P. (1995). Effects of predictable versus unpredictable feeding schedules on chimpanzee behavior. *Applied Animal Behavioral Science* 44, 65–74.

Bodmer, R. E., Eisenberg, J. F. & Redford, K. H. (1997). Hunting and the likelihood of extinction of Amazonian mammals. *Conservation Biology* 11, 460–466.

Bodsworth, F. (1956). *Last of the curlews*. Museum Press, London.

Boecklen, W. J. (1986). Optimal design of nature reserves: consequences of genetic drift. *Biological Conservation* 38, 323–338.

Boehlert, G. W. & Mundy, B. C. (1988). Roles of behavioural and physical factors in larval and juvenile fish recruitment to estuarine nursery areas. *American Fisheries Society Symposium* 3, 51–67.

den Boer, P. J. (1968). Spreading of risk and stabilisation of animal numbers. *Acta Biotheoretica* 18, 165–194.

Bogoslovskaya, L. S. (1986). On the social behaviour of gray whales off Chukotka and Koryaka. In *Behaviour of whales in relation to management*, pp. 243–251. International Whaling Commission, Cambridge.

Bohler, E. & Bergstrom, S. (1995). Subsequent pregnancy effects morbidity of previous child. *Journal of Biosocial Science* 27, 431–442.

Boitani, L. (1992). Wolf research and conservation in Italy. *Biological Conservation* 61, 125–132.

Bolger, D. T., Alberts, A. C. & Soulé, M. E. (1991). Occurrence patterns of bird species in habitat fragments: sampling, extinction and nested species subsets. *American Naturalist* 137, 155–166.

Bonner, W. N. (1989). *The natural history of seals.* Christopher Helm, London.

Borgerhoff Mulder, M. (1987). On cultural and biological success: Kipsigis evidence. *American Anthropologist* **89**, 619–634.

Borgerhoff Mulder, M. (1991). Datoga pastoralists of Tanzania. *National Geographic Research and Exploration* **7**, 166–187.

Borgerhoff Mulder, M. & Sellen, D. W. (1994). Pastoralist decision making: a behavioral ecological perspective. In *African pastoralist systems: an integrated approach* (Eds E. Fratkin, K. A. Galvin & E. A. Roth), pp. 205–229. Lynne Reinner Publications, Boulder, CO.

Borgerhoff Mulder, M., Sieff, D. & Merus, M. (1989). Datoga history in Ngorongoro Crater. *Swara* **12**, 32–35.

Borison, H. L. & Wang, S. C. (1953). Physiology and pharmacology of vomiting. *Pharmacological Review* **5**, 193–230.

Botton, M. L., Loveland, R. E. & Jacobsen, T. R. (1994). Site selection by migratory shorebirds in Delaware Bay, and its relationship to beach characteristics and abundance of horseshoe crab (*Limulus polyphemus*) eggs. *Auk* **111**, 605–616.

Bottriell, L. G. (1987). *King cheetah: the story of the quest.* E. J. Brill, Leiden.

Bouman, J. (1977). The future of Przewalski horses in captivity. *International Zoo Yearbook* **17**, 62–68.

Bowles, D. (1996). Wildlife trade – a conserver or exploiter ? In *Behavioural approaches to conservation in the wild* (Eds J. R. Clemmons & R. Buchholz), pp. 266–288. Cambridge University Press, Cambridge.

Boyce, M. S. (1992). Population viability analysis. *Annual Review of Ecology and Systematics* **23**, 481–506.

Boyce, M. S. & Perrins, C. M. (1987). Optimising great tit clutch-size in a fluctuating environment. *Ecology* **68**, 142–153.

Boyd, D. K., Paquet, P. C., Donelon, S., Ream, R. R., Pletscher, D. H. & White, C. C. (1995). Transboundary movements of a recolonizing wolf population in the Rocky Mountains. In *Ecology and conservation of wolves in a changing world* (Eds L. N. Carbyn, S. H. Fritts & D. R. Seip). Canadian Circumpolar Institute, Edmonton.

Boyd, H. (1962). Mortality and fertility of European Charadrii. *Ibis* **104**, 368–387.

Boyd, R. & Richardson, P. J. (1985). *Culture and the evolutionary process.* University of Chicago Press, Chicago.

Bradburd, D. (1982). Volatility of animal wealth among southwest Asian pastoralists. *Human Ecology* **10**, 85–106.

Bradshaw, E. L. & Bateson, P. (2000). Welfare implications of culling red deer (*Cervus elaphus*). *Animal welfare*, in press.

Brambell, M. R. (1977). Reintroduction. *International Zoo Yearbook* **17**, 112–116.

Brander, K. (1981). Disappearance of the common skate *Raia batis* from Irish Sea. *Nature* **290**, 48–49.

Bretagnolle, V. (1996). Acoustic communication in a group of nonpasserine birds, the petrels. In *The ecology & evolution of acoustic communication in birds* (Eds D. E. Kroodsma & E. H. Miller), pp. 160–177. Comstock, Ithaca, NY.

Brett, L. P., Hankins, W. G. & Garcia J. (1976). Prey-lithium aversions. III. Buteo hawks. *Behavioural Biology* **17**, 87–98.

Briggs, M. B. & Ott, R. L. (1986). Feline leukemia virus infection in a captive cheetah and the clinical and antibody response of six captive cheetahs to vaccination with a subunit feline leukemia virus vaccine. *Journal of the American Veterinary Medical Association* **189**, 1197–1199.

Britton, N. F., Partridge, L. W. & Franks, N. R. (1996). A mathematical model for the population dynamics of army ants. *Bulletin of Mathematical Biology* **58**, 471–492.

Brockmann, H. J. & Barnard, C. J. (1979). Kleptoparasitism in birds. *Animal Behaviour* **27**, 487–514.

Bronikowski, E. J., Jr, Beck, B. B. & Power, M. (1989). Innovation, exhibition and conservation: free-ranging tamarins at the National Zoological Park. In *American Association of Zoological Parks and Aquariums Annual Proceedings*, pp. 540–546. American Association of Zoological Parks and Aquariums, Wheeling, WV.

Broom, D. M. (1986). Indicators of poor welfare. *British Veterinary Journal* **142**, 524–526.

Broom, D. M. (1998). Welfare, stress and the evolution of feelings. *Advances in the Study of Behavior*, **27**, 371–403.

Broom, D. M. & Johnson, K. G. (1993). *Stress and animal welfare*. Chapman & Hall, London.

Brothers, N. (1991). Albatross mortality and associated bait loss in the Japanese longline fishery in the Southern Ocean. *Biological Conservation* **55**, 255–268.

Brower, L. P. & Malcolm, S. B. (1991). Animal migrations: endangered phenomena. *American Zoologist* **31**, 265–276.

Brower, T. P. (1969). Ecological chemistry. *Scientific American* **220**, 22–29.

Brown, I. L. & Ehrlich, P. R. (1980). Population biology of the checkerspot butterfly, *Euphydryas chalcedona*. Structure of the Jasper Ridge Colony. *Oecologia* **47**, 239–251.

Brown, J. H. (1971). Mammals on mountaintops: nonequilibrium insular biogeography. *American Naturalist* **105**, 467–478.

Brown, J. L. (1969). Territorial behaviour and population regulation in birds. *Wilson Bulletin* **81**, 293–329.

Bruford, M., Macdonald, D. W., McNutt, J. W., Mills, M. G. L., Sillero-Zubiri, C. & Woodroffe, R. (1997). Modeling/life history working group report. In *African wild dog* (Lycaon pictus) *population and habitat viability assessment report* (Eds M. G. L. Mills, S. Ellis, R. Woodroffe, D. W. Macdonald, P. Fletcher, M. Bruford, D. Wildt & U. S. Seal). IUCN/SSC Conservation Breeding Specialist Group, Apple Valley, MN.

Brunton, C. F. A., Macdonald, D. W. & Buckle, A. P. (1993). Behavioural resistance towards poison baits in brown rats, *Rattus norvegicus*. *Applied Animal Behavioural Science* **38**, 159–174.

Bruton, M. N. (1995). Have fishes had their chips? The dilemma of threatened fishes. *Environmental Biology of Fishes* **43**, 1–27.

Bucher, E. H. (1992). The causes of extinction of the passenger pigeon. In *Current ornithology*, vol. 9 (Ed. D. M. Power), pp. 1–36. Plenum Press, New York.

Buchholz, R. & Clemmons, J. R. (1997). Behavioral variation: a valuable but neglected biodiversity. In *Behavioural approaches to conservation in the wild* (Eds J. R. Clemmons & R. Buchholz), pp. 181–208. Cambridge University Press, Cambridge.

Buechner, M. (1987). Conservation of insular parks: simulation models of factors affecting the movements of animals across park boundaries. *Biological Conservation* **41**, 57–76.

Bulmer, R. N. H. (1982). Traditional conservation practices in Papua New Guinea. In *Traditional conservation in Papua New Guinea: implications for today* (Eds L. Morauta, J. Pernetta & W. Heaney), pp. 59–78. Institute of Applied Social and Economic Research, Boroko, Papua New Guinea.

Bures, J. & Buresova, O. (1977). Physiological mechanisms of conditioned food aversion. In *Food aversion learning* (Eds N. W. Milgram, L. Krames & T. M. Alloway), pp. 219–255. Plenum Press, New York and London.

Burger, J. (1981). The effect of human activity on birds at a coastal bay. *Biological Conservation* **21**, 231–241.

Burghardt, G. M., Wilcoxon, H. C. & Czaplicki, J. A. (1973). Conditioning in garter snakes: aversion to palatable prey induced by delayed illness. *Physiology and Behaviour* **11**, 435–439.

Burgman, M. A., Cantoni, D. & Vogel, P. (1992). Shrews in suburbia: an application of Goodman's extinction model. *Biological Conservation* **61**, 117–123.

Burgman, M. A., Ferson, S. & Akçakaya, H. R. (1993). *Risk assessment in conservation biology*. Chapman & Hall, London.

Burns, R. J. (1980). Evaluation of conditioned predation aversion for controlling coyote predation. *Journal of Wildlife Management* **44**, 938–942.

Butler, R. W., Williams, T. D., Warnock, N. & Bishop, M. A. (1997). Wind assistance: a requirement for migration of shorebirds? *Auk* **114**, 456–466.

Butynski, T. M., Chapman, C. A., Chapman, L. J. & Weary, D. M. (1992). Use of male blue monkey pyow calls for long-term individual identification. *American Journal of Primatology* **28**, 183–189.

Caddy, J. F. (1973). Underwater observations on tracks of dredges and trawls and some effects of dredging on a scallop ground. *Journal of the Fisheries Research Board of Canada* **30**, 173–180.

Cade, W. (1975). Acoustically orienting parasitoids: fly phonotaxis to cricket song. *Science* **190**, 1312–1313.

Cadée, N., Piersma, T. & Daan, S. (1996). Endogenous circannual rhythmicity in a non-passerine migrant, the knot *Calidris canutus*. *Ardea* **84**, 75–84.

Caillouet, C. W., Shaver, D. J., Teas, W. G., Nance, J. M., Revera, D. B. & Cannon, A. C. (1996). Relationship between sea-turtle stranding rates and shrimp fishing intensities in the northwestern Gulf of Mexico: 1986–1989 versus 1990–1993. *Fishery Bulletin* **94**, 237–249.

Caldecott, J. & Kavanagh, M. (1983). Can translocation help wild primates? *Oryx* **17**, 135–139.

Campbell, L. H. & Mudge, G. P. (1990). Conservation of black-throated divers in Scotland. *RSPB Conservation Review* **3**, 72–74.

Campbell, S. (1980). Is reintroduction a realistic goal? In *Conservation biology: an evolutionary–ecological perspective* (Eds M. E. Soulé & B. A. Wilcox), pp. 263–269. Sinauer, Sunderland, MA.

Camphuysen, C. J., Ensor, K., Furness, R. W., Garthe, S., Huppop, O., Leaper, G., Offringa, H. & Tasker, M. L. (1993). *Seabirds feeding on discards in winter in the*

North Sea. Netherlands Institute for Sea Research, Den Burg, Texel.

Camphuysen, C. J., Ens, B. J., Heg, D., Hulscher, J. B., van der Meer, J. & Smit, C. J. (1996). Oystercatcher *Haematopus ostralegus* winter mortality in The Netherlands: the effect of severe weather and food supply. *Ardea* **84A**, 469–492.

Carlton, J. T. (1996). Marine bioinvasions: the alternation of marine ecosystems by nonindegenous species. *Oceanography* **9**, 36–43.

Carlton, J. T. & Geller, J. B. (1993). Ecological roulette: the global transport of nonindigenous marine organisms. *Science* **261**, 78–82.

Caro, T. M. (1976). Observations on the ranging behaviour and daily activity of lone silver back mountain gorillas (*Gorilla gorilla beringei*). *Animal Behaviour* **24**, 889–897.

Caro, T. M. (1994). *Cheetahs of the Serengeti Plains: group living in an asocial species*. University of Chicago Press, Chicago.

Caro, T. (1999). *Behavioural ecology & conservation biology*. Oxford University Press, Oxford.

Caro, T. M. & Durant, S. M. (1991). Use of quantitative analyses of pelage characteristics to reveal family resemblances in genetically monomorphic cheetahs. *Journal of Heredity* **82**, 8–14.

Caro, T. M. & Durant, S. M. (1995). The importance of behavioural ecology for conservation biology: examples from studies of Serengeti carnivores. In *Serengeti. II. Dynamics, management and conservation of an ecosystem* (Eds A. R. E. Sinclair & P. Arcese), pp. 451–472. University of Chicago Press, Chicago.

Caro, T. M. & Laurenson, M. K. (1994). Ecological and genetic factors in conservation: a cautionary tail. *Science* **263**, 485–486.

Carpenter, F. L., Hixon, M. A., Russell, R.W., Paton, D.C. & Temeles, E. J. (1993). Interference asymmetries among age–sex classes of rufous hummingbirds during migratory stopovers. *Behavioural Ecology & Sociobiology* **33**, 305–312.

Carpenter, L. F. (1987). The study of territoriality: complexities and future directions. *American Zoologist* **27**, 401–409.

Cary, J. R., Small, R. J. & Rush, D. H. (1992). Dispersal of ruffed grouse: a large-scale individual-based model. In *Wildlife 2001: populations* (Eds D. R. McCullough & R. H. Barrett), pp. 727–737. Elsevier Applied Science, London.

Caswell, H. (1989). *Matrix population models*. Sinauer, Sunderland, MA.

Caswell, H. & Etter, R. J. (1993). Ecological interactions in patchy environments: from patch-occupancy models to cellular automata. In *Patch dynamics, volume 96 of lecture notes in biomathematics* (Eds S. A. Levin, T. M. Powell & J. Steele), pp. 176–183. Springer, New York.

Caughley, G. (1977). *Analysis of vertebrate populations*. John Wiley & Sons, London.

Caughley, G. (1994). Directions in conservation biology. *Journal of Animal Ecology* **63**, 215–244.

Caughley, G. & Gunn, A. (1996). *Conservation biology in theory and practice*. Blackwell Science, Cambridge, MA.

Caughley, G. & Sinclair, A. R. E. (1994). *Wildlife ecology and management*. Blackwell Science, Oxford, and Cambridge, MA.

Cavalli-Svorza, L. L. & Feldman, M. W. (1981). *Cultural transmission and evolution*. Princeton University Press, Princeton.

Chambers, L. K., Singleton, G. R. & Hood, G. M. (1997). Immunocontraception as a potential control method of wild rodent populations. *Belgian Journal of Zoology* **127**, 145–156.

Charnov, E. L. (1976). Optimal foraging: attack strategy of a mantid. *American Naturalist* **110**, 141–151.

Charnov, E. L. (1986). Life history evolution in a 'recruitment population': why are adult mortality rates constant? *Oikos* **47**, 129–134.

Charnov, E. L. & Krebs, J. R. (1974). On clutch-size and fitness. *Ibis* **116**, 217–219.

Cheeseman, C. L., Mallinson, P. J., Ryan, J. & Wilesmith, J. W. (1993). Recolonisation by badgers in Gloucestershire. In *The badger* (Ed. T. J. Hayden), pp. 78–93. Royal Irish Academy, Dublin.

Chitty, D. (1954). The study of the brown rat and its control by poison. In *The control of rats and mice*, vol. 1 (Ed. D. Chitty), pp. 160–305. Clarendon Press, Oxford.

Chivers, D. J. (1991). Guidelines for re-introduction: procedures and problems. *Zoological Society of London Symposium* No. 62, 89–99.

Christian, J. J. (1970). Social subordination, population density, and evolution. *Science* **168**, 84–90.

Christy, J. H. (1988). Pillar function in the fiddler crab *Uca beebei* (II): competitive courtship signalling. *Ethology* **78**, 113–128.

Ciriacy-Wantrup, S. V. & Bishop, R. C. (1975). 'Common property' as a concept in natural resources policy. *Natural Resources Journal* **15**, 713–727.

Clark, C. W. (1973). The economics of overexploitation. *Science* **181**, 630–634.

Clark, C. W. & Mangel, M. (1986). The evolutionary advantages of group foraging. *Theoretical Population Biology* **30**, 45–75.

Clark, N. A. (1993). Wash oystercatchers starving. *BTO News* **185**, 1 & 24.

Clark, W. R. (1990). Compensation in furbearer populations: current data compared with a review of concepts. *Transactions of the North American Wildlife and Natural Resources Conference* **55**, 491–500.

Clarke, R. T. & Goss-Custard, J. D. (1996). The Exe-estuary oystercatcher–mussel model. In *The oystercatcher: from individuals to populations* (Ed. J. D. Goss-Custard), pp. 390–393. Oxford University Press, Oxford.

Clayton, D. H. (1991). The influence of parasites on host sexual selection. *Parasitology Today* **7**, 329–334.

Cleland, J. (1995). Obstacles to fertility decline in developing countries. In *Human reproductive decisions* (Ed. R. Dunbar), pp. 207–229. Macmillan, London.

Clemmons, J. R. & Buchholz, R. (Eds) (1997). *Behavioral approaches to conservation in the wild.* Cambridge University Press, Cambridge.

Clinchy, M. & Krebs, C. J. (1997). !Viva Caughley! *Conservation Biology* **11**, 832–833.

Clout, M. N. & Barlow, N. D. (1982). Exploitation of brushtail possum populations in theory and practice. *New Zealand Journal of Ecology* **5**, 29–35.

Clout, M. N. & Efford, M. G. (1984). Sex differences in the dispersal and settlement of brushtail possums (*Trichosurus vulpecula*). *Journal of Animal Ecology* **53**, 737–749.

Clutton-Brock, T. H. (1988). Reproductive success. In *Reproductive success* (Ed. T. H. Clutton-Brock), pp. 472–485. University of Chicago Press, Chicago.

Clutton-Brock, T. H. & Parker, G. A. (1992). Potential reproductive rates and the operation of sexual selection. *Quarterly Review of Biology* **67**, 437–456.

Clutton-Brock, T. H. & Parker, G. A. (1995). Punishment in animal societies. *Nature*

373, 209–216.

Cohn, J. P. (1986). Surprising cheetah genetics. *BioScience* **36**, 358–362.

Colinvaux, P. (1980). *Why big fierce animals are rare.* Allen & Unwin, London.

Collie, J. S., Escanero, G. A. & Valentine, P. C. (1997). Effect of bottom fishing on the benthic megafauna of Georges Bank. *Marine Ecology Progress Series* **155**, 159–172.

Conner, D. A. (1982a). Geographic variation in short calls of pika (*Ochotona princeps*). *Journal of Mammalogy* **63**, 48–52.

Conner, D. A. (1982b). Dialects versus geographic variation in mammalian vocalizations. *Animal Behaviour* **30**, 297–298.

Connor, R. C. (1995). The benefits of mutualism: a conceptual framework. *Biological Review* **70**, 427–457.

Conover, M. R. (1989). Potential compounds for establishing conditioned food aversion in raccoons. *Wildlife Society Bulletin* **17**, 430–435.

Conover, M. R., Francik, J. G. & Miller, D. E. (1977). An experimental evaluation of aversive conditioning for controlling coyote predation. *Journal of Wildlife Management* **41**, 775–779.

Conover, M. R., Francik, J. G. & Miller, D. E. (1979). Aversive conditioning in coyotes: a reply. *Journal of Wildlife Management* **43**, 209–211.

Conrad, C., Lechner, M. & Werner, W. (1996). East German fertility after unification: crisis or adaptation. *Population Development Review* **22**, 331–358.

Conway, W. G. (1980). An overview of captive propagation. In *Conservation biology: an evolutionary–ecological perspective* (Eds M. E. Soulé & B. A. Wilcox), pp. 199–208. Sinauer, Sunderland, MA.

Cooper, J. E. & Knowler, C. (1991). Snails and snail farming – an introduction for the veterinary profession. *Veterinary Record* **129**, 541–549.

Cope, E. D. (1896). *The primary factors of organic evolution.* Open Court, Chicago.

Copeland, G. L. (1980). Antelope buck breeding behavior, habitat selection and hunting impact. *Idaho Department of Fish and Game Wildlife Bulletin* **8**, 1–45.

Corbet, C. B. & Harris, S. (1991). *The handbook of British mammals*, 3rd edn. Blackwell Scientific Publications, Oxford.

Corkeron, P. J. (1995). Humpback whales (*Megaptera novaeangliae*) in Hervey Bay, Queensland – behavior and responses to whale-watching vessels. *Canadian Journal of Zoology* **73**, 1290–1299.

Côté, I. M. & Sutherland, W. J. (1997). The effectiveness of removing predators to protect bird populations. *Conservation Biology* **11**, 395–405.

Cowan, P. E. (1996). Possum biocontrol – prospects for fertility regulation. *Reproduction Fertility and Development* **8**, 655–660.

Cox, D. R. (1970). *The analysis of binary data.* Methuen, London.

Cramp, S. & Simmons, K. E. L. (1980). *Handbook of the birds of Europe, the Middle East and North Africa*, vol. II. Oxford University Press, Oxford.

Crane, J. (1975). *Fiddler crabs of the world.* Princeton University Press, New Jersey.

Creel, S. R. & Creel, N. M. (1996). Limitation of African wild dogs by competition with larger carnivores. *Conservation Biology* **10**, 1–15.

Creel, S. & Creel, N. M. (1997). Lion density and population structure in the Selous Game Reserve: evaluation of hunting quotas and offtake. *African Journal of Ecology* **35**, 83–93.

Creel, S. & Waser, P. M. (1994). Inclusive fitness and reproductive strategies in dwarf mongooses. *Behavioral Ecology* 5, 339–348.

Creel, S., Creel, N. M., Mills, M. G. L. & Monfort, S. L. (1997). Rank and reproduction in cooperatively breeding African wild dogs: behavioral and endocrine correlates. *Behavioral Ecology* 8, 298–306.

Cresswell, W. (1994). Flocking is an effective anti-predation strategy in redshanks, *Tringa totanus*. *Animal Behaviour* 47, 433–442.

Crocker, D. R., Perry, S. M., Wilson, M., Bishop, J. D. & Scanlon, C. B. (1993). Repellency of cinnamic acid derivatives to rock doves. *Journal of Wildlife Management* 57, 113–122.

Crouse, D. T., Crowder, L. B. & Caswell, H. (1987). A stage-based population model for loggerhead sea turtles and implications for conservation. *Ecology* 68, 1412–1423.

Crow, J. F. & Kimura, M. (1970). *An introduction to population genetics theory*. Harper & Row, New York.

Cushing, D. H. (1992). A short history of the Downs stock of herring. *ICES Journal of Marine Science* 49, 437–443.

Dadswell, M. J., Rulifson, R. A. & Daborn, G. R. (1986). Potential impact of large scale tidal power development in the Upper Bay of Fundy on fisheries resources of the northwest Atlantic. *Fisheries* 11, 26–35.

Davidson, N. C. & Evans, P. R. (1982). Mortality of redshanks and oystercatchers from starvation during severe weather. *Bird Study* 29, 183–188.

Davidson, N. C. & Piersma, T. (1992). The migration of knots: conservation needs and implications. *Wader Study Group Bulletin* 64 (suppl.), 198–209.

Davidson, N. & Rothwell, P. (1993). Disturbance to waterfowl on estuaries. *Wader Study Group Bulletin* (Special Issue) 68.

Davies, G. P., Dare, P. J. & Edwards, D. B. (1980). Fenced enclosures for the protection of seed mussel (*Mytilus edulis*) from predation by shore crabs (*Carcinus maenas* (L).). *Fisheries Research Technical Report, Ministry of Fisheries and Food, Lowestoft* 56, 1–14.

Davies, J. K. (1988). A review of information relating to fish passage through turbines: implications to tidal power schemes. *Journal of Fish Biology* 33, 111–126.

Davies, N. B. (1978). Ecological questions about territorial behaviour. In *Behavioural ecology* (Eds J. R. Krebs & N. B. Davies), pp. 317–350. Sinauer, Sunderland, MA.

Dawkins, M. S. (1980). *Animal suffering: the science of animal welfare*. Chapman & Hall, London.

Dawkins, M. S. (1983). Battery hens name their price: consumer demand theory and the measurement of ethological 'needs'. *Animal Behaviour* 31, 1195–1205.

Dawkins, M. S. (1990). From an animal's point of view: motivation, fitness, and animal welfare. *Behaviour and Brain Sciences* 13, 1–61.

Dawkins, R. (1976). *The selfish gene*. Oxford University Press, Oxford.

Deblinger, R. D. & Alldredge, A. W. (1989). Management implications of variations in pronghorn social behavior. *Wildlife Society Bulletin* 17, 82–87.

Derrickson, S. R. (1986). A cooperative breeding program for the Guam rail (*Rallus owstoni*). In *American Association of Zoological Parks and Aquariums Annual Proceedings*, pp. 223–240. American Association of Zoological Parks and Aquariums, Wheeling, WV.

Derrickson, S. R. & Snyder, N. F. R. (1992). *Potentials and limits of captive breeding in parrots in crisis: solutions from conservation biology* (Eds S. R. Beissinger & N. F. R. Snyder). Smithsonian Institution Press, Washington.

Dethier, V. G. & MacArthur, R. H. (1964). A field's capacity to support a butterfly population. *Nature* **201**, 728–729.

Dhondt, A. A. (1966). A method to establish the boundaries of bird territories. *Le Gerfault* **56**, 404–408.

Dimmick, C. R. & Nicolaus, L. K. (1990). Efficiency of conditioned aversion in reducing depredation by crows. *Journal of Applied Ecology* **27**, 200–209.

Dinerstein, E., Wikramanayake, E., Robinson, J., Karanth, U., Rabinowitz, A., Olson, D., Mathew, T., Hedao, P., Connor, M., Hemley, G. & Bolze, D. (1997). *A framework for identifying high priority areas and actions for the conservation of tigers in the wild.* WWF, Washington, and WCS, New York.

Dingle, H. (1996). *Migration. The biology of life on the move.* Oxford University Press, New York.

Divyabhanusinh (1995). *The end of a trail: the cheetah in India.* Banyan Books, New Delhi.

Dobrott, S. R. (1993). *Masked bobwhite recovery plan. Region 2* (Ed. W. P. Kuvlesky, Jr). US Fish and Wildlife Service, Albuquerque, NM.

Dobzhansky, T. & Wallace, B. (1953). The genetics of homeostasis in *Drosophila. Proceedings of the National Academy of Science, USA* **39**, 162–171.

Dolman, P. M. (1995). The intensity of interference varies with resource density: evidence from a field study with snow buntings, *Plectrophenox nivalis. Oecologia* **102**, 511–514.

Dolman, P. M. & Sutherland, W. J. (1995a). The response of bird populations to habitat loss. *Ibis* **137**, S38–S46.

Dolman, P. M. & Sutherland, W. J. (1995b). Spatial patterns of depletion imposed by foraging vertebrates: theory, review and meta-analysis. *Journal of Animal Ecology* **66**, 481–494.

Domjan, M. (1973). Role of ingestion in odor-toxicosis learning in the rat. *Journal of Comparative and Physiological Psychology* **84**, 507–521.

Donald, P. F., Wilson, J. D. & Shepherd, M. (1994). The decline of the corn bunting. *British Birds* **87**, 106–132.Done, T. J. (1992). Phase-shifts in coral-reef communities and their ecological significance. *Hydrobiologia* **247**, 121–132.

Dover, J. W. (1994). Arable field margins: factors affecting butterfly distribution and abundance. In *Field margins: integrating agriculture and conservation* (Ed. N. D. Boatman), pp. 59–66. BCPC, Farnham.

Dragoin, W. B. (1971). Conditioning and extinction of taste aversions with variations in intensity of the CS and US in two strains of rats. *Psychonomic Science* **22**, 303–304.

Draulans, D. & van Vessem, J. (1985). The effect of disturbance on nocturnal abundance and behaviour of grey herons, *Ardea cinerea*, at a fish farm in winter. *Journal of Animal Ecology* **22**, 19–27.

Drewien, R. C. & Bizeau, E. G. (1978). Cross-fostering whooping cranes to sandhill cranes foster parents. In *Endangered birds: management techniques for preserving threatened species* (Ed. S. A. Temple), pp. 201–222. University of Wisconsin Press, Madison.

Dulvy, N. K. & Reynolds, J. D. (1997). Evolutionary transitions among egg-laying, live-bearing and maternal inputs in sharks and rays. *Proceedings of the Royal Society of London B* **264**, 1309–1315.

Duncan, I. J. H. (1993). Welfare is to do with what animals feel. *Journal of Agricultural and Environmental Ethics* **6** (suppl. 2), 8–14.

Duncan, N. (1978). The effects of culling herring gulls (*Larus argentatus*) on recruitment and population dynamics. *Journal of Applied Ecology* **15**, 697–713.

Dunstone, N. & O'Sullivan, J. N. (1996). The impact of ecotourism development on rainforest mammals. In *The exploitation of mammalian populations* (Eds V. J. Taylor & N. Dunstone), pp. 243–261. Chapman & Hall, London.

Durant, S. M. (1998a). A minimum intervention approach to conservation: the influence of structure. In *Behavioural ecology and conservation biology* (Ed. T. M. Caro). Oxford University Press, Oxford.

Durant, S. M. (1998b). Competition refuges and coexistence: an example from Serengeti carnivores. *Journal of Animal Ecology* **67**, 370–386.

Durant, S. M. & Harwood, J. (1992). Assessment of monitoring and management strategies for local populations of the Mediterranean monk seal *Monachus monachus*. *Biological Conservation* **61**, 81–92.

Durant, S. M. & Mace, G. M. (1994). Species differences and population structure in population viability analysis. In *Creative conservation: interactive management of wild and captive animals* (Eds P. J. S. Olney, G. M. Mace & A. T. C. Feistner), pp. 67–91. Chapman & Hall, London.

Durning, A. T. (1993). *Guardians of the land: indigenous peoples and the health of the earth*. Worldwatch Paper 12, Washington, DC.

van Dyk, G., Schoeman, F., Somers, M., van Staden, P., du Toit, K., Verbekmoes, W. & Wambua, J. (1997). Reintroduction and translocation working group report. In *African wild dog* (Lycaon pictus) *population and habitat viability assessment report* (Eds M. G. L. Mills, S. Ellis, R. Woodroffe, D. W. Macdonald, P. Fletcher, M. Bruford, D. Wildt & U. S. Seal). IUCN/SSC Conservation Breeding Specialist Group, Apple Valley, MN.

Dyson-Hudson, R. & Dyson-Hudson, N. (1969). Subsistence herding in Uganda. *Scientific American* **220**, 76–89.

Eaton, R. L. (1974). *The cheetah: biology, ecology and behaviour of an endangered species*. Van Nostrand Reinhold, New York.

Eisenberg, J. F. (1981). *The mammalian radiations*. University of Chicago Press, Chicago.

Ellis, D. H., Dobrott, S. J. & Goodwin, J. G., Jr. (1978). Reintroduction techniques for masked bobwhite. In *Endangered birds: management techniques for preserving threatened species* (Ed. S. A. Temple), pp. 345–354. University of Wisconsin Press, Madison.

Ellis, J. E. & Swift, D. M. (1988). Stability of African pastoral ecosystems: alternate paradigms and implications for development. *Journal of Range Management* **41**, 450–459.

Ellis, R. J. (1964). Tracking raccoons by radio. *Journal of Wildlife Management* **28**, 363–368.

Emlen J. T. & DeJong, M. J. (1992). Counting birds, the problems of variable hearing abilities. *Journal of Field Ornithology* **63**, 26–31.

Endler, J. A. (1980). Natural selection on colour patterns in *Poecilia reticulata*. *Evolution* **34**, 76–91.

Ens, B. J. & Cayford, J. T. (1996). Feeding with other oystercatchers. In *The oystercatcher: from individuals to populations* (Ed. J. D. Goss-Custard), pp. 77–104. Oxford University Press, Oxford.

Ens, B. J. & Goss-Custard, J. D. (1984). Interference among oystercatchers, *Haematopus ostralegus*, feeding on mussels, *Mytilus edulis*, on the Exe estuary. *Journal of Animal Ecology* **53**, 217–231.

Ens, B. J., Kersten, M., Brenninkmeijer, A. & Hulscher, J. B. (1992). Territory quality, parental effort and reproductive success of oystercatchers (*Haematopus ostralegus*). *Journal of Animal Ecology* **61**, 703–715.

Ens, B. J., Piersma, T. & Drent, R. H. (1994). The dependence of waders and waterfowl migrating along the East Atlantic Flyway on their coastal food supplies: what is the most profitable research programme? *Ophelia* (suppl.) **6**, 127–151.

Ens, B. J., Weissing, F. J. & Drent, R. H. (1995). The despotic distribution and deferred maturity – two sides of the same coin. *American Naturalist* **146**, 625–650.

Ensminger, J. & Knight, J. (1997). Changing social norms: common property, bridewealth, and clan exogamy. *Current Anthropology* **38**, 1–24.

Erlinge, S., Hoogenboom, I., Agrell, J., Nelson, J. & Sandell, M. (1990). Density-related home-range size and overlap in adult field voles (*Microtus agrestis*) in Southern Sweden. *Journal of Mammalogy* **71**, 597–603.

Erwin, T. L. (1983). Beetles and other insects of tropical forest canopies at Manaus, Brazil, sampled by insecticidal fogging. In *The tropical rain forest. British Ecological Society Symposium* (Eds S. L. Sutton, A. C. Chadwick & T. C. Whitmore), pp. 59–75. Blackwell Scientific Publications, Oxford.

Erwin, T. L. (1991). How many species are there – revisited. *Conservation Biology* **5**, 330–333.

Essen, L. (1991). A note on the lesser white-fronted goose *Anser erythropus* in Sweden and the result of a re-introduction scheme. *Ardea* **79**, 305–306.

Estep, D. Q. & Dewsbury, D. A. (1996). Mammalian reproductive behavior. In *Wild mammals in captivity* (Eds D. G. Kleiman, M. E. Allen, K. V. Thompson & S. Lumpkin), pp. 379–389. University of Chicago Press, Chicago.

Etheridge, B., Summers, R. W. & Green R. E. (1997). The effects of illegal killing and destruction of nests by humans on the population dynamics of the hen harrier *Circus cyaneus* in Scotland. *Journal of Applied Ecology* **34**, 1081–1105.

Evans, M. R. & Thomas, A. L. R. (1992). The aerodynamic and mechanical effects of elongated tails in the scarlet-tufted malachite-sunbird: measuring the cost of a handicap. *Animal Behaviour* **43**, 337–347.

Evans, P. R. (1991). Seasonal and annual patterns of mortality in migratory shorebirds: some conservation implications. In *Bird population studies. Relevance to conservation and management* (Eds C. M. Perrins, J.-D. Lebreton & G. J. M. Hirons), pp. 346–359. Oxford University Press, Oxford.

Evans, P. R. & Pienkowski, M. W. (1984). Population dynamics of shorebirds. In *Shorebirds. Breeding behavior and populations* (Eds J. Burger & B. L. Olla), pp. 83–123. Plenum Press, New York.

Evermann, J. F., Heeney, J. L., Roelke, M. E., McKeirnan, A. J. & O'Brien, S. J.

(1988). Biological and pathological consequences of feline infectious peritonitis virus infection in the cheetah. *Archives of Virology* **102**, 155–171.

Evermann, J. F., Laurenson, M. K., McKeirnan, A. J. & Caro, T. M. (1993). Infectious disease surveillance in captive and free-living cheetahs: an integral part of the Species Survival Plan. *Zoo Biology* **12**, 125–133.

Falconer, D. S. (1989). *Introduction to quantitative genetics*, 3rd edn. Longman, Harlow.

Falls, J. B. (1981). Mapping territories with playback: an accurate census method for songbirds. In *Estimating the numbers of terrestrial birds* (Eds C. J. Ralph & J. M. Scott), pp. 86–91. Allen Press, Lawrence, KS.

Falls, J. B. (1982). Individual recognition by sound in birds. In *Acoustic communication in birds*, vol. 2 (Eds D. E. Kroodsma & E. H. Miller), pp. 237–278. Academic Press, New York.

Fanshawe, J. H., Ginsberg, J. R., Sillero-Zubiri, C. & Woodroffe, R. (1997). The status and distribution of remaining wild dog populations. In *The African wild dog: status survey and conservation action plan* (Eds R. Woodroffe, J. R. Ginsberg & D. W. Macdonald). IUCN, Gland.

Felsenstein, J. (1985). Phylogenies and the comparative method. *American Naturalist* **125**, 1–15.

Fenn, M. G. P. & Macdonald, D. W. (1995). Use of middens by red foxes: risk reverses rhythms of rats. *Journal of Mammalogy* **76**, 130–136.

Ferreras, P., Aldama, J. J., Beltrán, J. F. & Delibes, M. (1992). Rates and causes of mortality in a fragmented population of Iberian lynx *Felis pardina* Temminck, 1824. *Biological Conservation* **61**, 197–202.

Ferrière, R., Sarrazin, F., Legendre, S. & Baron, J.-P. (1996). Matrix population models applied to viability analysis and conservation: theory and practice using ULM software. *Acta Oecologica* **17**, 629–656.

Ferson, S. & Akçakaya, H. R. (1990). *Modeling fluctuations in age-structured populations: Ramas/Age use manual*. Applied Biomathematics, Setauket, New York.

Fisher, R. A. (1930). *The genetical theory of natural selection*. Clarendon, Oxford.

FitzSimmons, N. N., Buskirk, S. W. & Smith, M. H. (1995). Population history, genetic variability, and horn growth in bighorn sheep. *Conservation Biology* **9**, 314–323.

Fleming, I. A. (1994). Captive breeding and the conservation of wild salmon populations. *Conservation Biology* **8**, 886–888.

Fleming, I. A. (1995). Reproductive success and the genetic threat of cultured fish to wild populations. In *Protection of aquatic biodiversity. Proceedings of the world fisheries congress, Theme 3* (Eds D. P. Philipp, J. M. Epifanio, J. E. Marsden & J. E. Claussen), pp. 117–135. IBH, New Delhi and Oxford.

Fleming, I. A. & Gross, M. R. (1989) Evolution of adult female life history and morphology in a Pacific salmon (coho: *Oncorhynchus kisutch*). *Evolution* **43**, 141–157.

Fleming, I. A., Jonsson, B., Gross, M. R. & Lamberg, A. (1996). An experimental study of the reproductive behaviour and success of farmed and wild Atlantic salmon (*Salmo salar*). *Journal of Applied Ecology* **33**, 893–905.

Flickinger, E. L., King, K. A., Stout, W. F. & Mohn, M. M. (1980). Wildlife hazards from Furadan 3G applications to rice in Texas. *Journal of Wildlife Management*

44, 190–197.

Fogarty, M. J., Sissenwine, M. P. & Cohen, E. B. (1991). Recruitment variability and the dynamics of exploited marine populations. *Trends in Ecology and Evolution* **6**, 241–246.

Folstad, I. & Karter, A. J. (1992). Parasites, bright males, and the immunocompetence handicap. *American Naturalist* **139**, 603–622.

Foose, T. J., Lande, R., Flesness, N. R., Rabb, G. & Read, B. (1986). Propagation plans. *Zoo Biology* **5**, 139–146.

Forbes, G. J. & Theberge, J. B. (1995). Influences of a migratory deer herd on wolf movements and mortality in and near Algonquin Park, Ontario. In *Ecology and conservation of wolves in a changing world* (Eds L. N. Carbyn, S. H. Fritts & D. R. Seip). Canadian Circumpolar Institute, Edmonton.

Ford, J. K. B. (1991). Vocal traditions among resident killer whales (*Orcinus orca*) in coastal waters of British Columbia. *Canadian Journal of Zoology* **69**, 1454–1483.

Forney, K. A. & Gilpin, M. E. (1989). Spatial structure and population extinction: a study with *Drosophila* flies. *Conservation Biology* **3**, 45–51.

Forthman-Quick, D., Gustavson, C. R. & Rusiniak, K. W. (1985). Coyote control and taste aversion. *Appetite* **6**, 253–264.

Frank, L. G. (1986). Social organization of the spotted hyena (*Crocuta crocuta*). 1. Demography. *Animal Behaviour* **34**, 1500–1509.

Frank, L. G., Holekamp, K. E. & Smale, L. (1995). Dominance, demography and reproductive success of female spotted hyenas. In *Serengeti. II. Dynamics, management, and conservation of an ecosystem* (Eds A. R. E. Sinclair & P. Arcese). Chicago University Press, Chicago.

Frankham, R. (1995a). Conservation genetics. *Annual Review of Genetics* **29**, 305–327.

Frankham, R. (1995b). Inbreeding and extinction: a threshold effect. *Conservation Biology* **9**, 792–799.

Frankham, R. (1995c). Effective population size/adult population size ratios in wildlife: a review. *Genetical Research, Cambridge* **66**, 95–107.

Franklin, A. & Pickett, G. D. (1978). Studies on the indirect effects of fishing on stocks of cockles, *Cardium edule*, in the Thames Estuary and Wash. *Fishery Research Technical Report, MAFF, Directorate of Fisheries Research Lowestoft* **42**, 1–9.

Franklin, I. R. (1980). Evolutionary change in small populations. In *Conservation biology: an evolutionary–ecological perspective* (Eds M. E. Soulé & B. A. Wilcox), pp. 135–149. Sinauer, Sunderland, MA.

Franks, N. R. (1982a). Ecology and population regulation in the army ant *Eciton burchelli*. In *The ecology of a tropical forest, seasonal rhythms and long-term changes* (Eds E. G. Leigh, A. Stanley & D. M. Windsor), pp. 389–395. Smithsonian Institution Press, Washington, DC.

Franks, N. R. (1982b). A new method for censusing animal populations – the number of *Eciton burchelli* army ant colonies on Barro-Colorado Island, Panama. *Oecologia* **52**, 266–268.

Franks, N. R. (1985). Reproduction, foraging efficiency and worker polymorphism in army ants. In *Experimental behavioural ecology* (Eds B. Hölldobler & M. Lindauer), pp. 91–107. G. Fisher Verlag, Stuttgart.

Franks, N. R. (1989). Army ants: a collective intelligence. *American Scientist* 77, 138–145.

Franks, N. R. & Bossert, W. H. (1983). Swarm raiding army ants and the patchiness and diversity of a tropical leaf litter ant community. In *The tropical rain forest. British Ecological Society Symposium* (Eds S. L. Sutton, A. C. Chadwick & T. C. Whitmore), pp. 151–163. Blackwell Scientific Publications, Oxford.

Franks, N. R. & Fletcher, C. R. (1983). Spatial patterns in army ant foraging and migration. *Eciton burchelli* on Barro Colorado Island, Panama. *Behavioural Ecology and Sociobiology* 12, 261–270.

Franks, N. R. & Hölldobler, B. (1987). Sexual competition during colony reproduction in army ants. *Biological Journal of the Linnean Society* 30, 229–243.

Franks, N. R. & Partridge, L. W. (1993). Lanchester battles and the evolution of combat in ants. *Animal Behaviour* 45, 197–199.

Fraser, D., Weary, D. M., Pajor, E. A. & Milligan, B. N. (1997). A scientific conception of animal welfare that reflects ethical concerns. *Animal Welfare* 6, 187–205.

Fratkin, E. (1997). Pastoralism: governance and development issues. *Annual Reviews in Anthropology* 26, 235–261.

Fratkin, E. & Roth, E. A. (1990). Drought and economic differentiation among Ariaal pastoralists of Kenya. *Human Ecology* 18, 385–402.

Fratkin, E. & Smith, K. (1994). Labor, livestock and land: the organization of pastoral production. In *African pastoralist systems: an integrated approach* (Eds E. Frakin, K. A. Galvin & E. A. Roth), pp. 91–112. Lynne Reinner Publications, Boulder, CO.

Fretwell, S. D. & Lucas, H. L. (1970). On territorial behaviour and other factors influencing habitat distribution in birds. *Acta Biotheoretica* 19, 16–36.

Friedlander, A., Beets, J. & Tobias, W. (1994). Effects of fish aggregating device design and location on fishing success in the US Virgin Islands. *Bulletin of Marine Science* 55, 592–601.

Fry, G. L. A. & W. J. Robson (1994). The effects of field margins on butterfly movement. In *Field margins: integrating agriculture and conservation* (Ed. N. D. Boatman), pp. 111–116. BCPC, Farnham.

Fuller, R. J., Gregory, R.D., Gibbons, D.W., Marchant, J. H., Wilson, J. D., Baillie, S. R. & Carter, N. (1995). Population declines and range contractions among lowland farmland birds in Britain. *Conservation Biology* 9, 1425–1441.

Fuller, T. K. (1989). Population dynamics of wolves in North-Central Minnesota. *Wildlife Monographs* 105, 1–41.

Fuller, T. K. & Keith, L. B. (1980). Wolf population dynamics and prey relationships in Northeastern Alberta. *Journal of Wildlife Management* 44, 583–602.

Furness, R. W. (1996). A review of seabird responses to natural or fisheries-induced changes in food supply. In *Aquatic predators and their prey* (Eds S. P. R. Greenstreet & M. L. Tasker), pp. 168–173. Blackwell Scientific Publications, Oxford.

Furness, R. W. & Camphuysen, C. J. (1997). Seabirds as monitors of the marine environment. *ICES Journal of Marine Science* 54, 726–737.

Gabriel, W. & Burger, R. (1992). Survival of small populations under demographic stochasticity. *Theoretical Population Biology* 41, 44–71.

Gadgil, M. & Berkes, F. (1991). Traditional resource management systems. *Resource*

Management and Optimization 8, 127–141.

Gaines, M. S. & McClenaghan, L. R. (1980). Dispersal in small mammals. *Annual Review of Ecology and Systematics* 11, 163–196.

Galaty, J. G. (1994). Rangeland tenure and pastoralism in Africa. In *African pastoralist systems: an integrated approach* (Eds E. Fratkin, K. A. Galvin & E. A. Roth), pp. 185–204. Lynne Reinner Publications, Boulder, CO.

Galef, B. G. & Osborne, B. (1978). Novel taste facilitation of the association of visual cues with toxicosis in rats. *Journal of Comparative and Physiological Psychology* 92, 907–916.

Galeotti, P. & Pavan, G. (1991). Individual recognition of male tawny owls (*Strix aluco*) using spectrograms of their territorial calls. *Ethology, Ecology & Evolution* 3, 113–126.

Galeotti, P., Saino, N., Sacchi, R. & Mller, A. P. (1997). Song correlates with social context, testosterone and body condition in male barn swallows. *Animal Behaviour* 53, 687–700.

Game, M. (1980). Best shapes for nature reserves. *Nature* 287, 630–632.

Gao, Y. & Short, R. V. (1993). The control of rodent populations. *Oxford Review of Reproductive Biology* 15, 265–310.

Garcia, J. & Koelling, R. A. (1967). A comparison of aversions induced by X-rays, toxins and drugs in the rat. *Radiation Research Supplement* 7, 439–450.

Garcia, J., Kimeldorf, D. J. & Koelling, R. A. (1955). A conditioned aversion to saccharin resulting from exposure to gamma radiation. *Science* 122, 157–159.

Garcia, J., Hankins, W. G. & Rusiniak, W. (1974). Behavioral regulation of the milieu interne in man and rat. *Science* 185, 824–831.

Garner, R. (Ed.) (1996). *Animal rights: the changing debate*. Macmillan, Basingstoke.

Garshelis, D. L. & Pelton, M. R. (1981). Movements of black bears in the Great Smoky Mountains National Park. *Journal of Wildlife Management* 41, 912–925.

Garthe, S., Camphuysen, C. J. & Furness, R. W. (1996). Amounts of discards by commercial fisheries and their significance as food for seabirds in the North Sea. *Marine Ecology Progress Series* 136, 1–11.

Gausen, D. & Moen, V. (1991). Large-scale escapes of farmed Atlantic salmon (*Salmo salar*) into Norwegian rivers threaten natural populations. *Canadian Journal of Fisheries and Aquatic Sciences* 48, 426–428.

Gerritsen, A. F. C. & Meijboom, A. (1986). The role of touch in prey density estimation by *Calidris alba*. *Netherlands Journal of Zoology* 36, 530–562.

Gese, E. M., Rongstad, O. J. & Mytton, W. R. (1989). Population dynamics of coyotes in southeastern Colorado. *Journal of Wildlife Management* 53, 174–181.

Gibbons, D.W., Avery, M. I. & Brown, A. F. (1996). Population trends of breeding birds in the United Kingdom since 1800. *British Birds* 89, 291–305.

Gibbs, J. P. & Wenny, D. G. (1993). Song output as a population estimator: effect of male pairing status. *Journal of Field Ornithology* 64, 316–322.

Gibson, C. C. & Marks, S. A. (1995). Transforming rural hunters into conservationists: an assessment of community-based wildlife management programs in Africa. *World Development* 23, 941–956.

Gilbert, G. (1993). Vocal individuality as a census and monitoring tool: practical considerations illustrated by a study of the bittern *Botaurus stellaris* and the black-throated diver *Gavia arctica*. Unpublished PhD thesis, University of Nottingham, UK.

Gilbert, G., McGregor, P. K. & Tyler, G. (1994). Vocal individuality as a census tool: practical considerations illustrated by a study of two rare species. *Journal of Field Ornithology* 65, 335–348.

Gilbert, G., Gibbons, D. W. & Evans, J. (1998). *Bird monitoring manual: a compendium of monitoring and survey techniques for birds of conservation concern and others in the UK.* RSPB/BTO/WWT/JNCC/ITE/Seabird Group, Sandy, UK.

Gilbert, L. E. & Singer, M. C. (1973). Dispersal and gene flow in a butterfly species. *American Naturalist* 107, 58–72.

Gill, E. L., Serra, M. B., Canavelli, S. B., Feare, C. J., Zaccagnini, M. E., Nadian, A. K., Heffernan, M. L. & Watkins, R. W. (1994). Cinnamamide prevents captive chestnut-capped blackbirds from eating rice. *International Journal of Pest Management* 40, 195–198.

Gill, J. A. (1996). Habitat choice in wintering pink-footed geese: quantifying the constraints determining winter site use. *Journal of Applied Ecology* 33, 884–892.

Gill, J. A., Sutherland, W. J. & Watkinson, A. R. (1996). A method to quantify the effects of human disturbance on animal populations. *Journal of Applied Ecology* 33, 786–792.

Gill, R. E., Canevari, P. & Iversen, E. H. (1998). Eskimo curlew (*Numenius borealis*). In *The birds of North America* (Eds A. Poole & F. Gill), 347, 1–28. The Academy of Sciences, Philadelphia, and The American Ornithologists' Union, Washington, DC.

Gilpin, M. & Hanski, I. (Eds) (1991). *Metapopulation dynamics: empirical and theoretical investigations.* Academic Press, London.

Gilpin, M. E. & Soulé, M. E. (1986). Minimum viable populations: processes of species extinction. In *Conservation biology: the science of scarcity and diversity* (Ed. M. E. Soulé), pp. 19–34. Sinauer, Sunderland, MA.

Ginsberg, J. R. & Macdonald, D. W. (1990). *Foxes, wolves, jackals and dogs: an action plan for the conservation of canids.* IUCN, Gland.

Ginzburg, L. R., Slobodkin, L. B., Johnson, K. & Bindman, A. G. (1982). Quasiextinction probabilities as a measure of impact on population growth. *Risk Analysis* 2, 171–181.

Gipps, J. H. W. (Ed.) (1991). Beyond captive breeding: re-introducing endangered mammals to the wild. *Zoological Society of London Symposium* No. 62, 1–278.

Gittleman, J. L. & Harvey, P. H. (1982). Carnivore home range size, metabolic needs and ecology. *Behavioural Ecology and Sociobiology* 10, 57–63.

Given, D. R. (1993). What is conservation biology and why is it so important? *Journal of the Royal Society of New Zealand* 23, 55–60.

Gollop, J. B., Barry, T. W. & Iversen, E. H. (1986). *Eskimo curlew: a vanishing species?* Saskatchewan Natural History Society, Regina.

Gompper, M. E. & Gittleman, J. L. (1991). Home range scaling: intraspecific and comparative trends. *Oecologia* 87, 343–348.

Goodman, D. (1987). The demography of chance extinction. In *Viable populations for conservation* (Ed. M. E. Soulé), pp. 11–34. Cambridge University Press, Cambridge.

Gordon, H. S. (1954). The economic theory of a common property resource: the fishery. *Journal of Political Economy* 62, 124–142.

Gordon, J. & Moscrop, A. (1996). Underwater noise pollution and its significance for whales and dolphins. In *The conservation of whales and dolphins: science and practice* (Eds M. P. Simmonds & J. Hutchinson), pp. 281–319. Wiley, Chichester.

Gosling, L. M. & Baker, S. J. (1989a). Demographic consequences of differences in the ranging behaviour of male and female coypus. In *Mammals as pests* (Ed. R. J. Putman), pp. 155–167. Chapman & Hall, London.

Gosling, L. M. & Baker, S. J. (1989b). The eradication of muskrats and coypus from Britain. *Biological Journal of the Linnean Society* **38**, 39–51.

Goss-Custard, J. D. (1977). The ecology of the Wash. III. Density-related behaviour and the possible effects of a loss of feeding grounds on wading birds (Charadrii). *Journal of Applied Ecology* **14**, 721–739.

Goss-Custard, J. D. (1980). Competition for food and interference among waders. *Ardea* **68**, 31–52.

Goss-Custard, J. D. (1993). The effect of migration and scale on the study of bird populations: 1991 Witherby Lecture. *Bird Study* **40**, 81–96.

Goss-Custard, J. D. (1996) Introduction. In *The oystercatcher: from individuals to populations* (Ed. J. D. Goss-Custard), pp. 1–4. Oxford University Press, Oxford.

Goss-Custard, J.D. & Durell, S.E.A. le V. dit (1984). Feeding ecology, winter mortality and the population dynamics of oystercatchers, *Haematopus ostralegus*, on the Exe estuary. In *Coastal waders and wildfowl in winter* (Eds P. R. Evans, J. D. Goss-Custard & W. G. Hale), pp. 190–208. Cambridge University Press, Cambridge.

Goss-Custard, J. D. & Durell, S. E. A. le V. dit (1987). Age-related effects in oystercatchers, *Haematopus ostralegus*, feeding on mussels, *Mytilus edulis*. III. The effect of interference on overall intake rates. *Journal of Animal Ecology* **56**, 549–558.

Goss-Custard, J. D. & Sutherland, W. J. (1997). Individual behaviour, populations and conservation. In *Behavioural ecology: an evolutionary approach*, 4th edn (Eds J. R. Krebs & N. B. Davies), pp. 373–395. Blackwell Science, Oxford.

Goss-Custard, J. D. & Willows, R. (1996). Modelling the responses of mussel *Mytilus edulis* and oystercatcher *Haematopus ostralegus* populations to environmental change. In *Aquatic predators and their prey* (Eds S. P. R. Greenstreet & M. L. Tasker), pp. 73–85. Blackwell Science, Oxford.

Goss-Custard, J. D., Caldow, R. W. G., Clarke, R. T., Durell, S. E. A. le V. dit, Urfi, A. J. & West, A. D. (1995a). Consequences of habitat loss and change to populations of wintering migratory birds: predicting the local and global effects from studies of individuals. *Ibis* **137** (suppl.), S56–S66.

Goss-Custard, J. D., Caldow, R. W. G., Clarke, R. T., Durell, S. E. A. le V. dit & Sutherland, W. J. (1995b). Deriving population parameters from individual variation in foraging behaviour. I. Empirical game theory distribution model of oystercatchers *Haematopus ostralegus* feeding on mussels *Mytlius edulis*. *Journal of Animal Ecology* **64**, 265–276.

Goss-Custard, J. D., Caldow, R. W. G., Clarke, R. T. & West, A. D. (1995c). Deriving population parameters from individual variation in foraging behaviour. II. Model tests and population parameters. *Journal of Animal Ecology* **64**, 277–289.

Goss-Custard, J. D., Clarke, R. T., Briggs, K. B., Ens, B. J., Exo, K.-M., Smit, C. J., Beintema, A. J., Caldow, R. W. G., Catt, D. C., Clark, N. A., Durell, S. E. A. le V. dit, Harris, M. P., Hulscher, J. B., Meininger, P. L., Picozzi, N., Prys-Jones, R. P., Safriel, U. N. & West, A. D. (1995d). Population consequences of winter habitat loss in a migratory shorebird. I. Estimating model parameters. *Journal of Applied Ecology* **32**, 317–333.

Goss-Custard, J. D., Clarke, R. T. , Durell, S. E. A. le V. dit, Caldow, R. W. G. & Ens, B. J. (1995e). Population consequences of habitat loss and change in wintering migratory birds. II. Model predictions. *Journal of Applied Ecology* **32**, 334–348.

Goss-Custard, J. D., Durell, S.E.A. le V. dit, Clarke, R. T., Beintema, A. J., Caldow, R. W. G., Meininger, P. L. & Smit, C. (1996a). Population dynamics of the oyster-catcher. In *The oystercatcher: from individuals to populations* (Ed. J. D. Goss-Custard), pp. 352–383. Oxford University Press, Oxford.

Goss-Custard, J. D., McGrorty, S. & Durell, S.E.A. le V. dit (1996b). The effect of oystercatchers *Haematopus ostralegus* on shellfish populations. *Ardea* **84**, 453–468.

Goss-Custard, J. D., West, A. D. & Sutherland, W. J. (1996c). Where to feed. In *The oystercatcher: from individuals to populations* (Ed. J. D. Goss-Custard), pp. 105–132. Oxford University Press, Oxford.

Gossow, H. (1970). Vergleichende verhaltensstudien an Marderartigen. I. Uber Lautau Berungen und zum Beuteurhalten. *Zeitschrift fur. Tierpsychologie* **27**, 405–480.

Gotwald, W. H. (1995). Army ants: the biology of social predation. *Oecologia* **52**, 266–268.

Gould, S. J. (1997). Cope's rule as psychological artefact. *Nature* **385**, 199–200.

Grabowski, R. (1988). Theory of induced institutional innovation: a critique. *World Development* **16**, 385–394.

Gradwohl, J. & Greenberg, R. (1988). *Saving the tropical forests*. Earthscan Publications Ltd, London.

Green, K. G. & Churchill, P. A. (1970). An effect of flavor on strength of conditioned aversions. *Psychonomic Science* **21**, 19–20.

Green, M. J. B. (1994). Protecting the mountains of Central Asia and their snow leopard populations. In *Proceedings of the seventh international snow leopard symposium in Xining 1992* (Eds J. L. Fox & D. Jizeng), pp. 223–239. International Snow Leopard Trust, Seattle, and Chicago Zoological Society, Chicago.

Green, R. E. (1995). Diagnosing causes of bird population declines. *Ibis* **137** (suppl.), S75–S84.

Green, R. E. & Hirons, G. J. M. (1991). The relevance of population studies to the conservation of threatened birds. In *Bird population studies* (Eds C. M. Perrins, J. D. Lebreton & G. J. M. Hirons), pp. 594–633. Oxford University Press, Oxford.

Green, R. E., Pienkowski, M. W. & Love, J. A. (1996). Long-term viability of the re-introduced population of the white-tailed eagle *Haliaeetus albicilla* in Scotland. *Journal of Applied Ecology* **33**, 357–368.

Green, W. Q. (1984). A review of ecological studies relevant to management of the common brushtail possum. In *Possums and gliders* (Eds A. P. Smith & I. D. Hume), pp. 483–499. Australian Mammal Society, Sydney.

Greenspan, B. N. (1980). Male size and reproductive success in the communal courtship system of the fiddler crab *Uca rapax*. *Animal Behaviour* **28**, 387–392.

Greenwood, J. J. D., Baillie, S.R., Gregory, R. D., Peach, W. J. & Fuller, R. J. (1995). Some new approaches to conservation monitoring of British breeding birds. *Ibis* **137** (suppl.), S16–S28.

Greenwood, P. J. (1980). Mating systems, philopatry, and dispersal in birds and mammals. *Animal Behaviour* **28**, 1140–1162.

Greenwood, P. J. & Harvey, P. H. (1978). Inbreeding and dispersal in the great tit. *Nature* **271**, 52–54.

Grenfell, B. & Harwood, J. (1997). (Meta)population dynamics of infectious diseases. *Trends in Ecology and Evolution* **12**, 395–399.

Gretton, A. (1991). *The ecology and conservation of the slender-billed curlew* (Numenius tenuirostris). International Council for Bird Preservation, Cambridge.

Griffin, J. F. T., Thomson, A. J., Cross, J. P., Buchan, G. S. & Mackintosh C. G. (1992). The impact of domestication on red deer immunity and disease resistance. In *The biology of deer* (Ed. R. D. Brown). Springer, New York.

Griffith, B., Scott, J. M., Carpenter, J. W. & Reed, C. (1989). Translocation as a conservation tool: status and strategy. *Science* **245**, 477–480.

Groom, M. J. & Pascual, M. J. (1998). The analysis of population persistence: an outlook on the practice of viability analysis. In *Conservation biology for the coming decade* (Eds P. L. Fielder & P. Kariva), pp. 4–27. Chapman & Hall, New York.

Gros, P. M. (1996). The status of the cheetah in Malawi. *Nyala* **19**, 33–38.

Gros, P. M. (1997). Conservation status of cheetahs in East Africa: estimation, determinants and projections. PhD thesis, University of California, Davis, USA.

Gros, P. M. (1998). Status of the cheetah *Acinonyx jubatus* in Kenya: a field-interview assessment. *Biological Conservation* **85**, 137–149.

Gurney, J. E., Watkins, R. W., Gill, E. L. & Cowan, D. P. (1996). Non-lethal mouse repellents: evaluation of cinnamamide as a repellent against commensal and field rodents. *Applied Animal Behaviour Science* **49**, 353–363.

Gustavson, C. R. (1977). Comparative and field aspects of learned food aversions. In *Learning mechanisms in food selection* (Eds L. M. Barker, M. R. Best & M. Domjan), pp. 23–43. Baylor University Press, USA.

Gustavson, C. R., Garcia, J., Hankins, W. G. & Rusiniak, K. W. (1974). Coyote predation: control by aversive conditioning. *Science* **184**, 581–583.

Gustavson, C. R., Kelly, D. J., Sweeney, M. & Garcia, J. (1976). Prey-lithium aversions. I. Coyotes and wolves. *Behavioural Biology* **17**, 61–72.

Gustavson, C. R., Gustavson, J. C. & Holzer, G. A. (1983). Thiabendazole-based taste aversion in dingoes and New Guinea wild dogs. *Applied Animal Ethology* **10**, 385–388.

Gwinner, E. (1986). *Circannual rhythms. Endogenous annual clocks in the organization of seasonal processes.* Springer-Verlag, Berlin.

Haig, S. M. & Avise, J. C. (1996). Avian conservation genetics. In *Conservation genetics. Case histories from nature* (Eds J. C. Avise & J. L. Hamrick), pp. 160–189. Chapman & Hall, New York.

Hamer, K. C., Furness, R. W. & Caldow, R. W. G. (1991). The effects of changes in food availablity on the breeding ecology of great skuas *Catharacta skua* in Shetland. *Journal of Zoology* **223**, 175–188.

Hames, R. (1987). Game conservation or efficient hunting. In *The question of the commons. The culture and ecology of communal resources* (Eds B. J. McCay & J. M. Acheson), pp. 92–107. University of Arizona Press, Tucson.

Hames, R. (1991). Wildlife conservation in tribal societies. In *Biodiversity: culture, conservation and ecodevelopment* (Eds M. L. Oldfield & J. B. Alcorn), pp. 172–199. Westview Press, Boulder, CO.

Hamilton, W. J. (1936). Rats and their control. *Bulletin of New York State Agricultural Experimental Station, Cornell* No. 353, 1–32.

Hamilton, W. H. & Howard, J. (Eds) (1997). *Infection, polymorphism and evolution.* Chapman & Hall, London.

Hammond, P. S., Mizroch, S. A. & Donovan, G. P. (1990). *Individual recognition of cetaceans, use of photo-identification and other techniques to estimate population parameters. Report of IWC*, Special Issue 12. International Whaling Commission, Cambridge.

Hamre, J. (1991). Interrelation between environmental changes and fluctuating fish populations in the Barents Sea. In *Long term variability of pelagic fish populations and their environment* (Eds T. Kawasaki, S. Tanaka, Y. Toba & A. Taniguchi), pp. 259–270. Pergamon Press, Oxford.

Hanna, S., Folke, C. & Maler, K.-G. (Eds) (1996). *Rights to nature: ecological, economic, cultural, and political principles of institutions for the environment.* Island Press, Washington, DC.

Hanski, I. (1991). Single-species metapopulation dynamics: concepts, models and observations. *Biological Journal of the Linnean Society* **42**, 17–38.

Hanski, I. (1994). A practical model of metapopulation dynamics. *Journal of Animal Ecology* **63**, 151–162.

Hanski, I. & Gilpin, M. (1991). Metapopulation dynamics: brief history and conceptual domain. *Biological Journal of the Linnean Society* **42**, 3–16.

Hanski, I. A. & Gilpin, M. E. (Eds) (1997). *Metapopulation dynamics: ecology, genetics, and evolution.* Academic Press, London and San Diego.

Hanski, I. & Thomas, C. D. (1994). Metapopulation dynamics and conservation: a spatially explicit model applied to butterflies. *Biological Conservation* **68**, 167–180.

Hanski, I. & Zhang, D.-Y. (1993). Migration, metapopulation dynamics and fugitive co-existence. *Journal of Theoretical Biology* **163**, 491–504.

Hanski, I., Kuussaari, M. & Nieminen, M. (1994). Metapopulation structure and migration in the butterfly *Melitaea cinxia*. *Ecology* **75**, 747–762.

Harcourt, A. H. (1996). Is the gorilla a threatened species – how should we judge? *Biological Conservation* **75**, 165–176.

Harcourt, A. H., Stewart, K. J. & Fossey, D. (1976). Male emigration and female transfer in wild mountain gorilla. *Nature, London* **263**, 226–227.

Harcourt, A. H., Fossey, D., Stewart, K. J. & Watts, D. P. (1980). Reproduction of wild gorillas and some comparisons with chimpanzees. *Journal of Reproduction and Fertility* (suppl.) **28**, 59–70.

Harcourt, A. H., Fossey, D. & Sabater-Pi, J. (1981). Demography of *Gorilla gorilla*. *Journal of Zoology* **195**, 215–233.

Hardin, G. (1968). The tragedy of the commons. *Science* **162**, 1243–1248.

Harmelin-Vivien, M. L., Harmelin, J. G., Chauvet, C., Duval, C., Galzin, R., Lejeune, P., Barnabe, G., Blanc, F., Chevalier, R., Duclerc, J. & Lassere, G. (1985). Evaluation visuelle des peuplements et populations de poissons: methods et problemes. *Revue d'Ecologie la Terre et la Vie* **40**, 467–539.

Harrington, B. (1996). *The flight of the red knot.* Norton, New York.

Harris, R. B., Metzgar, L. H. & Bevin, C. D. (1986). *GAPPS: generalized animal population projection system.* Version 3.0. *User's manual.* Montana Co-operative Wildlife Research Unit, University of Montana, Missoula.

Harrison, S. (1989). Long-distance dispersal and colonization in the bay checkerspot butterfly, *Euphydryas editha bayensis. Ecology* **70**, 1236–1243.

Harrison, S. (1991). Local extinction in a metapopulation context: an empirical evaluation. *Biological Journal of the Linnean Society* **42**, 73–88.

Harrison, S. (1994). Metapopulations and conservation. In *Large scale ecology and conservation biology* (Eds P. J. Edwards, R. M. May & N. Webb), pp. 111–128. Blackwell Scientific Publications, Oxford.

Hartley, I. R., Shepherd, M., Robson, T. & Burke, T. (1993). Reproductive success of polgynous male corn buntings (*Miliaria calandra*) as confirmed by DNA fingerprinting. *Behavioral Ecology* **4**, 310–317.

Hartshorn, G. (1989). Application of gap theory to tropical forest management – natural regeneration on strip clear cuts in the Peruvian Amazon. *Ecology* **70**, 567–569.

Hartshorn, G. (1995). Ecological basis for suitable development in tropical forests. *Annual Review of Ecology and Systematics* **26**, 155–175.

Harwood, J. & Hall, A. (1990). Mass mortality in marine mammals: its implications for population dynamics and genetics. *Trends in Ecology and Evolution* **5**, 254–257.

Hasselquist, D., Bensch, S. & von Schantz, T. (1996). Correlation between male song repertoire, extra-pair paternity and offspring survival in the great reed warbler. *Nature* **381**, 229–232.

Haug, T., Kroyer, A. B., Nilssen, K. T., Ugland, K. I. & Aspholm, P. E. (1991). Harp seal (*Phoca groenlandica*) invasions of Norwegian coastal waters: age composition and feeding habits. *ICES Journal of Marine Science* **48**, 363–371.

Hauser, M. D. (1996). *The evolution of communication.* MIT Press, Cambridge, MA.

Hayes, R. D. & Gunson, J. R. (1995). Status and management of wolves in Canada. In *Ecology and conservation of wolves in a changing world* (Eds L. N. Carbyn, S. H. Fritts & D. R. Seip). Canadian Circumpolar Institute, Edmonton.

Hedrick, P. W. (1983). *Genetics of populations.* Science Books International, Boston.

Hedrick, P. W. (1987). Genetic bottlenecks. *Science* **237**, 963.

Hedrick, P. W. (1996). Bottleneck(s) or metapopulation in cheetahs. *Conservation Biology* **10**, 897–899.

Hedrick, P. W., Lacy, R. C., Allendorf, F. W. & Soulé, M. E. (1996). Directions in conservation biology: comments on Caughley. *Conservation Biology* **10**, 1312–1320.

Heeney, J. L., Evermann, J. F., McKeirnan, A. J., Marker-Kraus, L., Roelke, M. E., Bush, M., Wildt, D. E., Meltzer, D. G., Colly, L., Lulas, J., Manton, V. J., Caro, T. M. & O'Brien, S. J. (1990). Prevalence and implications of feline coronavirus infections of captive and free-ranging cheetahs (*Acinonyx jubatus*). *Journal of*

Virology 64, 1964–1972.

Helbig, A. J., Martens, J., Seibold, I., Henning, F., Schottler, B., & Wink, M. (1996). Phylogeny and species limits in the Palearctic chiffchaff *Phylloscopus collybita* complex: mitochondrial genetic differentiation and bioacoustic evidence. *Ibis* 138, 650–666.

Henwood, T. A. & Stuntz. W. E. (1987). Analysis of sea turtle captures and mortalities during commercial shrimp trawling. *Fisheries Bulletin* 85, 813–817.

Herren, U. J. (1990). Socioeconomic stratification and small stock production in Mukogodo Division, Kenya. *Research in Economic Anthropology* 12, 111–148.

Hess, E. H. (1959). Imprinting. *Science* 130, 133–141.

Hess, E. H. (1972). The natural history of imprinting. *Annals of New York Academy of Science* 193, 124–136.

Heywood, J. S. (1989). Sexual selection by the handicap mechanism. *Evolution* 43, 1387–1397.

Hilborn, R. (1975). The effect of spatial heterogeneity on the persistence of predator–prey interactions. *Theoretical Population Biology* 8, 346–355.

Hilborn, R. & Walters, C. J. (1992). *Quantitative stock assessment: choice, dynamics and uncertainty.* Chapman & Hall, London.

Hill, D., Hockin, D., Price, D., Tucker, G., Morris, R. & Treweek, J. (1997). Bird disturbance: improving the quality and utility of disturbance research. *Journal of Applied Ecology* 34, 275–288.

Hill, J.K., Thomas, C. D. & Lewis, O. T. (1996). Effects of habitat patch size and isolation on dispersal by *Hesperia comma* butterflies: implications for metapopulation structure. *Journal of Animal Ecology* 65, 725–735.

Hill, K. (1995). Commentary. *Current Anthropology* 36, 789–818.

Hill, W. G. (1979). A note on effective population with overlapping generations. *Genetics* 92, 317–322.

Hills, D. M. & Smithers, R. H. N. (1980). The 'king cheetah': a historical review. *Arnoldia* 9, 1–23.

Hitchcock, C. L. & Gratto-Trevor, C. (1997). Diagnosing a shorebird local population decline with a stage-structured population model. *Ecology* 78, 522–534.

Hitchcock, C. L., Macdonald, D. W. & Cheeseman, C. L. (1999). A simulation of the role of behavioural perturbation in the control of bovine tuberculosis (*Mycobacterium bovis*) in badgers (*Meles meles*). *Journal of Applied Ecology*, in press.

Hobbs, R. J. (1992). The role of corridors in conservation: solution or bandwagon? *Trends in Ecology and Evolution* 7, 389–392.

Hockey, P. A. R. (1994). Man as a component of the littoral predator spectrum: a conceptual overview. In *Rocky shores: exploitation in Chile and South Africa* (Ed. W. R. Siegfried), pp. 17–31. Springer-Verlag, Berlin.

Hockin, D., Ounsted, M., Gorman, M., Hill, D., Keller, V. & Barker, M. A. (1992). Examination of the effects of disturbance on birds with reference to its importance in ecological assessments. *Journal of Environmental Management* 36, 253–286.

Hodgman, T. P., Harrison, D. J., Katnik, D. D. & Elowe, K. D. (1994). Survival in an intensively trapped marten population in Maine. *Journal of Wildlife Management* 58, 593–600.

Hoem, J. (1992). Public policy as the fuel of fertility – effects of a policy reform on the pace of childbearing in Sweden in the 1980s. *Acta Sociologica* **36**, 19–31.

Hofer, H. & East, M. L. (1993). The commuting system of Serengeti spotted hyaenas: how a predator copes with migratory prey. I. Social organization. *Animal Behaviour* **46**, 547–557.

Hofer, H., East, M. L. & Campbell, K. L. I. (1993). Snares, commuting hyaenas and migratory herbivores: humans as predators in the Serengeti. *Zoological Society of London Symposium* No. 65, 347–366.

Hoffmann, A. A. & Parsons, P. A. (1997). *Extreme environmental change and evolution.* Cambridge University Press, Cambridge.

Höglund, J. (1996). Can mating systems affect local extinction risks? Two examples of lek-breeding waders. *Oikos* **77**, 184–188.

Högstedt, G. (1980). Evolution of clutch-size in birds: adaptive variation in relation to territory quality. *Science* **210**, 1148–1150.

Holden, M. J. (1971). The rate of egg laying by three species of ray. *Journal du Conseil, Conseil International de l' Exploration de la Mer* **33**, 335–339.

Holland, J. & McGregor, P. K. (1997). Disappearing song dialects? The case of Cornish corn buntings. *UK Nature Conservation* **13**, 181–185.

Holland, J., McGregor, P. K. & Rowe, C. L. (1996). Microgeographic variation in the song of the corn bunting, *Miliaria calandra*: changes with time. *Journal of Avian Biology* **27**, 47–55.

Holland, K. N., Peterson, J. D., Lowe, C. G. & Wetherbee, B. M. (1993). Movements, distribution and growth rates of the white goatfish *Mulloides flavolineatus* in a fisheries conservation zone. *Bulletin of Marine Science* **52**, 982–992.

Hölldobler, B. & Wilson, E. O. (1990). *The ants.* Belknap Press of Harvard University Press, Cambridge, MA.

Holmgren, N. (1995). The ideal free distribution of unequal competitors: predictions from a behaviour-based functional response. *Journal of Animal Ecology* **64**, 197–212.

Holyoak, M. & Lawler, S. P. (1996). Persistence of an extinction-prone predator-prey interaction through metapopulation dynamics. *Ecology* **77**, 1867–1879.

Homewood, K. M. & Rodgers, W. A. (1991). *Maasailand ecology.* Cambridge University Press, Cambridge.

Hornocker, M. & Bailey, T. (1986). Natural regulation in three species of Felids. In *Cats of the world: biology, conservation, and management* (Eds S. D. Miller & D. D. Everett), pp. 211–220. National Wildlife Federation, Washington, DC.

Hornocker, M. G., Messick, J. P. & Melquist, W. E. (1983). Spatial strategies in three species of *Mustelidae. Acta Zoologica Fennica* **174**, 185–188.

Horwood, J.W. & Goss-Custard, J.D. (1977). Predation by the oystercatcher, *Haematopus ostralegus* (L.), in relation to the cockle, *Cerastoderma edule* (L.), fishery in the Burry Inlet, South Wales. *Journal of Applied Ecology* **14**, 139–158.

Houllier, F. & Lebreton, J. D. (1986). A renewal equation approach to the dynamics of stage-grouped populations. *Mathematical Biosciences* **79**, 185–197.

Houston, A. I., Clark, C. W., McNamara, J. & Mangel, M. (1988). Dynamic models in behavioural and evolutionary ecology. *Nature* **332**, 29–34.

Howell, R. (1985) The effect of bait-digging on the bioavailablity of heavy metals from surficial inter-tidal marine sediments. *Marine Pollution Bulletin* **16**, 292–295.

del Hoyo, J., Elliott, A. & Sargatal, J. (Eds) (1996). *Handbook of the birds of the world*, vol. 3. *Hoatzin to auks*. Lynx Edicions, Barcelona.

Hubback, J. (1957). *Wives who went to college*. Heinnemann, London.

Hubbell, S. P. & Foster, R. B. (1983). Diversity of canopy trees in a neotropical forest and implications for conservation. In *The tropical rain forest. British Ecological Society Symposium* (Eds S. L. Sutton, A. C. Chadwick & T. C. Whitmore), pp. 25–41. Blackwell Scientific Publications, Oxford.

Hubbell, S. P. & Foster, R. B. (1992). Short term dynamics of a neotropical forest: why ecological research matters to tropical conservation and management. *Oikos* 63, 48–61.

Hudson, A. V. & Furness, R. W. (1988). Utilization of discarded fish by scavenging seabirds behind whitefish trawlers in Shetland. *Journal of Zoology* 215, 151–166.

Hudson, A. V., Stowe, T. J. & Aspinall, S. J. (1990). Status and distribution of corncrakes in Britain in 1988. *British Birds* 83, 173–187.

Hudson, E. & Mace, G. (Eds) (1996). *Marine fish and the IUCN red list of threatened animals: report of a workshop held in collaboration with WWF and IUCN at the Zoological Society of London from April 29th–May 1st*. Zoological Society of London, London.

Hughes, T. P. (1994). Catastrophes, phase shifts and large-scale degradation of a Caribbean coral reef. *Science* 265, 1547–1551.

Hunn, E. (1982). Mobility as a factor limiting resource use in the Columbian Plateau of North America. In *Resource managers: North American and Australian hunter–gatherers* (Eds N. Williams & E. Hunn), pp. 17–43. Westview Press, Boulder, CO.

Hunt, G. L., Barrett, R. T., Joiris, C. & Montevecchi, W. A. (1996). Seabird/fish interactions: an introduction. In *Seabird/fish interactions, with particular reference to seabirds in the North Sea ICES Cooperative Research Report* No. 216. (Eds G. L. Hunt & R. W. Furness), pp. 2–5. International Council for the Exploration of the Sea, Copenhagen.

Hutchings, J. A. & Myers, R. A. (1994) What can be learned from the collapse of a renewable resource? Atlantic cod, *Gadus morhua*, of Newfoundland and Labrador. *Canadian Journal of Fisheries and Aquatic Sciences* 51, 2126–2146.

Hutchings, J. A., Walters, C. & Haedrich, R. L. (1997). Is scientific inquiry incompatible with government information control? *Canadian Journal of Fisheries and Aquatic Sciences* 54, 1198–1210.

Hutchinson, R. E., Stevenson, J. G. & Thorpe, W. H. (1968). The basis for individual recognition by voice in the sandwich tern (*Sterna sandwichensis*). *Behaviour* 32, 150–157.

Hyatt, G. W. & Salmon, M. (1978). Combat in the fiddler crab *Uca pugilator* and *U. pugnax*: a quantitative analysis. *Behaviour* 65, 182–211.

ICCAT (1991). *Recommendations for enhanced management of western Atlantic bluefin tuna, Annex 7. Report of the 12th regular meeting of the Commission (provisional), 11–15 November*. ICCAT, Madrid.

IIED (1994). *Whose Eden? An overview of community approaches to wildlife management*. International Institute for Environment and Development, London.

Immelmenn, K. (1975). Ecological significance of imprinting and early learning.

Annual Review of Ecology and Systematics **6**, 15–37.

Inglis, I. R. & Lazarus, J. (1981). Vigilance and flock size in brent geese: the edge effect. *Zeitschrift fuer Tierpsychologie* **57**, 193–200.

Irish, K. E. & Norse, E. A. (1996). Scant emphasis on marine biodiversity. *Conservation Biology* **10**, 680.

Irons, W. (1979). Cultural and biological success. In *Evolutionary biology and human social behaviour: an anthropological perspective* (Eds N. Chagnon & W. Irons), pp. 257–272. Duxbury, North Scituate.

IUCN (1982). *IUCN directory of neotropical protected areas*. Tycooly International Publishing Limited, Dublin, Ireland.

IUCN (1992a). *Protected areas of the world: a review of national systems*, vol. 1. *Indomalaya, Oceania, Australia and Antarctic*. IUCN, Gland, Switzerland.

IUCN (1992b). *Protected areas of the world: a review of national systems*, vol. 3. *Afrotropical*. IUCN, Gland, Switzerland.

IUCN (1992c). *Protected areas of the world: a review of national systems*, vol. 4. *Neoarctic and Neotropical*. IUCN, Gland, Switzerland.

IUCN (1996). *1996 IUCN red list of threatened animals*. IUCN, Gland, Switzerland.

Iwasa, Y., Pomiankowski, A. & Nee, S. (1991). The evolution of costly mate preferences. II. The 'handicap' principle. *Evolution* **45**, 1431–1442.

IWIGIA (1992). Declaration by the indigenous peoples. *IWIGIA Yearbook* **1992**, 157–163.

Jablonski, D. (1987). Heritability at the species level: analysis of geographic ranges of Cretaceous mollusks. *Science* **238**, 360–363.

Jablonski, D. (1997). Body-size evolution in Cretaceous molluscs and the status of Cope's rule. *Nature* **385**, 250–252.

Jackson, M. J. & James, R. (1979). The influence of bait digging on cockle, *Cerastoderma edule*, populations in north Norfolk. *Journal of Animal Ecology* **16**, 671–679.

Jackson, R. & Ahlborn, G. (1989). Snow leopards (*Panthera uncia*) in Nepal – home ranges and movements. *National Geographic Research* **5**, 161–175.

Janik, V. M., Dehnhardt, G. & Todt, D. (1994). Signature whistle variations in a bottlenosed dolphin, *Tursiops truncatus*. *Behavioural Ecology and Sociobiology* **35**, 243–248.

Jaquette, D. L. (1970). A stochastic model for the optimal control of epidemics and pest populations. *Mathematical Biosciences* **8**, 343–354.

Jarrett, W. F. H., Jennings, F. W., Murray, M. & Harthoorn, A. M. (1964). Muscular dystrophy in a wild Hunter's antelope. *East African Wildlife Journal* **2**, 158–159.

Jasper, J. M. (1996). The American animal rights movement. In *Animal rights: the changing debate* (Ed. R. Garner), pp. 129–142. Macmillan Press, Basingstoke.

Jenni, L. & Jenni-Eiermann, S. (1999). Fuel supply and metabolic constraints in migrating birds. *Journal of Avian Biology* **29**, 521–528.

Jennings, S. (1992). Potential effects of estuarine development on the success of management strategies for the British bass fishery. *Ambio* **21**, 468–470.

Jennings, S. & Kaiser, M. J. (1998). The effects of fishing on marine ecosystems. *Advances in Marine Biology* **34**, 201–352.

Jennings, S., Lancaster, J. E., Ryland, J. S. & Shackley, S. E. (1991). The age structure and growth dynamics of young-of-the-year bass, *Dicentrarchus labrax*, populations. *Journal of the Marine Biological Association of the United Kingdom* 71, 799–810.

Jennings, S., Reynolds, J. D. & Mills, S. C. (1998). Life history correlates of responses to fisheries exploitation. *Proceedings of the Royal Society of London B* **265**, 333–339.

Jeppesen, J. L. (1987). Impact of human disturbance on home range, movements and activity of red deer (*Cervus elaphus*) in a Danish environment. *Danish Review of Game Biology* **13**(2).

Jimenez, J. A., Hughes, K. A., Alaks, G., Graham, L. & Lacy, R. C. (1994). An experimental study of inbreeding depression in a natural habitat. *Science* **266**, 271–273.

Johannes, R. E. (1980). Using knowledge of reproductive behaviour of reef and lagoon fishes to improve yields. In *Fish behaviour and fisheries management* (Eds J. Bardach, J. Magnuson, R. May & J. Reinhart), pp. 247–270. ICLARM, Manila.

Johannes, R. E. (1981). *Words of the lagoon.* University of California Press, Berkeley.

Johansson, S. (1987). Status anxiety and demographic contraction of privileged populations. *Population Development Review* **13**, 439–470.

Johnsingh, A. J. T. (1985). Distribution and status of dhole *Cuon alpinus* Pallas 1811 in South Asia. *Mammalia* **49**, 203–208.

Johnson, A. M., Delong, R. L., Fiscus, C. H. & Kenyon, K. W. (1982). Population status of the Hawaiian monk seal (*Monachus schauinslandi*), 1978. *Journal of Mammalogy* **63**, 413–421.

Johnson, M. L. & Gaines, M. S. (1990). Evolution of dispersal: theoretical models and empirical tests using birds and mammals. *Annual Review of Ecology and Systematics* **21**, 449–480.

Jones, J. C. & Reynolds, J. D. (1997). Effects of pollution on reproductive behaviour of fishes. *Reviews in Fish Biology and Fisheries* **7**, 463–491.

Joubert, S. C. T. (1974) The social organisation of the roan antelope (*Hippotragus equinus*) and its influence on the spatial distribution of herds in the Kruger National Park. In *The behaviour of ungulates and its relation to management* (Eds V. Geist & F. Walther), pp. 661–675. IUCN, Morges.

Jouventin, P. & Cornet, A. (1980). The sociobiology of pinnipeds. *Advances in the Study of Behavior* **11**, 121–141.

Judge, D. S. & Hrdy, S. B. (1992). Allocation of accumulated resources among close kin – inheritance in Sacramento, California 1980–1984. *Ethology & Sociobiology* **15**, 495–522.

Jukema, J. (1987). Were lesser golden plovers *Pluvialis fulva* regular winter visitors to Friesland, The Netherlands, in the first half of the 20th century? *Wader Study Group Bulletin* **51**, 56–58.

Jukema, J. & Hulscher, J. B. (1988). Recovery rate of ringed golden plovers *Pluvialis apricaria* in relation to the severity of the winter. *Limosa* **61**, 85–90.

Jukema, J., Piersma, T., Louwsma, L., Monkel, C., Rijpma, U., Visser, K. & van der Zee, D. (1995). Moult and mass changes of northward migrating ruffs in Friesland, March–April 1993–1994. *Vanellus* **48**, 55–61.

Kaiser, M. J. & Spencer, B. E. (1994). Fish scavenging behaviour in recently trawled areas. *Marine Ecology Progress Series* **112**, 41–49.

Kaiser, M. J. & Spencer, B. E. (1996). The behavioural response of scavengers to beam-trawl disturbance. In *Aquatic predators and their prey* (Eds S. P. R. Greenstreet & M. L. Tasker), pp. 117–123. Blackwell Scientific Publications, Oxford.

Kalat, J. W. (1977). Biological significance of food aversion learning. In *Food aversion learning* (Eds N. W. Milgram, L. Krames & T. M. Alloway), pp. 73–103. Plenum Press, New York and London.

Kalat, J. W. & Rozin, P. (1973). 'Learned safety' as a mechanism in long-delay taste-aversion learning in rats. *Journal of Comparative and Physiological Psychology* 83, 198–207.

Kaplan, H. & Hill, K. (1992). The evolutionary ecology of food acquisition. In *Evolutionary ecology and human behavior* (Eds E. A. Smith & B. Winterhalder), pp. 167–201. Aldine de Gruyter, New York.

Kaplan, H. & Kopishke, K. (1992). Resource use, traditional technology, and change among native peoples of lowland South America. In *Conservation of neotropical forests* (Eds K. H. Redford & C. Padoch), pp. 83–107. Columbia University Press, New York.

Kaplan, H., Lancaster, J., Bock, J. A. & Johnson, S. E. (1995). Does observed fertility maximise fitness among New Mexican men? A test of an optimality model and a new theory of parental investment in the embodied capital of offspring. *Human Nature* 6, 325–360.

Karban, R. (1997). Neighbourhood affects a plant's risk of herbivory and subsequent success. *Ecological Entomology* 22, 433–439.

Katnik, D. D., Harrison, D. J. & Hodgman, T. P. (1994). Spatial relations in a harvested population of marten in Maine. *Journal of Wildlife Management* 58, 600–607.

Kay, J. (1985). Native Americans in the fur trade and wildlife depletion. *Environmental Review* 9, 118–130.

Keller, L. F., Arcese, P., Smith, J. N. M., Hochachka, W. M. & Stearns, S. C. (1994). Selection against inbred song sparrows during a natural population bottleneck. *Nature* 372, 356–357.

Keller, V. (1991). The effect of disturbance from roads on the distribution of feeding sites of geese wintering in north-east Scotland. *Ardea* 79, 229–231.

Kelly, M. J. & Durant, S. M. (1999). Viability of the Serengeti cheetah population. *Conservation Biology*, in press.

Kelly, M. J., Laurenson, M. K., FitzGibbon, C. D., Collins, D. A., Durant, S. M., Frame, G. W., Bertram, B. C. R. & Caro, T. M. (1998). Demography of the Serengeti cheetah (*Acinonyx jubatus*) population: the first 25 years. *Journal of Zoology* 244, 473–488.

Kenyon, K. W. (1981). Monk seals *Monachus* (Fleming 1822). In *Handbook of marine mammals*, vol. 2. *Seals* (Eds S. H. Ridgeway & R. J. Harrison), pp. 195–220. Academic Press, London.

Kerle, J. A. (1984). Variation in the ecology of *Trichosurus*: its adaptive significance. In *Possums and gliders* (Eds A. P. Smith & I. D. Hume), pp. 115–128. Australian Mammal Society, Sydney.

Kestin, C. (1995). Welfare aspects of the commercial slaughter of whales. *Animal Welfare* 4, 11–27.

Kimura, M. (1953). 'Stepping-stone' model of population. *Annual Report of the Na-*

tional Institute of Genetics, Japan **3**, 62–63.

King, C. M. & Moors, P. J. (1979). The life-history tactics of mustelids, and their significance for predator control and conservation in New Zealand. *New Zealand Journal of Zoology* **6**, 619–622.

Kirby, J. S. & Bell, M. C. (1996). Surveillance of non-breeding waterfowl populations: methods to trigger conservation action. *Gibier Faune Sauvage (Game and Wildlife)* **13**, 493–512.

Kirby, J. S., Clee, C. & Seager, V. (1993). Impact and extent of recreational disturbance to wader roosts on the Dee estuary: some preliminary results. *Wader Study Group Bulletin* **68**, 53–58.

Kirkpatrick, J. F., Turner, J. W., Liu, I. K. M. & Fayrerhosken, R. (1996). Applications of pig zona-pellucida immunocontraception to wildlife fertility control. *Journal of Reproduction and Fertility* **S50**, 183–189.

Kirkpatrick, J. F., Turner, J. W., Liu, I. K. M., Fayrerhosken, R. & Rutberg, A. T. (1997). Case studies in wildlife immunocontraception: wild and feral equids and white-tailed deer. *Reproduction Fertility and Development* **9**, 105–110.

Kirkwood, J. K. (1993). Interventions for wildlife health, conservation and welfare. *Veterinary Record* **132**, 235–238.

Kleiman, D. G. (1977). Monogamy in mammals. *Quarterly Review of Biology* **52**, 39–69.

Kleiman, D. G. (1980). The sociobiology of captive propagation. In *Conservation biology: an evolutionary–ecological perspective* (Eds M. E. Soulé & B. A. Wilcox), pp. 243–261. Sinauer, Sunderland, MA.

Kleiman, D. G. (1989). Reintroduction of captive mammals for conservation. *Bioscience* **39**, 152–161.

Kleiman, D. G., Beck, B. B., Dietz, J. M., Dietz, L. A., Ballou, J. B. & Coimbra-Filho, A. C. (1986). Conservation program for the golden lion tamarins: captive rearing and management, ecological studies, education strategies and reintroduction. In *Primates: the road to self-sustaining populations* (Ed. K. Benirschke), pp. 959–979. Springer-Verlag, New York.

Klepper, O., van der Tol, M. W. M., Scholten, H. & Herman, P. M. J. (1994). SMOES: a simulation model for the Oosterschelde ecosystem. I. Description and uncertainty analysis. *Hydrobiologia* **282/283**, 437–452.

Klima, G. (1964). Jural relations between the sexes among the Barabaig. *Africa* **34**, 9–19.

Klima, G. (1965). Kinship, property and jural relations among the Barabaig. PhD dissertation, University of California, USA.

Klomp, N. I. & Furness, R. W. (1992). Non-breeders as a buffer against environmental stress: declines in numbers of great skuas on Foula, Shetland, and prediction of future recruitment. *Journal of Applied Ecology* **29**, 341–348.

Klump, G. M. (1996). Bird communication in the noisy world. In *The ecology & evolution of acoustic communication in birds* (Eds D. E. Kroodsma & E. H. Miller), pp. 321–338. Comstock, Ithaca, NY.

Knick, S. T. (1990). Ecology of bobcats relative to exploitation and a prey decline in southeastern Idaho. *Wildlife Monographs* **108**, 1–42.

Kock, M. D. (1996). Zimbabwe: a model for the sustainable use of wildlife and the development of innovative wildlife management practices. In *The exploitation of mammal populations* (Eds V. J. Taylor & N. Dunstone), pp. 229–249. Chapman & Hall, London.

Kokko, H. & Ebenhard, T. (1996). Measuring the strength of demographic stochasticity. *Journal of Theoretical Biology* **183**, 169–178.

Komdeur, J. & Deerenberg, C. (1997). The importance of social behavior studies for conservation. In *Behavioral approaches to conservation in the wild* (Eds J. R. Clemmons & R. Buchholz), pp. 262–276. Cambridge University Press, Cambridge.

Kotrschal, K., Hemetsberger, J. & Dittami, J. (1993). Food exploitation by a winter flock of greylag geese: behavioural dynamics, competition and social status. *Behavioural, Ecology & Sociobiology* **33**, 289–295.

Krasnov, J. V. & Barrett, R. T. (1995). Large-scale interactions among seabirds, their prey and humans in the southern Barents Sea. In *Ecology of fjords and coastal waters* (Eds H. R. Skjoldal, C. Hopkins, K. E. Erikstad & H. P. Leinaas), pp. 443–456. Elsevier Science BV, Amsterdam.

Krebs, C. J. (1985). Do changes in spacing behaviour drive population cycles in small mammals? In *Behavioural ecology* (Eds R. M. Sibley & R. H. Smith), pp. 295–312. Blackwell Scientific Publications, Oxford.

Krebs, C. J. & Chitty, D. (1995). Can changes in social behaviour help to explain house mouse plagues in Australia. *Oikos* **73**, 429–434.

Krebs, C. J., Redfield, J. A. & Taitt, M. J. (1978). A pulsed-removal experiment on the vole *Microtus townsendii*. *Canadian Journal of Zoology* **56**, 2253–2262.

Kroodsma, D. E. (1988). Contrasting styles of song development and their consequences among the Passeriformes. In *Evolution and learning* (Eds R. C. Bolles & M. D. Beecher), pp. 157–184. Erlbaum, Hillsdale, NJ.

Kruuk, H. (1972). *The spotted hyaena*. Chicago University Press, Chicago.

Kruuk, H. (1989). *The social badger*. Oxford University Press, Oxford.

Kruuk, H. (1995). *Wild otters: predation and populations*. Oxford University Press, Oxford.

Kruuk, H. & Macdonald, D. W. (1985). Group territories of carnivores: empires and enclaves. In *Behavioural ecology: ecological consequences of adaptive behaviour* (Eds R. M. Sibly & R. H. Smith), pp. 521–536. Blackwell Scientific Publications, Oxford.

Kuussaari, M., Nieminen, M. & Hanski, I. (1996). An experimental study of migration in the Glanville fritillary butterfly *Melitaea cinxia*. *Journal of Animal Ecology* **65**, 791–801.

L'Abée-Lund, J. H., Langeland, A., Jonsson, B. & Ugedal, O. (1993). Spatial segregation by age and size in Arctic charr: a trade-off between feeding possibility and risk of predation. *Journal of Animal Ecology* **62**, 160–168.

Lack, D. (1954). *The regulation of animal numbers*. Izd-vo Inostrannoi Literatury, Moskva.

Lack, D. (1968). *Ecological adaptations for breeding in birds*. Methuen, London.

Lacy, R. C. (1987). Loss of genetic diversity from managed populations: interacting effects of drift, mutation, immigration, selection and population subdivision. *Conservation Biology* **1**, 143–158.

Lacy, R. C. (1997). Importance of genetic variation to the viability of mammalian populations. *Journal of Mammalogy* **78**, 320–335.

Lambeck, R.H.D., Goss-Custard, J.D. & Triplet, P. (1996). Oystercatchers and man in the coastal zone. In *The oystercatcher: from individuals to populations* (Ed. J. D. Goss-Custard), pp. 289–326. Oxford University Press, Oxford.

Lambin, X. & Krebs, C. J. (1993). Influence of female relatedness on the demography of Townsend's vole populations in spring. *Journal of Animal Ecology* 62, 536–550.

Lamprey, H. F. (1983). Pastoralism yesterday and today: the overgrazing problem. In *Tropical savannas* (Ed. F. Bourliere), pp. 643–666. Elsevier, Amsterdam.

Lande, R. (1981). Models of speciation by sexual selection on polygenic characters. *Proceedings of the National Academy of Sciences of the USA* 78, 3721–3725.

Lande, R. (1987). Extinction thresholds in demographic models of territorial populations. *American Naturalist* 130, 624–635.

Lande, R. (1988a) Genetics and demography in biological conservation. *Science* 241, 1455–1460.

Lande, R. (1988b). Demographic models of the northern spotted owl (*Strix occidentalis caurina*). *Oecologia* 75, 601–607.

Lane, C. (1990). *Barabaig natural resource management: sustainable land use under threat of destruction.* United Nations Research Institute for Social Development, Discussion Paper 12, June 1990.

Lane, C. (1996). *Pastures lost: Barabaig economy, resource tenure, and the alienation of their land in Tanzania.* Initiatives Publishers, Nairobi, Kenya.

Lang, A., Houston, A. I., Black, J. M., Pettifor, R. A. & Prop, J. (1999). From individual feeding performance to predicting population dynamics in barnacle geese: the spring staging model. In *Research on Arctic geese. Proceedings of the Svalbard goose symposium, Oslo, Norway, 23–26 September 1997* (Eds F. Mehlum, J. M. Black & J. Madsen). Norsk Polarinstitutt Skrifter 200, Oslolufhavn.

Langbein, J. & Putman, R. (1996). Studies of English red deer populations subject to hunting-to-hounds. In *The exploitation of mammal populations* (Eds V. J. Taylor & N. Dunstone), pp. 208–225. Chapman & Hall, London.

LaRoe, E. T., Farris, G. S., Puckett, C. E., Doran, P. D. & Mac, M. J. (1995). *Our living resources: a report to the nation on the distribution, abundance and health of U.S. plants, animals and ecosystems.* US Department of the Interior, Washington, DC.

Latter, B. D. H. (1973). The island model of population differentiation: a general solution. *Genetics* 73, 147–157.

Laurance, W. F. (1991). Ecological correlates of extinction proneness in Australian tropical forest mammals. *Conservation Biology* 5, 79–89.

Laurance, W. F. & Bierregaard, R. O. (Eds) (1997). *Tropical rainforest remnants: ecology, management and conservation of fragmented communities.* The University of Chicago Press, Chicago.

Laurenson, M. K. (1993). Early maternal behavior of wild cheetahs: implications for captive husbandry. *Zoo Biology* 12, 31–43.

Laurenson, M. K. (1994). High juvenile mortality in cheetahs (*Acinonyx jubatus*) and its consequences for maternal care. *Journal of Zoology* 234, 387–408.

Laurenson, M. K. (1995). Implications of high offspring mortality for cheetah population dynamics. In *Serengeti. II. Dynamics, management and conservation of an ecosystem* (Eds A. R. E. Sinclair & P. Arcese), pp. 385–399. University of Chicago Press, Chicago.

Laurenson, M. K. & Caro, T. M. (1994). Monitoring the effects of non-trivial handling in free-living cheetahs. *Animal Behaviour* **47**, 547–557.

Laurenson, M. K., Caro, T. M. & Borner, M. (1992). Female cheetah reproduction. *National Geographic Research and Exploration* **8**, 64–75.

Laurenson, M. K., Caro, T. M., Gros, P. & Wielebnowski, N. (1995a). Controversial cheetahs? *Nature* **377**, 392.

Laurenson, M. K., Wielebnowski, N. & Caro, T. M. (1995b). Extrinsic factors and juvenile mortality in cheetahs. *Conservation Biology* **9**, 1329–1331.

Lawton, J. H. & May, R. M. (Eds) (1995). *Assessing extinction rates*. Oxford University Press, Oxford.

Lazarus, J. (1978). Vigilance, flock size and domain of danger size in the white-fronted goose. *Wildfowl* **29**, 855–865.

Le Boeuf, B. J. & Reiter, J. (1988). Lifetime reproductive success in northern elephant seals. In *Reproductive success* (Ed. T. H. Clutton-Brock), pp. 344–362. University of Chicago Press, Chicago.

Lederhouse, R. C. (1982). Territorial defense and lek behavior of the black swallowtail butterfly *Papilio polyxenes. Behavioural Ecology and Sociobiology* **10**, 109–118.

Ledig, F. T. (1986). Heterozygosity, heterosis and fitness in outbreeding plants. In *Conservation biology: the science of scarcity and diversity* (Ed. M. E. Soulé), pp. 77–104. Sinauer, Sunderland, MA.

Legendre, S. & Clobert, J. (1995). ULM, a software for conservation and evolutionary biologists. *Journal of Applied Statistics* **22**, 817–834.

Legendre, S., Clobert, J., Møller, A. P. & Sorci, G. (1999). Demographic stochasticity and social mating system in the process of extinction of small populations: the case of passerines introduced to New Zealand. *American Naturalist*, in press.

LeGrand, T. & Phillips, J. F. (1996). The effect of fertility reductions on infant and child mortality: evidence from Matlab in Rural Bangladesh. *Population Studies* **50**, 51–68.

Leonard, M. L. & Horn, A. G. (1995). Crowing in relation to status: talk is cheap. *Animal Behaviour* **49**, 1283–1290.

Lessells, C. M., Rowe, C. L. & McGregor, P. K. (1995). Individual and sex differences in the provisioning calls of European bee-eaters. *Animal Behaviour* **49**, 244–247.

Levins, R. (1969). Some demographic and genetic consequences of environmental heterogeneity for biological control. *Bulletin of Entomological Society of America* **15**, 237–240.

Levitan, D. R., Sewell, M. A. & Chia, F. S. (1992). How distribution and abundance influence fertilization success in the sea-urchin *Strongylocentrotus franciscanus. Ecology* **73**, 248–254.

Levy, C. J., Carroll, M. E., Smith, J. C. & Hofer, K. G. (1974). Antihistamines block radiation-induced taste aversions. *Science* **186**, 1044–1046.

Lewis, A. R., Pinchin, A. M. & Kestin, S. C. (1997). Welfare implications of the night shooting of wild impala (*Aepyceros melampus*). *Animal Welfare* **6**, 123–131.

Lewis, J. C. (1990). Captive propagation in the recovery of whooping cranes. *Endangered Species Update* **8**(1), 46–48.

Lewis, O. T., Thomas, C. D., Hill, J. K., Brookes, M. I., Crane, T. P. R., Graneau, Y. A., Mallet, J. L. B. & Rose, O. C. (1997). Three ways of assessing metapopulation structure in the butterfly *Plebejus argus. Ecological Entomology* **22**, 283–293.

Lima, S. L. & Dill, L. M. (1990). Behavioral decisions made under the risk of predation: a review and prospectus. *Canadian Journal of Zoology* **68**, 619–640.

Lindberg, D. R., Estes, J. A. & Warheit, K. I. (1998). Human influences on trophic cascades along rocky shores. *Ecological Applications* **8**, 880–890.

Lindburg, D. G., Durrant, B. S., Millard, S. E. & Oosterhuis, J. E. (1993). Fertility assessment of cheetah males with poor semen quality. *Zoo Biology* **12**, 97–103.

Lindenmayer, D. B. & Lacy, R. C. (1995). Metapopulation viability of arboreal marsupials in fragmented old-growth forests – comparison among species. *Ecological Applications* **5**, 183–199.

Lindley, A. (1994). Sustainable use of wildlife: assessment of animal welfare. Unpublished report to the IUCN Ethics Working Group.

Litvaitis, J. A., Major, J. T. & Sherburne, J. A. (1987). Influence of season and human-induced mortality on spatial organization of bobcats (*Felis rufus*) in Maine. *Journal of Mammalogy* **68**, 100–106.

Lloyd, E. R. (1990). *The staghunting controversy: some provocative questions and straight answers*. Masters of Deerhounds Association, UK.

Lockwood, S. J. (1988). *The mackerel: its biology, assessment and the management of a fishery*. Fishing News Books, Farnham, UK.

Lopez Pizarro, E. (1986). Estado actual del jaguar en Costa Rica. In *Proceedings of the conservation status of the jaguar* (Ed. I.C.f.G.a.W. Conservation).

Lorenz, K. Z. (1952). *King Solomon's ring*. Crowell, New York.

Lott, D. F. (1991). *Intraspecific variation in the social systems of wild vertebrates*. Cambridge University Press, Cambridge.

Loudon, A. S. I. (1985). Lactation and neonatal survival of mammals. *Zoological Society of London Symposium* No. 54, 183–207.

Low B. (1991). Reproductive life in 19th century Sweden: an evolutionary perspective on demographic phenomena. *Ethology & Sociobiology* **12**, 411–448.

Low, B. S. (1996). Behavioural ecology of conservation in traditional societies. *Human Nature* **7**, 353–379.

MacArthur, R. H. & Wilson, E. O. (1967). *The theory of island biogeography*. Princeton University Press, Princeton, NJ.

Macdonald, D. W. (1995). Wildlife rabies: the implications for Britain – unresolved questions for the control of wildlife rabies: social perturbation and interspecific interactions. In *Rabies in a changing world*. British Small Animal Veterinary Association, Cheltenham.

Macdonald, D. W. & Johnson, P. J. (1996). The impact of sport hunting: a case study. In *The exploitation of mammal populations* (Eds V. J. Taylor & N. Dunstone), pp. 160–207. Chapman & Hall, London.

Macdonald, D.W., Mace, G. & Rushton, S. R. (1998). *Proposals for monitoring British mammals*. Department of the Environment, Transport and the Regions, London.

Mace, G. M. & Balmford, A. (1999). Patterns and processes in mammalian extinction. In *Priorities for the conservation of mammalian biodiversity* (Eds A. Entwistle & N. Dunstone). Cambridge University Press, Cambridge.

Mace, R. (1993). Transitions between cultivation and pastoralism in sub-Saharan Africa. *Current Anthropology* **34**, 363–382.

Mace, R. (1996a). Biased parental investment and reproductive success in Gabbra pastoralists. *Behavioural Ecology & Sociobiology* **38**, 75–81.

Mace, R. (1996b). When to have another baby: a dynamic model of reproductive decision-making and evidence from Gabbra pastoralists. *Ethology & Sociobiology* **17**, 263–274.

Mace, R. (1998). The co-evolution of human fertility and wealth inheritance strategies. *Philosophical Transactions of the Royal Society of London Series B* **353**, 389–397.

Mace, R. & Sear R. (1996). Maternal mortality in a Kenyan, pastoralist population. *International Journal of Gynecology & Obstetrics* **54**, 137–141.

Mace, R. & Sear, R. (1997). Birth interval and the sex of children in a traditional African population: an evolutionary analysis. *Journal of Biosocial Sciences* **29**, 499–507.

Mace, R. & Sear, R (1998). Reproductive decision-making in the face of demographic risks in Gabbra pastoralists. *Nomadic Peoples*, **1**, 151–163.

Mackinson, S., Sumaila, U. R. & Pitcher, T. J. (1997). Bioeconomics and catchability: fish and fishers behaviour during stock collapse. *Fisheries Research* **31**, 11–17.

Madise, N. J. & Diamond, I. (1995). Determinants of infant mortality in Malawi: an analysis to control for death clustering within families. *Journal of Biosocial Sciences* **27**, 95–106.

Madsen, J. (1985). The impact of disturbance on field utilisation of pink-footed geese in W. Jutland, Denmark. *Biological Conservation* **33**, 53–63.

Madsen, J. & Fox, A. D. (1995). Impacts of hunting disturbance on waterbirds – a review. *Wildlife Biology* **1**, 193–207.

Madsen, T., Stille, B. & Shine, R. (1996). Inbreeding depression in an isolated population of adders *Vipera berus. Biological Conservation* **75**, 113–118.

Magnhagen, C. (1991). Predation risk as a cost of reproduction. *Trends in Ecology and Evolution* **6**, 183–186.

Maigret, J. (1986). The monk seal *Monachus monachus* Herman 1779 on the west African coasts. *Aquatic Mammals* **12**, 49–51.

Main, J. & Sangster, G. I. (1982a). A study of separating fish from *Nephrops norwegicus* L. in a bottom trawl. *Scottish Fisheries Research Reports* **24**, 1–9.

Main, J. & Sangster, G. I. (1982b). A study of a multi-level bottom trawl for species separation using direct observation techniques. *Scottish Fisheries Research Reports* **26**, 1–17.

Mallinson, J. J. C. (1995). Conservation breeding programs – an important ingredient for species survival. *Biodiversity and Conservation* **4**, 617–635.

Mangel, M. & Clark, C. W. (1988). *Dynamic modelling in behavioural ecology.* Princeton University Press, Princeton, NJ.

Mangel, M. & Tier, C. (1994). Four facts every conservation biologist should know about persistence. *Ecology* **75**, 607–614.

Marchessaux, D. (1989). *The biology, status and conservation of the monk seal* Monachus monachus. Council of Europe, Nature and Environment Series, Palais de l'Europe, Strasbourg.

Mares, M. A. & Lacher, T. E., Jr (1987). Social spacing in small mammals: patterns of individual variation. *American Zoologist* **27**, 293–306.

Marker, L. & O'Brien, S. J. (1989). Captive breeding of the cheetah (*Acinonyx jubatus*) in North American zoos 1871–1986. *Zoo Biology* **8**, 3–16.

Marks, C. A., Nijk, M., Gigliotti, F., Busanak, F. & Short, R. V. (1996). Preliminary field assessment of a cabergoline baiting campaign for reproductive control of the red fox (*Vulpes vulpes*). *Wildlife Research* **23**, 161–168.

Marks, J. S. & Redmond, R. L. (1994). Conservation problems and research needs for bristle-thighed curlews *Numenius tahitiensis* on their wintering grounds. *Bird Conservation International* **4**, 329–341.

Marks, J. S. & Redmond, R. L. (1996). Demography of bristle-thighed curlews *Numenius tahitiensis* wintering on Laysan Island. *Ibis* **138**, 438–447.

Marler, P. & Tamura, M. (1962). Song 'dialects' in three populations of White-crowned Sparrows. *Condor* **64**, 368–377.

Martens, J. (1996). Vocalizations and speciation of Palearctic birds. In *The ecology & evolution of acoustic communication in birds* (Eds D. E. Kroodsma & E. H. Miller), pp. 221–240. Comstock, Ithaca, NY.

Martin, L. T. & Lawrence, C. D. (1979). The importance of odor and texture cues in food aversion learning. *Behavioural and Neural Biology* **27**, 503–515.

Martin, P. S. (1984). Prehistoric overkill: the global model. In *Quaternary extinctions* (Eds P. S. Martin & R. G. Klein). University of Arizona Press, Tucson, AZ.

Martin, P. (1997). *The sickening mind*. HarperCollins, London.

Mason, G. J. & Mendl, M. (1993). Why is there no simple way of measuring animal welfare? *Animal Welfare* **2**, 301–319.

Mason, J. R. & Reidinger, R. F. (1983). Importance of colour for methiocarb-induced food aversions in Red-winged blackbirds. *Journal of Wildlife Management* **47**, 383–393.

Masters, W. M., Raver, K. A. S. & Kazial, K. A. (1995). Sonar signals of big brown bats, *Eptesicus fuscus*, contain information about individual identity, age and family affiliation. *Animal Behaviour* **50**, 143–160.

May, R. M. (1990). How many species? *Philosophical Transactions of the Royal Society London B* **330**, 293–304.

May, R. M. (1994). Biological diversity: differences between land and sea. *Philosophical Transactions of the Royal Society London B* **343**, 105–111.

May, R. M. (1995). The cheetah controversy. *Nature* **374**, 309–310.

Mayer, S. & Simmonds, M. (1996). Science and precaution in cetacean conservation. In *The conservation of whales and dolphins: science and practice* (Eds M. P. Simmonds & J. Hutchinson), pp. 391–406. Wiley, Chichester.

Maynard Smith, J. M. (1982). *Evolution and the theory of games*. Cambridge University Press, Cambridge.

McCabe, J. T. (1990). Turkana pastoralism: a case against the tragedy of the commons. *Human Ecology* **18**, 81–103.

McCarthy, M. A., Franklin, D. C. & Burgman, M. A. (1994). The importance of demographic uncertainty: an example from the helmeted honeyeater *Lichenostomus melanops cassidix*. *Biological Conservation* **67**, 137–142.

McCarthy, M. A., Burgman, M. A. & Ferson, S. (1995). Sensitivity analysis for models of population viability. *Biological Conservation* **73**, 93–100.

McCay, B. J. & Acheson, J. M. (1990). *The question of the commons*. University of Arizona Press, Tucson, AZ.

McComb, K. E. (1992). Playback as a tool for studying contests between social groups. In *Playback and studies of animal communication* (Ed. P. K. McGregor), pp. 111–119. Plenum Press, New York.

McCullough, D. R. (1996). Spatially structured populations and harvest theory. *Journal of Wildlife Management* **60**, 1–9.

McDonald, R. A., Hutchings, M. R. & Keeling, J. G. M. (1997). The status of ship rats *Rattus rattus* on the Shiant Islands, Outer Hebrides, Scotland. *Biological Conservation* **82**, 113–117.

McDowall, R. M. (1992). Particular problems for the conservation of diadromous fish. *Aquatic Conservation: Marine and Freshwater Ecosystems* **2**, 351–355.

McGregor, P. K. (1980). Song dialects in the corn bunting (*Emberiza calandra*). *Zeitschrift für Tierpsychologie* **54**, 285–297.

McGregor, P. K. (1994). Voice, social pattern and behaviour (of the corn bunting). In *The birds of the western Palearctic*, vol. IX (Eds S. Cramp & C. M. Perrins), pp. 329–335. Oxford University Press, Oxford.

McGregor, P. K. & Byle, P. (1992). Individually distinctive bittern booms: potential as a census tool. *Bioacoustics* **4**, 93–109.

McGregor, P. K. & Dabelsteen, T. (1996). Communication networks. In *The ecology & evolution of acoustic communication in birds* (Eds D. E. Kroodsma & E. H. Miller), pp. 409–425. Comstock, Ithaca, NY.

McGregor, P. K. & Peake, T. M. (1998). Individual identification and conservation biology. In *Behavioural ecology and conservation* (Ed. T. Caro), pp. 31–55. Oxford University Press, Oxford.

McGregor, P. K., Krebs, J. R. & Perrins, C. M. (1981). Song repertoires and lifetime reproductive success in the great tit (*Parus major*). *American Naturalist* **118**, 149–159.

McGregor, P. K., Walford, V. R. & Harper, D. G. C. (1988). Song inheritance and mating in a songbird with local dialects. *Bioacoustics* **1**, 107–129.

McGregor, P. K., Anderson, C. M., Harris, J., Seal, J. R. & Soulé, J. M. (1994). Individual differences in songs of fan-tailed Warblers *Cisticola juncidis* in Portugal. *Airo* **5**, 17–21.

McGregor, P. K., Holland, J. & Shepherd, M. (1997). The ecology of corn bunting *Miliaria calandra* song dialects and their potential use in conservation. *UK Nature Conservation* **13**, 76–87.

McGrorty, S. & Goss-Custard, J.D. (1991). Population dynamics of the mussel *Mytilus edulis*: spatial variations in age-class densities of an intertidal estuarine population along environmental gradients. *Marine Ecology Progress Series* **73**, 191–202.

McGrorty, S., Clarke, R.T., Reading, C.J. & Goss-Custard, J.D. (1990). Population dynamics of the mussel *Mytilus edulis*: density changes and regulation of the population in the Exe estuary, Devon. *Marine Ecology Progress Series* **67**, 157–169.

McKean, M. A. (1992). Success on the commons: a comparative examination of institutions for common property resource management. *Journal of Theoretical Politics* **4**, 243–281.

McLain, D. K. (1993). Cope's rule, sexual selection, and the loss of ecological plasticity. *Oikos* **68**, 490–500.

McLain, D. K., Boulton, M. P. & Redfearn, T. P. (1995). Sexual selection and the risk of extinction of introduced birds on oceanic islands. *Oikos* **74**, 27–34.

McLellan, B. N. (1989). Dynamics of a grizzly bear population during a period of industrial resource extraction. I. Density and age-sex composition. *Canadian Journal of Zoology* **67**, 1856–1860.

McLeod, B. J., Thompson, E. G., Crawford, J. L. & Shackell, G. H. (1997). Successful group housing of wild-caught brush-tailed possums (*Trichosurus vulpecula*). *Animal Welfare* **6**, 67–76.

McNamara, J. M. & Houston, A. I. (1986). The common currency for behavioural decisions. *American Naturalist* **127**, 358–378.

McNamara, J. M., Houston, A. I. & Lima, S. L. (1994). Foraging routines of small birds in winter: a theoretical investigation. *Journal of Avian Biology* **25**, 287–302.

McNeilage, A. (1996). Ecotourism and mountain gorillas in the Virunga Volcanoes. In *The exploitation of mammalian populations* (Eds V. J. Taylor & N. Dunstone), pp. 334–344. Chapman & Hall, London.

McVittie, R. (1979). Changes in the social behaviour of South West African cheetah. *Madoqua* **11**, 171–184.

Meffe, G. K. (1992). Techno-arrogance and halfway technologies: salmon hatcheries on the Pacific coast of North America. *Conservation Biology* **6**, 350–354.

Melquist, W. E. (1984). *Status survey of otters (Lutrinae) and spotted cats (Felidae) in Latin America. Completion Report.* IUCN, Gland, Switzerland.

Menotti-Raymond, M. & O'Brien, S. J. (1993). Dating the genetic bottleneck of the African cheetah. *Proceedings of the National Academy of Sciences USA* **90**, 3172–3176.

Menotti-Raymond, M. A. & O'Brien, S. J. (1995). Evolutionary conservation of ten microsatellite loci in four species of Felidae. *Journal of Heredity* **86**, 319–322.

Merola, M. (1994). A reassessment of homozygosity and the case for inbreeding depression in the cheetah, *Acinonyx jubatus*: implications for conservation. *Conservation Biology* **8**, 961–971.

Mestel, R. (1997). Feeling a little strange? *New Scientist* **154**(1086), 25–28.

Mesterton-Gibbons, M. & Dugatkin, L. A. (1992). Cooperation among unrelated individuals: evolutionary factors. *Quarterly Review of Biology* **67**, 267–281.

Metcalfe, J. D. & Arnold, G. P. (1997). Tracking fish with electronic tags. *Nature* **387**, 665–666.

Metcalfe, S. (1994). The Zimbabwe communal areas management programme for indigenous resources (CAMPFIRE). In *Natural connections: perspectives in community-based conservation* (Eds D. Western & M. Wright), pp. 161–192. Island Press, Washington, DC.

Milinski, M. (1985). Risk of predation taken by parasitised fish under competition for food. *Behaviour* **93**, 203–216.

Milinski, M. & Parker, G. A. (1991). Competition for resources. In *Behavioural ecology* (Eds J. R. Krebs & N. B. Davies), pp. 137–168. Blackwell Scientific Publications, Oxford.

Miller, B., Reading, R. & Forest, S. (1996). *Prairie night: black footed ferrets and the recovery of endangered species.* Smithsonian Institution Press, Washington, DC.

Miller, C. E. (1988). Collection of yodel calls for individual identification of male common loons. In *Papers from 1987 N. American conference on loon management*

(Ed. P. I. V. Strong), pp. 44–52. N. American Loon Fund, Meredith, NH.

Miller, C. E. & Dring, T. (1988). Territorial defense of multiple lakes by common loons. In *Papers from 1987 N. American conference on loon management* (Ed. P. I. V. Strong), pp. 1–14. N. American Loon Fund, Meredith, NH.

Miller, E. H. (1996). Acoustic differentiation and speciation in shorebirds. In *The ecology & evolution of acoustic communication in birds* (Eds D. E. Kroodsma & E. H. Miller), pp. 241–257. Comstock, Ithaca, NY.

Miller, K. V. & Ozoga, J. J. (1997). Density effects on deer sociobiology. In *The science of overabundance in deer ecology and population management* (Eds W. J. McShea, H. B. Underwood & J. H. Rappole), pp. 136–150. Smithsonian Institute Press, Washington, DC.

Miller, R. I. (1979). Conserving the genetic integrity of faunal populations and communities. *Environmental Conservation* 6, 197–304.

Miller-Edge, M. A. & Worley, M. B. (1992). In vitro responses of cheetah mononuclear cells to feline herpesvirus-1 and *Cryptococcus neoformans. Veterinary Immunology and Immunopathology* 30, 261–274.

Milner-Gulland, E. J. (1997). A stochastic dynamic programming model for the management of the saiga antelope. *Ecological Applications* 7, 130–142.

Milner-Gulland, E. J. & Mace, R. (1998). *Conservation of biological resources.* Blackwell Science, Oxford.

Mitchell, A. (1974). *A field guide to the trees of Britain and northern Europe.* Collins, London.

Mitra, S., Landel, H. & Pruett-Jones, S. (1996). Species richness covaries with mating system in birds. *Auk* 113, 544–551.

Mode, C. J. (1995). An extension of a Galton–Watson process to a two-sex density-dependent model. In *Branching processes. Proceedings of the first world congress* (Ed. C. C. Heyde), pp. 152–168. Springer-Verlag, New York.

Møller, A. P. (1989). Viability costs of male tail ornaments in a swallow. *Nature* 339, 132–135.

Møller, A. P. (1994). *Sexual selection and the barn swallow.* Oxford University Press, Oxford.

Møller, A. P. & Cuervo, J. J. (1998). Speciation and feather ornamentation in birds. *Evolution,* 52, 859–869.

Møller, A. P. & de Lope, F. (1994). Differential costs of a secondary sexual character: an experimental test of the handicap principle. *Evolution* 48, 1676–1683.

Møller, A. P. & Nielsen, J. T. (1997). Differential predation cost of a secondary sexual character: sparrowhawk predation on barn swallows. *Animal Behaviour* 54, 1545–1551.

Møller, A. P. & Saino, N. (1994). Parasites, immunology of hosts, and host sexual selection. *Journal of Parasitology* 80, 850–858.

Møller, A. P., de Lope, F. & López Caballero, J. M. (1995). Foraging costs of a tail ornament: experimental evidence from two populations of barn swallows *Hirundo rustica* with different degrees of sexual size dimorphism. *Behavioral Ecology and Sociobiology* 37, 289–295.

Møller, A. P., Christe, Ph. & Lux, E. (1999). Parasite-mediated sexual selection: effects of parasites and host immune function. *Quarterly Review of Biology,* 74, 3–20.

Mollison, D. (1985). Sensitivity analysis of simple endemic models. In *Population dynamics of rabies in wildlife* (Ed. P. J. Bacon), pp. 223–234. Academic Press, London.

Monaghan, P. (1992). Seabirds and sandeels: the conflict between exploitation and conservation in the North Sea. *Biodiversity and Conservation* 1, 98–111.

Monaghan, P., Walton, P., Wanless, S., Uttley, J. D. & Burns, M. D. (1994). Effects of prey abundance on the foraging behaviour, diving efficiency and time allocation of breeding guillemots *Uria aalge*. *Ibis* 136, 214–222.

Monaghan, P., Wright, P., Bailey, M. C., Uttley, J. D., Walton, P. & Burns, M. S. (1996). The influence of changes in food abundance on diving and surface feeding seabirds. In *Studies of high latitude seabirds. 4. Trophic relationships and energetics of endotherms in cold oceanic systems: Canadian Wildlife Service Occasional Paper No. 91.* (Ed. W. A. Montevecchi), pp. 10–19. Canadian Wildlife Service, Ottawa.

Moore, H. D. M., Jenkins, N. M. & Wong, C. (1997). Immunocontraception in rodents: a review of the development of a sperm-based immunocontraceptive vaccine for the grey squirrel (*Sciurus carolinensis*). *Reproduction Fertility and Development* 9, 125–129.

Moreiro, J. R. & Macdonald, D. W. (1996). Capybara use and conservation in South America. In *The exploitation of mammal populations* (Eds V. J. Taylor & N. Dunstone), pp. 88–101. Chapman & Hall, London.

Morris, P. A. (1993). *A red data book for British mammals.* The Mammal Society, London.

Morsbach, D. (1986). The behavioural ecology and movement of cheetahs on farmland in Southwest Africa/Namibia. *Annual progress reports to Directorate of Nature Conservation and Recreation Resorts.* Government of Namibia, Windhoek.

Mountford, M. D., Watson, A., Moss, R., Parr, R. & Rothery, P. (1990). Land inheritance and population cycles of red grouse. In *Red grouse population processes* (Eds A. N. Lance & J. H. Lawton), pp. 78–83. Royal Society for the Protection for Birds, Sandy, Beds.

Mulder, J. L. (1985). Spatial organization, movements and dispersal in a Dutch red fox (*Vulpes vulpes*) population: some preliminary results. *Revue d' Ecologie la Terre et la Vie* 40, 133–138.

Muller, L. I., Warren, R. J. & Evans, D. L. (1997). Theory and practice of immunocontraception in wild mammals. *Wildlife Society Bulletin* 25, 504–514.

Mundinger, P. C. (1982). Microgeographic and macrogeographic variation in acquired vocalizations of birds. In *Acoustic communication in birds*, vol. 2 (Eds D. E. Kroodsma & E. H. Miller), pp. 147–208. Academic Press, New York.

Murawski, S. A. & Idoine, J. S. (1992). Multispecies size composition: a conservative property of exploited fishery systems? *Journal of Norwestern Atlantic Fisheries Science* 14, 79–85.

Murphy, D. D. & White, R. R. (1984). Rainfall, resources, and dispersal in southern populations of *Euphydryas editha* (Lepidoptera: Nymphalidae). *Pan-Pacific Entomologist* 60, 350–354.

Murphy, D. D., Menninger, M. S. & Ehrlich, P. R. (1984). Nectar source distribution as a determinant of oviposition host species in *Euphydryas chalcedona*. *Oecologia* 62, 269–271.

Myers, J. H. & Krebs, C. J. (1971). Genetic, behavioural and reproductive attributes of dispersing field voles, *Microtus pennsylvanicus* and *Microtus ochrogaster*. *Ecological Monographs* **41**, 53–78.

Myers, J. P. (1986). Sex and gluttony on Delaware Bay. *Natural History* **95**, 68–77.

Myers, J. P. (1988). The sanderling. *Audubon Wildlife Report* **1988/1989**, 651–666.

Myers, J. P., Morrison, R. I. G., Antas, P. Z., Harrington, B. A., Lovejoy, T. E., Sallaberry, M., Senner, S. E. & Tarak, A. (1987). Conservation strategy for migratory species. *American Scientist* **75**, 18–26.

Myers, R. A, Hutchings, J. A. & Barrowman, N. J. (1997). Why do fish stocks collapse? The example of cod, *Gadus morhua*, in Atlantic Canada. *Ecological Applications* **7**, 91–106.

Myers, R. H. & De Castro, J. M. (1977). Learned aversion to intracerebral carbachol. *Physiology and Behaviour* **10**, 73–78.

Myers, S. A., Millan, J. R., Roudybush, T. E. & Grav, G. R. (1988). Reproductive success of hand-reared vs parent-reared cockatiels (*Nymphicus hollandicus*). *Auk* **105**, 536–542.

Nachman, M. & Hartley, P. L. (1975). Role of illness in producing learned taste aversions in rats: a comparison of several rodenticides. *Journal of Comparative and Physiological Psychology* **89**, 1010–1018.

Nachman, M., Rauschenberger, J. & Ashe, J. H. (1977). Stimulus characteristics in food aversion learning. In *Food aversion learning* (Eds N. W. Milgram, L. Krames & T. M. Alloway), pp. 105–131. Plenum Press, New York and London.

Nellis, D. W. & Small, V. (1983). Mongoose predation on sea turtle eggs and nests. *Biotropica* **15**, 159–160.

Nelson, R. (1982). A conservation ethic and environment: the Koyukon of Alaska. In *Resource managers: North American and Australian hunter-gatherers* (Eds N. Williams & E. Hunn). Westview Press, Boulder, CO.

Nesbitt, S. A. (1979). Notes on the suitability of captive reared sandhill cranes for release into the wild. In *Proceedings of the 1978 crane workshop* (Ed. J. C. Lewis), pp. 85–88. Colorado State University Printing Service, Fort Collins.

Netting, R. M. (1976). What alpine peasants have in common: observations on communal tenure in a Swiss village. *Human Ecology* **4**, 135–146.

Nettles, V. F. (1997). Potential consequences and problems with wildlife contraceptives. *Reproduction Fertility and Development* **9**, 137–143.

Newberry, R. C. (1995). Environmental enrichment – increasing the biological relevance of captive environments. *Applied Animal Behavioral Science* **44**, 229–243.

Newmark, W. D. (1995). Extinction of mammal populations in western North American national parks. *Conservation Biology* **9**, 512–526.

Newton, I. (1993). Predation and limitation of bird numbers. *Current Ornithology* **11**, 143–198.

Newton, I. (1995). The contribution of some recent research on birds to ecological understanding. *Journal of Animal Ecology* **64**, 675–696.

Nicholson, E. M. (1963). Preface to Atkinson-Willes, G. L. 1963 (Ed.). *Wildfowl in Great Britain*. HMSO, London.

Nicolaus, L. K. (1987). Conditioned aversions in a guild of predators: implications for aposematism and prey defense mimicry. *American Midland Naturalist* 117, 405–419.

Nicolaus, L. K & Nellis, D. W. (1987). The first evaluation of the use of conditioned taste aversion to control predation by mongooses upon eggs. *Applied Animal Behaviour Science* 17, 329–346.

Nicolaus, L. K., Hoffman, T. E. & Gustavson, C. R. (1982). Taste aversion conditioning in free ranging raccoons (*Procyon lotor*). *Northwest Science* 56, 165–169.

Nicolaus, L. K., Cassel, J. F., Carlson, R. B. & Gustavson, C. R. (1983). Taste-aversion conditioning of crows to control predation on eggs. *Science* 220, 212–214.

Nicolaus, L. K, Farmer, P. V., Gustavson, C. R. & Gustavson, J. C. (1989a). The potential of estrogen based conditioned aversion in controlling depredation: a step closer to the 'magic bullet'. *Applied Animal Behaviour Science* 23, 1–14.

Nicolaus, L. K., Herrera, J., Nicolaus, J. C. & Dimmick, C. R. (1989b). Carbachol as a conditioned taste aversion agent to control avian depredation. *Agricultural Ecosystems & Environment* 26, 13–21.

Nicolaus, L. K., Herrera, J., Nicolaus, J. C. & Gustavson, C. R. (1989c). Ethinyl estradiol and generalized aversions to eggs among free-ranging predators. *Applied Animal Behaviour Science*, 24, 313–324.

Nienhuis, P. H. & Smaal, A. C. (Eds) (1994). The Oosterschelde Estuary (The Netherlands): a case-study of a changing ecosystem. *Hydrobiologia* 282/283.

Norman, A. P., Jones, G. & Arlettaz, R. (1999). Noctuid moths show neural and behavioural responses to sounds made by some bat-marking rings. *Animal Behavior* 57, 829–835.

Norris, K. J. & Johnstone, I. G. (1998). The functional response of oystercatchers (*Haematopus ostralegus*) searching for cockles (*Cerastoderma edule*) by touch. *Journal of Animal Ecology*, 67, 329–346.

Norse, E. A. (1993). *Global marine biological diversity*. Island Press, Washington.

Novak, M., Baker, J. A., Obbard, M. E. & Malloch, B. (1987). *Wild furbearer management and conservation in North America*. Ontario Ministry of Natural Resources, Toronto.

Novaro, A. J. (1995). Sustainability of harvest of culpeo foxes in Patagonia. *Oryx* 29, 18–22.

Nowell, K. & Jackson, P. (1996). *Wild cats: status survey and conservation action plan*. IUCN/SSC Cat Specialist Group, Gland.

Nowlis, G. H. (1974). Conditioned stimulus intensity and acquired alimentary aversions in the rat. *Journal of Comparative and Physiological Psychology* 86, 1173–1184.

Nunney, L. (1991). The influence of age structure and fecundity on effective population size. *Proceedings of the Royal Society of London, Series B* 246, 71–76.

Nur, N. (1987). Alternative reproductive tactics in birds: individual variation in clutch-size. *Perspectives in ethology*, vol. 7 (Eds P. P. G. Bateson & P. H. Klopfer), pp. 49–77. Plenum Press, London.

O'Brien, S. J. (1994a). A role for molecular genetics in biological conservation. *Proceedings of the National Academy of Sciences USA* 91, 5748–5755.

O'Brien, S. J. (1994b). The cheetah's conservation controversy. *Conservation Biology* 8, 1153–1155.

O'Brien, S. J. & Evermann, J. F. (1988). Interactive influence of infectious disease and genetic diversity in natural populations. *Trends in Ecology and Evolution* 3, 254–259.

O'Brien, S. J., Wildt, D. E., Goldman, D., Merrill, C. R. & Bush, M. (1983). The cheetah is depauperate in genetic variation. *Science* 221, 459–462.

O'Brien, S. J., Roelke, M. E., Marker, L., Newman, A., Winkler, C. A., Meltzer, D., Colly, L., Bush, M., Evermann, J. F. & Wildt, D. E. (1985). Genetic basis for species vulnerability in the cheetah. *Science* 227, 1428–1434.

O'Brien, S. J., Wildt, D. E., Bush, M., Caro, T. M., FitzGibbon, C., Aggundey, I. & Leakey, R. E. (1987). East African cheetahs: evidence for two population bottlenecks. *Proceedings of the National Academy of Sciences USA* 84, 508–511.

Odendaal, F., Turchin, P. & Stermitz, F. R. (1989). Influence of host-plant density and male harassment on the distribution of female *Euphydryas anicia* (Nymphalidae). *Oecologia* 78, 283–288.

O'Donnell, C. F. (1996). Predators and the decline of New Zealand forest birds: an introduction to the hole-nesting bird and predator programme. *New Zealand Journal of Zoology* 23, 213–219.

Olive, P. J. W. (1993). Management of the exploitation of the lugworm *Arenicola marina* and the ragworm *Nereis virens* (Polychaeta) in conservation areas. *Aquatic Conservation: Marine and Freshwater Ecosystems* 3, 1–24.

Oliveira, R. F., Machado, J. L., Jordão, J. M., Burford, F. R. L., Latruffe, C. & McGregor, P. K. (1999). Human exploitation of male fiddler crab claws: behavioural consequences and implications for conservation. *Conservation Biology*, in press.

Olla, B. L., Davis, M. W. & Ryder, C. H. (1994). Behavioural deficits in hatchery-reared fish: potential effects on survival following release. *Aquaculture Fisheries Management* 25 (suppl.), 19–34.

Olla, B. L., Davis, M. W. & Schreck, C. B. (1997). Effects of simulated trawling on sablefish and walleye pollock: the role of light intensity, net velocity and towing duration. *Journal of Fish Biology* 50, 1181–1194.

Olney, P. J. S., Mace, G. M. & Feistner, A. T. C. (Eds) (1994). *Creative conservation: interactive management of wild and captive animals.* Chapman & Hall, London.

Olson, M. (1967). *The logic of collective action: public goods and the theory of groups.* Harvard University Press, Boston, MA.

Olsson, G. E., Willebrand, T. & Smith, A. (1996). The effects of hunting on willow grouse *Lagopus lagopus* movements. *Wildlife Biology* 2, 11–15.

van Ordsol, K. G., Hanby, J. P. & Bygott, J. D. (1985). Ecological correlates of lion social organization (*Panthera leo*). *Journal of Zoology* 206, 97–112.

O'Regan, F. M. (1997). Conservation and animal welfare. *Trends in Ecology and Evolution* 12, 318.

O'Reilly, K. M. & Wingfield, J. C. (1995). Spring and autumnal migration in arctic shorebirds: same distance, different strategies. *American Zoologist* 35, 222–233.

Orians, G. H. (1969). On the evolution of mating systems in birds and mammals. *American Naturalist* 103, 589–603.

Osterhaus, A., Groen, J., Niesters, H., Bildt, M. v. d., Vedder, B. M. L., Vos, J., Egmond, H. V., Sidi, B. A. & Barham, M. E. O. (1997). Morbillivirus in monk seal mass mortality. *Nature* 388, 838–839.

Ostfeld, R. S. (1986). Territoriality and mating system of California voles. *The Journal of Animal Ecology **55**, 691–706.

Ostrom, E. (1990). *Governing the commons: the evolution of institutions for collective action.* Cambridge University Press, Cambridge.

Ostrom, E., Walker, J. & Gardner, R. (1994). *Rules, games and common pool resources.* University of Michigan Press, Ann Arbor, MI.

OTA (1987). *Technologies to maintain biological diversity.* US Government Printing Office, Office of Technological Assessment, US Congress, Washington, DC.

Owens, I. P. F. & Bennett, P. M. (1994). Mortality costs of parental care and sexual dimorphism in birds. *Proceedings of the Royal Society of London B* **257**, 1–8.

Ozoga, J. J. & Verme, L. J. (1984). Effects of family-bond deprivation on reproductive performance in female white-tailed deer. *Journal of Wildlife Management* **48**, 1326–1334.

Ozoga, J. J. & Verme, L. J. (1985). Comparative breeding behavior and performance of yearling vs prime-age white-tailed bucks. *Journal of Wildlife Management* **49**, 364–372.

Packer, C. A. (1979). Inter-troop transfer and inbreeding avoidance in *Papio anubis*. *Animal Behaviour* **27**, 1–36.

Packer, C., Herbst, L., Pusey, A. E., Bygott, J. D., Hanby, J. P., Cairns, S. J. & Bogerhoff Mulder, M. (1988). Reproductive success of lions. In *Reproductive success* (Ed. T. H. Clutton-Brock). University of Chicago Press, Chicago.

Packer, C., Pusey, A. E., Rowley, H., Gilbert, D. A., Martenson, J. S. & O'Brien, S. J. (1991). A case study of a population bottleneck: lions in the Ngorongoro Crater. *Conservation Biology* **5**, 219–230.

Paine, R. T., Wootton, J. T. & Boersma, P. D. (1990). Direct and indirect effects of peregrine predation on seabird abundance. *Auk* **107**, 1–9.

Papastavrou, V. (1996). Sustainable use of whales: whaling or whale watching? In *The exploitation of mammal populations* (Eds V. J. Taylor & N. Dunstone), pp. 102–113. Chapman & Hall, London.

Paquet, P. C. & Hackman, A. (1993). *Large carnivore conservation in the Rocky Mountains.* WWF Canada, Toronto.

Parker, G. A. (1970). The reproductive behaviour and the nature of sexual selection in *Scatophagua stercoraria* L. II. The fertilisation rate and the spatial and temporal relationships of each sex around the site of mating and oviposition. *Journal of Animal Ecology* **39**, 205–228.

Parker, G. A. & Sutherland, W. J. (1986). Ideal free distributions when individuals differ in competitive ability: phenotype limited ideal free models. *Animal Behaviour* **34**, 1222–1242.

Parr, R. (1993). Nest predation and numbers of golden plovers *Pluviallis apricaria* and other moorland waders. *Bird Study* **40**, 223–231.

Parsons, P. A. (1993). Stress, extinctions and evolutionary change: from living organisms to fossils. *Biological Reviews* **68**, 313–333.

Parsons, P. A. (1995). Stress and limits to adaptation: sexual selection. *Journal of Evolutionary Biology* **8**, 445–461.

Partridge, L. W., Britton, N. F. & Franks, N. R. (1996). Army ant population dynamics: the effects of habitat quality and reserve size on population size and time to extinction. *Proceedings of the Royal Society London B* **263**, 735–741.

Patterson, I. J. (1977). The control of fox movement by electric fencing. *Biological*

Conservation 11, 267–278.

Pauly, D. & Palomares, M. L. (1989). New estimates of monthly biomass, recruitment and related statistics of anchoveta (*Engraulis ringens*) off Peru (4–14° S), 1953–1985. *ICLARM Conference Proceedings* 18, 189–206.

Pauly, D., Muck, P., Mendo, J. & Tsukayama, I. (Eds) (1987). *The Peruvian anchoveta and its upwelling ecosystem: three decades of change.* Instituto del Mar del Peru, Callao.

Pauly, D., Christensen, V., Dalsgaard, J., Froese, R. & Torres, F., Jr (1998). Fishing down marine food webs. *Science* 279, 860–863.

Pawson, M. G., Kelley, D. F. & Pickett, G. D. (1987). The distribution and migrations of bass *Dicentrarchus labrax* (L.) in waters around England and Wales as shown by tagging. *Journal of the Marine Biological Association of the United Kingdom* 67, 183–217.

Payne, R. & Webb, D. (1971). Orientation by means of long range acoustic signalling in baleen whales. *Annals of New York Academy of Science* 188, 110–141.

Peach, W. J., Thompson, P. S. & Coulson, J. C. (1994). Annual and long-term variation in the survival rates of British lapwings *Vanellus vanellus. Journal of Animal Ecology* 63, 60–70.

Peake, T. M. (1997). Variation in the vocal behaviour of the corncrake *Crex crex:* potential for conservation. Unpublished PhD thesis, University of Nottingham, UK.

Peake, T. M. & McGregor, P. K. (1999) Geographical variation in the vocalisation of the corncrake *Crex crex. Ethology, Ecology & Evolution,* in press.

Peake, T. M., McGregor, P. K., Smith, K. W., Tyler, G., Gilbert, G. & Green, R. E. (1998). Individuality in corncrake *Crex crex* vocalizations. *Ibis* 140, 121–127.

Peek, J. M., Pelton, M. R., Picton, H. D., Schoen, J. W. & Zager, P. (1987). Grizzly bear conservation and management: a review. *Wildlife Society Bulletin* 15, 160–169.

Penner, R. C. N. (1995). Oestrogen-based conditioned taste aversion as a method of reducing mammalian predation of duck eggs. MSc thesis, University of Saskatchewan, Saskatoon, Canada.

Percival, S. M., Sutherland, W. J. & Evans, P. R. (1996). A spatial depletion model of the responses of grazing wildfowl to the availability of intertidal vegetation. *Journal of Applied Ecology* 33, 979–992.

Peterson, C. H. & Summerson, H. C. (1992). Basin scale coherence of population dynamics of an exploited marine invertebrate, the bay scallop: implications of recruitment limitation. *Marine Ecology Progress Series* 90, 257–272.

Pettifor, R. A. (1993a). Brood manipulation experiments. I. The number of offspring surviving per nest in blue tits (*Parus caeruleus* L.). *Journal of Animal Ecology* 62, 131–144.

Pettifor, R. A. (1993b). Brood manipulation experiments. II. A cost of reproduction in blue tits (*Parus caeruleus* L.)? *Journal of Animal Ecology* 62, 145–159.

Pettifor, R. A. (1997). Population behaviour in response to anthropogenic change in wetland habitats: the use of long-term data-sets as tools in conservation. *Proceedings of the Xth international waterfowl congress,* pp. 103–115. IWRB/NERC Special Publication, NERC, Swindon.

Pettifor, R. A., Perrins, C. M. & McCleery, R. H. (1988). Individual optimisation of clutch-size in great tits. *Nature* **336**, 160–162.

Pettifor, R. A., Rowcliffe, J. M. & Mudge, G. P. (1997). *Population viability analysis of Icelandic/Greenlandic pink-footed geese. Scottish Natural Heritage Research, Survey and Monitoring Report* No. 65. SNH Publications, Perth.

Pettifor, R. A., Black, J. M., Owen, M., Rowcliffe, J. M. & Patterson, D. (1999). Growth of the Svalbard barnacle goose (*Branta leucopsis*) winter population 1958–1996: an initial review of temporal demographic changes. In *Research on Arctic geese. Proceedings of the Svalbard goose symposium, Oslo, Norway, 23–26 September 1997* (Eds F. Mehlum, J. M. Black & J. Madsen). Norsk Polarinstitutt Skrifter 200, Osloluthavn.

Pfister, C., Harrington, B. A. & Lavine, M. (1992). The impact of human disturbance on shorebirds at a migration staging area. *Biological Conservation* **60**, 115–126.

Phillips, M. K., Smith, R., Henry, V. G. & Lucash, C. (1995). Red wolf reintroduction program. In *Ecology and conservation of wolves in a changing world* (Eds L. N. Carbyn, S. H. Fritts & D. R. Seip). Canadian Circumpolar Institute, Edmonton.

Phillips, R. A., Caldow, R. W. G. & Furness, R. W. (1996a). The influence of food availability on the breeding effort and reproductive success of Arctic Skuas *Stercorarius parasiticus*. *Ibis* **138**, 410–419.

Phillips, R. A., Furness, R. W. & Caldow, R. W. G. (1996b). Behavioural responses of arctic skuas *Stercorarius parasiticus* to changes in sandeel availablity. In *Aquatic predators and their prey* (Eds S. P. R. Greenstreet & M. L. Tasker), pp. 17–25. Blackwell Science, Oxford.

Pianka, E. R. (1970). On r and k selection. *American Naturalist* **104**, 592–597.

Pickering, H. & Whitmarsh, D. (1997). Artificial reefs and fisheries exploitation: a review of the 'attraction versus production' debate, the influence of design and its significance for policy. *Fisheries Research* **31**, 39–59.

Picozzi, N. (1984). Breeding biology of polygynous hen harriers *Circus circus cyaneus* in Orkney. *Ornis Scandinavica* **15**, 1–10.

Pienaar, U. de V. (1969). Predator–prey relationships amongst the larger mammals of the Kruger National Park. *Koedoe* **12**, 108–176.

Pienkowski, M. W. (1984). Breeding biology and population dynamics of ringed plovers *Charadrius hiaticula* in Britain and Greenland: nest predation as a possible factor limiting distribution and timing of breeding. *Journal of Zoology* **202**, 83–114.

Piersma, T. (1987). Hop, skip or jump? Constraints on migration of arctic waders by feeding, fattening, and flight speed. *Limosa* **60**, 185–191.

Piersma, T. (1994). *Close to the edge: energetic bottlenecks and the evolution of migratory pathways in knots.* Uitgeverij Het Open Boek, Den Burg.

Piersma, T. (1997). Do global patterns of habitat use and migration strategies co-evolve with relative investments in immunocompetence due to spatial variation in parasite pressure? *Oikos* **71**, 623–631.

Piersma, T. (1998). Phenotypic flexibility during migration: physiological optimization of organ sizes contingent on the risks and rewards of fueling and flight? *Journal of Avian Biology* **29**, 511–520.

Piersma, T. & Davidson, N. C. (1992). The migrations and annual cycles of five subspecies of knots on perspective. *Wader Study Group Bulletin* **64** (suppl.),

187–197.

Piersma, T. & Gill, R. E., Jr (1998). Guts don't fly: small digestive organs in obese bar-tailed godwits. *Auk* **115**, 196–203.

Piersma, T. & Jukema, J. (1990). Budgeting the flight of a long-distance migrant: changes in nutrient reserve levels of bar-tailed godwits at successive spring staging sites. *Ardea* **78**, 315–337.

Piersma, T. & Jukema, J. (1993). Red breasts as honest signals of migratory quality in a long-distance migrant, the bar-tailed godwit. *Condor* **95**, 163–177.

Piersma, T. & Koolhaas, A. (1997). *Shorebirds, shellfish(eries) and sediments around Griend, western Wadden Sea, 1988–1996*. NIOZ-Report 1997-7, Texel.

Piersma, T. & Lindström, Å. (1997). Rapid reversible changes in organ size as a component of adaptive behaviour. *Trends in Ecology and Evolution* **12**, 134–138.

Piersma, T. & van de Sant, S. (1992). Pattern and predictability of potential wind assistance for waders and geese migrating from West Africa and the Wadden Sea to Siberia. *Ornis Svecica* **2**, 55–66.

Piersma, T. & Wiersma, P. (1996). Family Charadriidae (plovers). In *Handbook of birds of the world*, vol. 3. *Hoatzin to auks* (Eds J. del Hoyo, A. Elliott & J. Sargatal), pp. 384–442. Lynx Edicions, Barcelona.

Piersma, T., Zwarts, L. & Bruggemann, J. H. (1990). Behavioural aspects of the departure of waders before long-distance flights: flocking, vocalizations, flight paths and diurnal timing. *Ardea* **78**, 157–184.

Piersma, T., Hoekstra, R., Dekinga, A., Koolhaas, A., Wolf, P., Battley, P. & Wiersma, P. (1993a). Scale and intensity of intertidal habitat use by knots *Calidris canutus* in the western Wadden Sea in relation to food, friends and foes. *Netherlands Journal of Sea Research* **31**, 331–357.

Piersma, T., Koolhaas, A. & Dekinga, A. (1993b). Interactions between stomach structure and diet choice in shorebirds. *Auk* **110**, 552–564.

Piersma, T., Everaarts, J. M. & Jukema, J. (1996a). Build-up of red blood cells in refueling bar-tailed godwits in relation to individual migratory quality. *Condor* **98**, 363–370.

Piersma, T., van Gils, J. & Wiersma, P. (1996b). Family Scolopacidae (sandpipers, snipes and phalaropes). In *Handbook of birds of the world*, vol. 3. *Hoatzin to auks* (Eds J. del Hoyo, A. Elliott & J. Sargatal), pp. 444–533. Lynx Edicions, Barcelona.

Piersma, T., Wiersma, P. & van Gils, J. (1997). The many unknowns about plovers and sandpipers of the world: introduction to a wealth of research opportunities highly relevant for shorebird conservation. *Wader Study Group Bulletin* **82**, 23–33.

Piersma, T., van Aelst, R., Kurk, K., Berkhoudt, H. & Maas, L. R. M. (1998). A new pressure sensory mechanism for prey detection in birds: the use of seabed-dynamic principles? *Proceedings of the Royal Society of London, Series B*, **265**, 1377–1383.

Pison, G., Trape, J. F., Lefebvre, M. & Enel, C. (1993). Rapid decline in child mortality in a rural area of Senegal. *International Journal of Epidemiology* **22**, 72–80.

Pitcher, T. J. (1995). The impact of pelagic fish behaviour on fisheries. *Scientia Marina* **59**, 295–306.

Pitcher, T. J. & Parrish, J. K. (1993). Functions of shoaling behaviour in teleosts. In

Behaviour of teleost fishes, 2nd edn (Eds T. J. Pitcher), pp. 363–439. Chapman & Hall, London.

Pitelka, F. A., Holmes, R. T. & MacLean, S. F. (1974). Ecology and evolution of social organization in arctic sandpipers. *American Zoologist* **14**, 185–204.

Polsky, R. H. (1975). Hunger, prey, feeding and predatory aggression. *Behavioral Biology* **13**, 81–93.

Polson, J. E. (1983). *Application of aversion techniques for the reduction of losses to beehives by black bears in N. E. Sakatchewan*. Report to the Department of Supply & Services, Ottawa, SRC Publication C-805-13-E-83.

Pomiankowski, A., Iwasa, Y. & Nee, S. (1991). The evolution of costly mate preferences. I. Fisher and biased mutation. *Evolution* **45**, 1422–1430.

Poole, J. H. & Thomsen, J. B. (1989). Elephants are not beetles: implications of the ivory trade for the survival of the African elephant. *Oryx* **23**, 188–198.

Posey, D. A. & Balee, W. (Eds) (1989). *Resource management in Amazonia: indigenous and folk strategies*. New York Botanical Garden, The Bronx, NY.

Possingham, H. (1996). Decision theory and biodiversity management: how to manage a metapopulation. In *Frontiers of population ecology* (Eds R. B. Floyd, A. W. Sheppard & P. J. de Barro). CSIRO Publishing, Melbourne, Australia.

Powell, R. A. (1979). Mustelid spacing patterns: variations on a theme by *Mustela*. *Zeitschrift für Tierpsychologie* **50**, 153–165.

Powell, R. A., Zimmerman, J. W., Seaman, D. E. & Gilliam, J. F. (1996). Demographic analyses of a hunted black bear population with access to a refuge. *Conservation Biology* **10**, 224–234.

Prendergast, M. A., Hendricks, S. E., Yells, D. P. & Balogh, S. (1996). Conditioned taste aversion induced by fluoxetine. *Physiology & Behaviour* **60**, 311–315.

Price, E. C., Ashmore, L. A. & McGivern, A. M. (1994). Reactions of zoo visitors to free-range monkeys. *Zoo Biology* **13**, 355–373.

Price, M. V. & Kelly, P. A. (1994). An age structured demographic model for the endangered Stephen's kangaroo rat. *Conservation Biology* **8**, 810–821.

Prins, H. H. T., van der Jeugd, H. P. & Beekman, J. H. (1994). Elephant decline in Lake Manyara National Park, Tanzania. *Journal of African Ecology* **32**, 185–191.

Prins, T. C. & Smaal, A. C. (1990). Benthic–pelagic coupling: the release of inorganic nutrients by an intertidal bed of *Mytilus edulis*. In *Proceedings of the 24th European marine biology symposium* (Eds M. Barnes & R. N. Gibson), pp. 89–103. Aberdeen University Press, Aberdeen.

Promislow, D. E. L., Montgomerie, R. D. & Martin, T. E. (1992). Mortality costs of sexual dimorphism in birds. *Proceedings of the Royal Society of London B* **250**, 143–150.

Promislow, D. E. L., Montgomerie, R. D. & Martin, T. E. (1994). Sexual selection and survival in North American waterfowl. *Evolution* **48**, 2045–2050.

Pulliam, H. R. (1988). Sources, sinks and population regulation. *American Naturalist* **123**, 652–661.

Puri, R.K. (1995). Commentary. *Current Anthropology* **36**, 789–818.

Purvis, A. & Rambaut, A. (1994). *Comparative analysis by independent contrasts (CAIC)*, version 2. Oxford University, Oxford.

Pusey, A. E. & Packer, C. (1993). Infanticide in lions: consequences and counter-strategies. In *Infanticide and parental care* (Eds S. Parmigiani & F. S. v. Saal).

Harwood Academic Press, London.

Putman, R. J. (1990). The care and rehabilitation of injured wild deer. *Deer* **8**, 31–35.

Quattro, J. M. & Vrijenhoek, R. C. (1989). Fitness differences among remnant populations of the endangered Sonoran topminnow. *Science* **245**, 976–978.

Rabinowitz, A. R. & Nottingham, B. G., Jr (1986). Ecology and behaviour of the jaguar (*Panthera onca*) in Belize, Central America. *Journal of Zoology* **210**, 149–159.

Radford, M. (1996). Partial protection: animal welfare and the law. In *Animal rights: the changing debate* (Ed. R. Garner), pp. 67–91. Macmillan Press, Basingstoke.

Ralls, K. & Ballou, J. D. (1986). Captive breeding programs for populations with a small number of founders. *Trends in Ecology and Evolution* **1**, 19–22.

Ralls, K., Harvey, P. H. & Lyles, A. M. (1986). Seven forms of rarity and their frequency in the flora of the British Isles. In *Conservation biology: the science of scarcity and diversity* (Ed. M. E. Soulé), pp. 35–56. Sinauer, Sunderland, MA.

Ralls, K., Ballou, J. D. & Templeton, A. (1988). Estimates of lethal equivalents and the cost of inbreeding in mammals. *Conservation Biology* **2**, 185–193.

Ramenofsky, M., Piersma, T. & Jukema, J. (1995). Plasma corticosterone in bar-tailed godwits at a major stopover site during spring migration. *Condor* **97**, 580–585.

Ramsay, K., Kaiser, M. J. & Hughes, R. N. (1997). A field study of intraspecific competition for food in hermit crabs (*Pagurus bernhardus*). *Estuarine Coastal and Shelf Science* **44**, 213–220.

Ratcliffe, N., Vaughan, D., Whyte, C. & Shepherd, M. (1998). Development of playback census methods for storm petrels *Hydrobates pelagicus*. *Bird Study* **45**, 302–312.

Rawles, K. (1996). *Conservation and animal welfare*. Department of Philosophy, Lancaster University, Lancaster, UK.

Redford, K. (1991). The ecologically noble savage. *Orion Nature Quarterly* **9**, 25–29.

Redford, K. H. (1992). The empty forest. *BioScience* **42**, 412–422.

Reed, W. J. (1974). A stochastic model for the economic management of a renewable animal resource. *Mathematical Biosciences* **22**, 313–337.

Regan, T. (1983). *The case for animal rights*. Routledge & Kegan Paul, London.

Regehr, H. & Montevecchi, W. A. (1997). Interactive effects of food shortage and predation on breeding failure of black-legged kittiwakes: indirect effects of fisheries activities and implications for indicator species. *Marine Ecology Progress Series* **155**, 249–260.

Reijnen, R. & Foppen, R. (1991). Effect of road traffic on the breeding site-tenacity of male willow warblers (*Phylloscopus trochilus*). *Journal für Ornithologie* **312**, 291–295.

Reijnen, R., Foppen, R., Terbraak, C. & Thissen, J. (1995). The effects of car traffic on breeding bird populations in woodland. 3. Reduction of density in relation to the proximity of main roads. *Journal of Applied Ecology* **32**, 187–202.

Reijnen, R., Foppen, R. & Meeuwsen, H. (1996). The effects of traffic on the density of breeding birds in dutch agricultural grasslands. *Biological Conservation* **75**, 255–260.

Reijnen, R., Foppen, R. & Veenbaas, G. (1997). Disturbance by traffic of breeding birds: evaluation of the effect and considerations in planning and managing road corridors. *Biodiversity and Conservation* 6, 567–581.

Reinhardt, V. & Roberts, A. (1997). Effective feeding enrichment for non-human primates: a brief review. *Animal Welfare* 6, 265–272.

Rettenmeyer, C. W. (1963). Behavioural studies of army ants. *Kansas University Science Bulletin* 44, 281–465.

Revusky, S. & Gorry, T. (1973). Flavour aversions produced by contingent drug infection: relative effectiveness of apomorphine, emetine and lithium. *Behavioural Research & Therapy* 11, 403–409.

Reynolds, J. C. & Nicolaus, L. (1994). Learning to hate gamebirds! *The Game Conservancy Review* 25, 97–99.

Reynolds, J. C. & Tapper, S. C. (1995). The ecology of the red fox (*Vulpes vulpes*) in relation to small game in rural southern England. *Wildlife Biology* 1, 105–119.

Reynolds, J. C. & Tapper, S. C. (1996). Control of mammalian predators in game management and conservation. *Mammal Review* 26, 127–156.

Reynolds, J. D. (1996). Animal breeding systems. *Trends in Ecology and Evolution* 11, 68–72.

Reynolds, J. D. & Szekely, T. (1997). The evolution of parental care in shorebirds: life histories, ecology, and sexual selection. *Behavioral Ecology* 8, 126–134.

Rice, A. G., Lopez, A. & Garcia, J. (1987). Oestrogen produces conditioned taste aversions in rats which are blocked by antihistamine. *Society for Neuroscience Abstracts, 153* 14, 556.

Richards, R. A., Cobb, J. S. & Fogarty, M. J. (1983). Effects of behavioral interactions on the catchability of American lobster, *Homarus americanus*, and two species of *Cancer* crab. *Fishery Bulletin* 81, 51–60.

Ridley, M. (1986). *Animal behaviour: a concise introduction.* Blackwell Scientific Publications, Oxford.

Riley, A. L. & Tuck, D. L. (1985). Conditioned food aversions: a bibliography. In *Experimental assessments and clinical applications of conditioned food aversions* (Eds N. S. Braveman & P. Bronstein). *Annals of the New York Academy of Science* 200, 212–214.

Robinson, J. A. & Warnock, S. E. (1997). The staging paradigm and wetland conservation in arid environments: shorebirds and wetlands of the North American Great Basin. *International Wader Studies* 9, 37–44.

Robinson, J. G. & Redford, K. H. (1991). Sustainable harvest of neotropical forest mammals. In *Neotropical wildlife use and conservation* (Eds J. G. Robinson & K. H. Redford), pp. 415–429. University of Chicago Press, Chicago.

Robinson, S. K., Thompson III, F. R., Donovan, T. M., Whitehead, D. R. & Faaborg, J. (1995). Regional forest fragmentation and the nesting success of migratory birds. *Science* 267, 1987–1990.

Robins-Troeger, J. B. (1994). Evaluation of the Morrison soft turtle excluder device: prawn and bycatch variation in Moreton Bay, Queensland. *Fisheries Research* 19, 205–217.

Rodriguez, J. P. (1997). Conservation and animal welfare – reply. *Trends in Ecology and Evolution* 12, 318.

Roelke, M. E., Martenson, J. S. & O'Brien, S. J. (1993). The consequences of demographic reduction and genetic depletion in the endangered florida panther. *Current Biology* 3, 340–350.

Roelke-Parker, M. E., Munson, L., Packer, C., Kock, R., Cleaveland, S., Carpenter, M., O'Brien, S. J., Popischil, A., Hofmann-Lehmann, R., Lutz, H., Mwamengele, G. L. M., Mgasa, M. N., Machange, G. A., Summers, B. A. & Appel, M. J. G. (1996). A canine distemper virus epidemic in Serengeti lions (*Panthera leo*). *Nature* **379**, 441–445.

Rogers, A. R. (1991). Conserving resources for children. *Human Nature* **2**, 73–82.

Rogers, A. (1994). Evolution of time preference by natural selection. *American Economic Review* **84**, 460–481.

Rogers, A. (1995). For love or money: the evolution of reproductive and material motivations. In *Human reproductive decisions* (Ed. R. Dunbar), pp. 76–95. Macmillan, London.

Rogers, J. G. (1978). Some characteristics of conditioned aversion in red-winged blackbirds. *Auk* **95**, 362–369.

Roll, D. L. & Smith, J. C. (1972). Conditioned taste aversion in anaesthetized rats. In *The biological boundaries of learning* (Eds M. E. P. Seligman & J. L. Hager), pp. 98–102. Appleton-Centuary-Crofts, New York.

Rolley, R. E. (1985). Dynamics of a harvested bobcat population in Oklahoma. *Journal of Wildlife Management* **49**, 283–292.

Ronsmans, C. (1995). Patterns of clustering of child mortality in rural area of Senegal. *Population Studies* **49**, 443–461.

Rose, P. M. & Scott, D. A. (1997). *Waterfowl population estimates*, 2nd edn. Wetlands International, Wageningen.

Roselaar, C. S. (1979). Fluctuaties in aantallen krombekstrandlopers *Calidris ferruginea*. *Watervogels* **4**, 202–210.

Ross, R. M. (1990). The evolution of sex change mechanisms in fishes. *Environmental Biology of Fishes* **29**, 81–93.

Rowcliffe, J. M., Pettifor, R. A., Black, J. M. & Owen, M. (1995). *Population viability analysis of Svalbard barnacle geese*. WWT Report to Scottish Natural Heritage.

Rowcliffe, J. M., Pettifor, R. A. & Black, J. M. (1999). Modelling the dynamics of winter barnacle goose flocks: a progress report. In *Research on Arctic geese. Proceedings of the Svalbard goose symposium, Oslo, Norway, 23–26 September 1997*. (Eds F. Mehlum, J. M. Black & J. Madsen). Norsk Polarinstitutt Skrifter 200, Oslolufhavn.

Rozenberg, M. S. (1997). Evolution of shape differences between the major and minor chelipeds of *Uca pugnax* (Decapoda: Ocypodidae). *Journal of Crustacean Biology* **17**, 52–59.

Rozin, P. & Kalat, J. W. (1971). Specific hungers and poison avoidance as adaptive specialisations of learning. *Psychological Review* **78**, 459–486.

RSPB (1997). Black-throated diver raft project newsletter. RSPB/SNH/Forest Enterprise, unpublished.

Rudnai, J. (1979). Ecology of lions in Nairobi National Park and the adjoining Kitengela Conservation Unit in Kenya. *African Journal of Ecology* **17**, 85–95.

Rulifson, R. A., Murray, J. D. & Bahen, J. J. (1992). Finfish catch reduction in south-Atlantic shrimp trawls using 3 designs of by-catch reduction devices. *Fisheries* **17**, 9–20.

Rusiniak, K. W., Gustavson, C. R., Hankins, W. G. & Garcia, J. (1976a). Prey-lithium aversions. II. Laboratory rats and ferrets. *Behavioural Biology* **17**, 73–86.

Rusiniak, K. W., Hankins, W. G., Garcia, J. & Brett L. P. (1976b). Flavor-illness aversions: potentiation of odor by taste in rats. *Behavioural and Neural Biology* **25**, 1–17.

Russell, E. S. (1939). An elementary treatment of the overfishing problem. *Journal du Conseil, Conseil International pour l'Exploration de la Mer* **110**, 5–14.

Ruttan, L. M. (1998). Closing the commons: cooperation for gain or restraint? *Human Ecology* **26**, 25–49.

Rutton, L. M. & Borgerhoff Mulder, M. (1999). Are East African pastoralists truly conservationists? *Current Anthropology*, **40**, in press.

Ruxton, G. D., Guerney, W. S. C. & De Roos, A. M. (1992). Interference and generation cycles. *Theoretical Population Biology* **42**, 235–253.

Ruzzante, D. E. (1994). Domestication effects on aggressive and schooling behavior in fish. *Aquaculture* **120**, 1–24.

Sabelis, M. W., Diekmann, O. & Jansen, V. A. A. (1991). Metapopulation persistence despite local extinction: predator–prey patch models of the Lotka–Voltera type. *Biological Journal of the Linnean Society* **42**, 267–283.

Saccheri, I., Kuussaari, M., Kankare, M., Vikman, P., Fortelius, W. & Hanski, I. (1998). Inbreeding and extinction in a butterfly population. *Nature* **392**, 491–494.

Sadovy, Y. J. (1996). Reproduction of reef fishery species. In *Reef fisheries* (Eds N. V. C. Polunin & C. M. Roberts), pp. 15–59. Chapman & Hall, London.

Saether, B.-E., Ringsby, T. H. & Roskaft, E. (1996). Life history variation, population processes and priorities in species conservation: towards a reunion of research paradigms. *Oikos* **77**, 217–226.

Saino, N. & Møller, A. P. (1996). Sexual ornamentation and immunocompetence in the barn swallow. *Behavioral Ecology* **7**, 227–232.

Saino, N., Bolzern, A. M. & Møller, A. P. (1997a). Immunocompetence, ornamentation and viability of male barn swallows (*Hirundo rustica*). *Proceedings of the National Academy of Sciences of the USA* **94**, 549–552.

Saino, N., Cuervo, J. J., Krivacek, M., de Lope, F. & Møller, A. P. (1997b). Experimental manipulation of tail ornament size affects haematocrit of male barn swallows (*Hirundo rustica*). *Oecologia* **110**, 186–190.

Saino, N., Cuervo, J. J., Ninni, P., de Lope, F. & Møller, A. P. (1997c). Haematocrit correlates with tail ornament size in three populations of the barn swallow (*Hirundo rustica*). *Functional Ecology* **11**, 604–610.

Sainsbury, A. W., Bennett, P. M. & Kirkwood, J. K. (1995). The welfare of free-living wild animals in Europe: harm caused by human activities. *Animal Welfare* **4**, 183–206.

Salvioni, M. & Lidicker, W. Z. (1995). Social organization and space-use in California voles: seasonal, sexual, and age-specific strategies. *Oecologia* **101**, 526–538.

Samoilys, M. A. (1997). Movement in a large predatory fish: coral trout, *Plectropomus leopardus* (Pisces: Serranidae), on Heron Reef, Australia. *Coral Reefs* **16**, 151–158.

Sandberg, R. & Gudmundsson, G. A. (1996). Orientation cage experiments with dunlins during autumn migration in Iceland. *Journal of Avian Biology* **27**, 183–188.

Sandford, S. (1983). *Management of pastoral development in the third world.* John

Wiley and Sons, London.

Sanjayan, M. A. & Crooks, K. (1996). Skin grafts and cheetahs. *Nature* **381**, 566.

Sanjayan, M. A., Crooks, K., Zegers, G. & Foran, D. (1996). Genetic variation and the immune response in natural populations of pocket gophers. *Conservation Biology* **10**, 1519–1527.

Savage, R. J. G. (1993). *The conservation and management of red deer in the West Country.* National Trust, London.

Savidge, J. A. (1987). Extinction of an island forest avifauna by an introduced snake. *Ecology* **68**, 660–668.

Schaefer, M. B. (1954). Some aspects of the dynamics of populations important to the management of commercial marine fisheries. *Bulletin of the Inter-American Tropical Tuna Commission* **1**, 27–56.

Schaeffer, A. A. (1911). Habit formation in frogs. *Journal of Animal Behaviour* **1**, 309–335.

Schaller, G. B. (1963). *The mountain gorilla, ecology and behaviour.* University of Chicago Press, Chicago.

Schaller, G. B. (1972). *The Serengeti lion: a study of predator–prey relations.* University of Chicago Press, Chicago.

Scheepers, J. L. & Venze, K. A. E. (1995). Attempts to reintroduce African wild dog *Lycaon pictus* into Etosha National Park, Namibia. *South African Journal of Wildlife Research* **25**, 138–140.

Schluter, D. & Price, T. (1993). Honesty, perception and population divergence in sexually selected traits. *Proceedings of the Royal Society of London B* **253**, 117–122.

Schneirla, T. C. (1971). *Army ants: a study in social organization* (Ed. H. R. Topoff). Freeman, San Francisco.

Schofield, R. D. (1960). A thousand miles of fox trails in Michigan's ruffed grouse range. *Journal of Wildlife Management* **24**, 432–434.

Scholten, H. & van der Tol, M. W. M. (1994). SMOES: a simulation model for the Oosterschelde ecosystem. II. Calibration and validation. *Hydrobiologia* **282/283**, 453–474.

Scott, A. (1955). The fishery: the objectives of sole ownership. *Journal of Political Economy* **63**, 116–124.

Scott, J. A. (1975). Flight patterns among eleven species of diurnal Lepidoptera. *Ecology* **56**, 1367–1377.

Seal, U. S. (1991). Life after extinction. *Zoological Society of London Symposium* No. 62, 39–55.

Searcy, W. A. & Yasukawa, K. (1996). Song and female choice. In *The ecology & evolution of acoustic communication in birds* (Eds D. E. Kroodsma & E. H. Miller), pp. 454–473. Comstock, Ithaca, NY.

Segal, E. F. (Ed.) (1989). *Housing care and psychological well-being of captive and laboratory primates.* Noyes, Park Ridge, NJ.

Seidensticker, J. (1986). Large carnivores and the consequences of habitat insuralisation: ecology and conservation of tigers in Indonesia and Bangladesh. In *Cats of the world: biology, conservation and management* (Eds D. Miller & D. D. Everett). National Wildlife Federation, Washington, DC.

Semel, B. & Nicolaus, L. K. (1992). Estrogen-based aversion to eggs among free-ranging raccoons. *Ecological Applications* **2**, 439–449.

Sergeant, D., Ronald, K., Boulva, J. & Berkes, F. (1978). The recent status of *Monachus monachus*, the Mediterranean monk seal. *Biological Conservation* 14, 259–287.

Servheen, C. (1983). Grizzly bear food habits, movements and habitat selection in the Mission Mountains, Montana. *Journal of Wildlife Management* 47, 1026–1035.

Shaffer, M. L. (1983). Determining minimum viable population sizes for the grizzly bear. *International Conference on Bear Research and Management* 5, 133–139.

Shaffer, M. L. & Samson, F. B. (1985). Population size and extinction: a note on determining critical population sizes. *The American Naturalist* 125, 144–152.

Shapiro, A. M. (1970). The role of sexual behavior in density related dispersal in Pierid butterflies. *American Naturalist* 104, 367–372.

Sharp, B. E. (1996). Post-release survival of oiled cleaned sea birds in North America. *Ibis* 138, 222–228.

Sharrock, J. T. R. (1976). *The atlas of breeding birds in Britain*. BTO & IWC, Tring.

Sheffer, R. J., Hedrick, P. W., Minckley, W. L. & Velasco, A. L. (1996). Fitness in the endangered Gila topminnow. *Conservation Biology* 11, 162–171.

Shepherdson, D. (1994). The role of environmental enrichment in the captive breeding and reintroduction of endangered species. In *Creative conservation: interactive management of wild and captive animals* (Eds P. J. S. Olney, G. M. Mace & A. T. C. Feistner), pp. 167–177. Chapman & Hall, London.

Sherrod, S. K., Heinrich, W. R., Burnham, W. A., Baroley, K. H. & Cade, T. J. (1982). *Hacking, a method for releasing peregrine falcons and other birds of prey*. The Peregrine Fund, Inc.

Shields, O. (1967). Hilltopping. *Journal of Research on the Lepidoptera* 6, 69–178.

Shoemaker, A. H. (1982). The effect of inbreeding and management on propagation of pedigree leopards *Panthera pardus* spp. *International Zoo Yearbook* 22, 198–206.

Shoemaker, C. A. (1982). Optimal integrated control of univoltine pest populations with age structure. *Operations Research* 30, 40–61.

Siebels, R. E. (1993). The Bali connection. In *ZooView (Summer)*. Los Angeles Zoo, Los Angeles, CA.

Sieff, D. F. (1995). The effects of resource availability on the subsistence strategies of Datoga pastoralists of north west Tanzania. PhD, University of Oxford, UK.

Sillero-Zubiri, C. & Gottelli, D. (1992). Population ecology of spotted hyaena in an equatorial mountain forest. *African Journal of Ecology* 30, 292–300.

Simmonds, M. P. & Hutchinson, J. D. (Eds) (1996). *The conservation of whales and dolphins*. John Wiley and Sons, Chichester.

Singer, M. C. (1982). Quantification of oviposition behavior in the butterfly *Euphydryas editha*. *Oecologia* 52, 224–229.

Singer, M. C. & Thomas, C. D. (1992). The difficulty of deducing behavior from resource use: an example from hilltopping in checkerspot butterflies. *American Naturalist* 140, 654–664.

Singer, M. C. & Thomas, C. D. (1996). Evolutionary responses of a butterfly metapopulation to human and climate-caused environmental variation. *American Naturalist* 148, S9–S39.

Singer, M. C., Ng, D. & Thomas, C. D. (1988). Heritability of oviposition preference and its relationship to offspring performance within a single insect population. *Evolution* **42**, 977–985.

Singer, P. (1990). *Animal liberation*, 2nd edn. Cape, London.

Slagsvold, T. (1980). Habitat selection in birds: on the presence of other bird species with special regard to *Turdus pilaris*. *Journal of Animal Ecology* **49**, 523–536.

Slater, P. J. B. (1989). Bird song learning: causes and consequences. *Ethology, Ecology & Evolution* **1**, 19–46.

Slatkin, M. (1977). Gene flow and genetic drift in a species subject to frequent local extinction. *Theoretical Population Biology* **12**, 253–262.

Slatkin, M. (1985). Gene flow in natural populations. *Annual Review of Ecology and Systematics* **16**, 393–430.

Slatkin, M. (1987). Gene flow and the geographic structure of natural populations. *Science* **236**, 787–792.

Slobodkin, L. (1968). How to be a predator. *American Zoologist* **8**, 43–51.

Slough, B. G. & Mowat, G. (1996). Lynx population dynamics in an untrapped refugium. *Journal of Wildlife Management* **60**, 946–961.

Smal, C. M. (1991). Population studies on feral American mink *Mustela vison* in Ireland. *Journal of Zoology* **224**, 233–249.

Small, G. L. (1971). *The blue whale*. Columbia University Press, New York.

Smit, C. J. & Visser, G. J. M. (1993). Effects of disturbance on shorebirds: a summary of existing knowledge from the Dutch Wadden Sea and Delta area. *Wader Study Group Bulletin* **68**, 6–19.

Smith, A. T. & Peacock, M. M. (1990). Conspecific attraction and the determination of metapopulation colonization rates. *Conservation Biology* **4**, 320–323.

Smith, B. D. & Jamieson, G. S. (1991). Possible consequences of intensive fishing for males on the mating opportunities of Dungeness crabs. *Transactions of the American Fisheries Society* **120**, 650–653.

Smith, E. A. (1983). Anthropological applications of optimal foraging theory: a critical review. *Current Anthropology* **24**, 625–651.

Smith, E. A. (1995). Commentary. *Current Anthropology* **36**, 789–818.

Smith, J. C. (1971). Radiation: its detection and its effects on taste preferences. In *Progress in physiological psychology*, vol. 4 (Eds E. Stellar & J. M. Sprague), pp. 53–118. Academic Press, New York.

Smith, P., Inglis, I. R., Cowan, D. P., Kerins, G. M. & Bull, D. S. (1994). Symptom-dependent taste aversion induced by an anticoagulant rodenticide in the brown rat (*Rattus norvegicus*). *Journal of Comparative Psychology* **108**, 282–290.

Smith, T. D. (1983). Changes in size of three dolphin (*Stenella* sp.) populations in the eastern tropical Pacific. *Fishery Bulletin* **81**, 1–13.

Smithers, R. H. N. (1983). *Mammals of the Southern African subregion*. University of Pretoria, Pretoria.

Smuts, G. L. (1978). Effects of population reduction on the travels and reproduction of lions in Kruger National Park. *Carnivore* **1**, 61–72.

Snyder, N. F. R., Wiley, J. W. & Kepler, C. B. (1987). The parrots of Loquillo. In *Natural history and conservation of the Puerto Rican parrot*. Western Foundation of Vertebrate Zoology, Los Angeles.

Sorci, G., Møller, A. P. & Clobert, J. (1998). Plumage dichromatism of birds predicts introduction success to New Zealand. *Journal of Animal Ecology* **67**, 263–269.

Soulé, M. E. (1987). *Viable populations for conservation*. Cambridge University Press, Cambridge.

Soulé, M. E. & Wilcox, B. A. (Eds) (1980). *Conservation biology: an evolutionary-ecological perspective*. Sinauer, Sunderland, MA.

Soulé, M. E., Gilpin, M., Conway, W. & Foose, T. (1986). The millennium ark: how long a voyage, how many staterooms, how many passengers? *Zoo Biology* **5**, 101–113.

Spencer, J. A. & Burroughs, R. (1991). Antibody response of captive cheetahs to modified-live feline virus vaccine. *Journal of Wildlife Diseases* **27**, 578–583.

Sperling, L. & Galaty, J. G. (1990). Cattle, culture, and economy. In *The world of pastoralism: herding systems in comparative perspective* (Eds J. G. Galaty & D. L. Johnson), pp. 69–98. Guildford Press, New York.

Stalmaster, M. V. & Newman, J. R. (1978). Behavioral responses of wintering bald eagles to human activity. *Journal of Wildlife Management* **42**, 506–513.

Stamp Dawkins, M. S. (1995). *Unravelling animal behaviour*, 2nd edn. Longmans, Harlow.

Stamps, J. A. (1988). Conspecific attraction and aggregation in territorial species. *American Naturalist* **131**, 329–347.

Stamps, J. A. (1994). Territorial behavior: testing the assumptions. *Advances in the Study of Behavior* **23**, 173–232.

Stamps, J. A., Buechner, M. & Krishnan, V. V. (1987). The effects of edge-permeability and habitat geometry on emigration from patches of habitat. *American Naturalist* **124**, 533–552.

Stander, P. E. (1991). Demography of lions in Etosha National Park, Namibia. *Madoqua* **18**, 1–9.

Stanley Price, M. (1989). *Animal re-introductions: the Arabian oryx in Oman*. Cambridge University Press, Cambridge.

Starfield, A. M. (1997). A pragmatic approach to modelling for wildlife management. *Journal of Wildlife Management* **61**, 261–270.

Starfield, A. M. & Beloch, A. L. (1986). *Building models for conservation and wildlife management*. Macmillan Publishing Company, New York.

Stauffer, D. (1985). *Introduction to percolation theory*. Taylor & Francis Ltd, London.

Steadman, D. R. (1995). Prehistoric extinctions of Pacific Island birds: biodiversity meets zooarchaeology. *Science* **267**, 1123–1131.

Stearman, A. M. (1994). Only slaves climb trees. *Human Nature* **5**, 339–357.

Stearman, A. M. (1995). Commentary. *Current Anthropology* **36**, 789–818.

Steinhart, P. (1992). In the blood of the cheetah. *Audobon* **94**, 40–46.

Stelter, C., Reich, M., Grimm, V. & Wissel, C. (1997). Modelling persistence in dynamic landscapes: lessons from a metapopulation of the grasshopper *Bryodema tuberculata*. *Journal of Animal Ecology* **66**, 508–518.

Stephens, D. W. & Krebs, J. R. (1986). *Foraging theory*. Princeton University Press, Princeton, NJ.

Stevens, S. (1997). Consultation, co-management, and conflict in Sagarmatha (Mount Everest) National Park, Nepal. In *Conservation through cultural survival: indigenous peoples and protected areas* (Ed. S. Stevens), pp. 63–97. Island Press, Washington, DC.

Stewart, P. D. (1990). Mapping the dhole. *Canid News* **1**, 18–21.

Stillman, R. A., Goss-Custard, J. D. & Caldow, R. W. G. (1997). Modelling interference from basic foraging behaviour. *Journal of Animal Ecology* **66**, 692–703.

Stock, M. (1993). Studies on the effects of disturbances on staging brent geese: a progress report. *Wader Study Group Bulletin* **68**, 29–35.

Stokesbury, K. D. E. & Himmelman, J. H. (1993). Spatial distribution of the giant scallop *Placopecten magellanicus* in unharvested beds in the Baie des Chaleurs, Quebec. *Marine Ecology Progress Series* **96**, 159–168.

Stork, N. E. (1988). Insect diversity – facts, fiction and speculation. *Biological Journal of the Linnean Society* **35**, 321–337.

Stork, N. E. (1993). How many species are there? *Biodiversity and Conservation* **2**, 215–232.

Stork, N. E. (1994). How many species are there? (addendum to vol. 2, p. 215, 1993). *Biodiversity and Conservation* **3**, 204–205.

Stouffer, P. C. & Bierregaard, R. O. (1995). Use of Amazonian forest fragments by understory insectivorous birds. *Ecology* **76**, 2429–2445.

Strachan, C., Jefferies, D. J., Barreto, G. R., Macdonald, D. W. & Strachan, R. (1998). The rapid impact of resident American mink on water voles: case studies in lowland England. *Zoological Society of London Symposium* No. 71, 339–357.

van Stralen, M. R. & Dijkema, R. D. (1994). Mussel culture in a changing environment: the effects of a coastal engineering project on mussel culture (*Mytilus edulis*(L.)) in the Oosterschelde estuary (SW Netherlands). *Hydrobiologia* **282/283**, 359–381.

Strickberger, M. W. (1985). *Genetics*. MacMillan, New York.

Suarez, F. M., Yanes, M., Herranz, J. & Manrique, J. (1993). Nature reserves and the conservation of Iberian shrub steppe passerines – the paradox of nest predation. *Biological Conservation* **64**, 77–81.

Sukumar, R. (1989). *The Asian elephant: ecology and management*. Cambridge University Press, Cambridge.

Sukumar, R. (1991). The management of large mammals in relation to male strategies and conflict with people. *Biological Conservation* **55**, 93–102.

Sunquist, F. (1992). Cheetahs, closer than kissing cousins. *Wildlife Conservation* **95**(3), 39–43.

Sunquist, M. E. (1981). The social organisation of tigers (*Panthera tigris*) in Royal Chitwan National Park, Nepal. *Smithsonian Contributions to Zoology* **336**, 1–98.

Sutcliffe, O. L. & Thomas, C. D. (1996). Open corridors appear to facilitate dispersal by ringlet butterflies (*Aphantopus hyperantus*) between woodland clearings. *Conservation Biology* **10**, 1359–1365.

Sutcliffe, O. L., Thomas, C. D. & Peggie, D. (1997). Area-dependent migration by ringlet butterflies generates a mixture of patchy population and metapopulation attributes. *Oecologia* **109**, 229–234.

Sutherland, W. J. (1996a). *From individual behaviour to population ecology. Oxford series in ecology and evolution*. Oxford University Press, Oxford.

Sutherland, W. J. (1996b). Predicting the consequences of habitat loss for migratory populations. *Proceedings of the Royal Society of London B* **263**, 1325–1327.

Sutherland, W. J. (1998a). Managing habits and species. In *Conservation science and action* (Ed. W. J. Sutherland), pp. 202–218. Blackwell Science, Oxford.

Sutherland, W. J. (1998b). The effect of change in habitat quality on populations of migratory species. *Journal of Applied Ecology* **35**, 418–421.

Sutherland, W. J. & Allport, G. A. (1994). A spatial depletion model of the interaction between bean geese and wigeon with the consequences for habitat management. *Journal of Animal Ecology* **63**, 51–59.

Sutherland, W. J. & Anderson, G. W. (1993). Predicting the distribution of individuals and the consequences of habitat loss: the role of prey depletion. *Journal of Theoretical Biology* **160**, 223–230.

Sutherland, W. J. & Crockford, N. J. (1993). Factors affecting the feeding distribution of red-breasted geese *Branta ruficollis* wintering in Romania. *Biological Conservation* **63**, 61–65.

Sutherland, W. J. & Dolman, P. (1994). Combining behaviour and population dynamics with applications for predicting consequences of habitat loss. *Proceedings of the Royal Society of London, Series B* **255**, 133–138.

Sutherland, W. J. & Koene, P. (1982). Field estimates of the strength of interference between oystercatchers *Haematopus ostralegus*. *Oecologia* **55**, 108–109.

Sutherland, W. J. & Parker, G. A. (1985). Distribution of unequal competitors. In *Behavioural ecology* (Eds R. M. Sibly & R. H. Smith), pp. 255–273. Blackwell Scientific Publications, Oxford.

Swennen, C., Leopold, M. F. & de Bruijn, L. L. M. (1989). Time-stressed oystercatchers, *Haematopus ostralegus*, can increase their intake rate. *Animal Behaviour* **38**, 8–22.

Swinton, J., Tuyttens, F. A. M., Macdonald, D. W. & Cheeseman, C. L. (1997). Social perturbation and bovine tuberculosis in badgers: fertility control and lethal control compared. *Philosophical Transactions of the Royal Society of London, Series B* **352**, 619–631.

Tanaka, Y. (1996). Sexual selection enhances population extinction in a changing environment. *Journal of Theoretical Biology* **180**, 197–206.

Tapper, S. C., Potts, G. R. & Brockless, M. H. (1996). The effect of an experimental reduction in predation pressure on the breeding success and population density of grey partridges (*Perdix perdix*). *Journal of Applied Ecology* **33**, 965–978.

Taylor, B. L. (1995). The reliability of using population viability analysis for risk classification of species. *Conservation Biology* **9**, 551–558.

Taylor, V. J. & Dunstone, N. (1996). The exploitation, sustainable use and welfare of wild mammals. In *The exploitation of mammal populations* (Eds V. J. Taylor & N. Dunstone), pp. 3–15. Chapman & Hall, London.

Temple, S. A. (1978). Reintroducing birds of prey to the wild. In *Endangered birds: management techniques for preserving threatened species* (Ed. S. A. Temple), pp. 355–363. University of Wisconsin, Madison.

Templeton, A. R. (1987). Inferences on natural population structure from genetic studies on captive mammalian populations. In *Mammalian dispersal patterns: the effects of social structure on population genetics* (Eds B. D. Chepko-Sade & Z. Tang Halpin), pp. 257–272. University of Chicago Press, Chicago.

Templeton, A. R. & Read, B. (1984). Factors eliminating inbreeding depression in a captive herd of Speke's gazelle (*Gazella spekei*). *Zoo Biology* **3**, 177–199.

Tewes, M. E. & Schmidly, D. J. (1987). The neotropical felids: jaguar, ocelot, margay and jaguarundi. In *Wild furbearer management and conservation in North America* (Eds M. Novak, J. A. Baker, M. E. Obbard & B. Malloch). Ontario Ministry of

Natural Resources, Toronto.

Thomas, C. D. (1983). An effect of the colony edge on gatekeeper butterflies, *Pyronia tithonus* L. (Satyridae). *Journal of Research on the Lepidoptera* **21**, 206–207.

Thomas, C. D. (1992). The establishment of rare insects in vacant habitat. *Antenna* **16**, 89–93.

Thomas, C. D. (1994). Local extinctions, colonizations and distributions: habitat tracking by British butterflies. In *Individuals, populations and patterns in ecology* (Eds S. R. Leather, A. D. Watt, N. J. Mills & K. F. A. Walters), pp. 319–336. Intercept Ltd, Andover.

Thomas, C. D. (1995). Ecology and conservation of butterfly metapopulations in the fragmented British landscape. In *Ecology and conservation of butterflies* (Ed. A. S. Pullin), pp. 46–63. Chapman & Hall, London.

Thomas, C. D. & Hanski, I. (1997). Butterfly metapopulations. In *Metapopulation dynamics: ecology, genetics and evolution* (Eds I. A. Hanski & M. E. Gilpin), pp. 359–386. Academic Press, London.

Thomas, C. D. & Jones, T. M. (1993). Partial recovery of a skipper butterfly (*Hesperia comma*) from population refuges: lessons for conservation in a fragmented landscape. *Journal of Animal Ecology* **62**, 472–481.

Thomas, C. D. & Singer, M. C. (1987). Variation in host preference affects movement patterns in a butterfly population. *Ecology* **68**, 1262–1267.

Thomas, C. D. & Singer, M. C. (1996). Catastrophic extinction of population sources in a butterfly metapopulation. *The American Naturalist* **148**, 957–975.

Thomas, C. D., Jordano, D., Lewis, O. T., Hill, J. K., Sutcliffe, O. L. & Thomas, J. A. (1999). Butterfly distributional patterns, processes and conservation. In *Conservation in a changing world: integrating processes into priorities for action* (Eds G. M. Mace, A. Balmford & J. R. Ginsberg). Cambridge University Press, Cambridge.

Thomas, D. R., Nelson, D. R. & Johnson, A. M. (1987). Biochemical effects of the anti-depressant paroxetine, a specific hydroxytryptamine uptake inhibitor. *Psychopharmacology* **93**, 193–200.

Thomas, J. A., Thomas, C. D., Simcox, D. J. & Clarke, R. T. (1986). The ecology and declining status of the silver-spotted skipper butterfly (*Hesperia comma*) in Britain. *Journal of Applied Ecology* **23**, 365–380.

Thompson, K. V. (1996). Behavioral development and play. In *Wild mammals in captivity* (Eds D. G. Kleiman, M. E. Allen, K. V. Thompson & S. Lumpkin), pp. 352–371. University of Chicago Press, Chicago.

Tillman, M. F. & Donovan, G. P. (Eds) (1986). Report of the workshop. In *Behaviour of whales in relation to management*, pp. 1–56. International Whaling Commission, Cambridge.

Tilman, D., May, R. M., Lehman, C. L. & Nowak, M. A. (1994). Habitat destruction and the extinction debt. *Nature* **371**, 65–66.

Timberlake, W. (1997). An animal-centred, causal-system approach to the understanding and control of behaviour. *Applied Animal Behaviour Science* **53**, 107–129.

Tishendorf, L. & Wissel, C. (1997). Corridors as conduits for small animals: attainable distances depending on movement pattern, boundary reaction and corridor width. *Oikos* **79**, 603–611.

du Toit, J. T., Provenza, F. D. & Nastis, A. (1991). Conditioned taste aversions: how sick must a ruminant get before it learns about toxicity in foods. *Applied Animal Behaviour Science* **30**, 35–46.

Tomikawa, M. (1979). The migrations and inter-tribal relations of the pastoral Datoga. *Senri Ethnological Studies* **5**, 1–46.

Tomkovich, P. S. (1992). An analysis of the geographic variability in knots *Calidris canutus* based on museum skins. *Wader Study Group Bulletin* **64** (suppl.), 17–23.

Tomkovich, P. S. & Soloviev, M.Y. (1994). Site fidelity in high arctic breeding waders. *Ostrich* **65**, 174–180.

Townshend, D. J. & O'Connor, D. A. (1993). Some effects of disturbance to waterfowl from bait-digging and wildfowling at Lindisfarne National Nature Reserve, north-east England. *Wader Study Group Bulletin* **68**, 47–52.

Trewhella, W. J., Harris, S., Smith, G. C. & Nadian, A. K. (1991). A field trial evaluating bait uptake by an urban fox (*Vulpes vulpes*) population. *Journal of Applied Ecology* **28**, 454–466.

Triplet, P. & Etienne, P. (1991). L'huitrier-pie (*Haematopus ostralegus*) face à une diminution de sa principale resource alimentaire la cocque (*Cerastoderma edule*) en baie de Somme. *Bulletin Mensuel Office National de la Chasse* **153**, 21–28.

Tuite, C. H., Hanson, P. R. & Owen, M. (1984). Some ecological factors affecting winter wildfowl distribution on inland waters in England and Wales, and the influence of water-based recreation. *Journal of Applied Ecology* **21**, 41–62.

Tuljapurkar, S. & Caswell, H. (Eds) (1997). *Structured-population models in marine, terrestrial, and freshwater systems.* Chapman & Hall, London.

Turing, A. M. (1952). The chemical basis of morphogenesis. *Philosophical Transactions of the Royal Society B* **237**, 37–72.

Turnpenny, A. W. H., Langford, T. E. & Aston, R. J. (1985). Power stations and fish. *Central Electricity Generating Board Research* **17**, 27–39.

Tuyttens, F. A. M. (1999). The consequences of perturbation caused by badger removal for the control of bovine tuberculosis in cattle: a study of behaviour, population dynamics and epidemiology. D. Phil. thesis, University of Oxford.

Tuyttens, F. A. M. & Macdonald, D. W. (1998a). Fertility control: an option for non-lethal control of wild carnivores? *Animal Welfare* **7**, 339–364.

Tuyttens, F. A. M. & Macdonald, D. W. (1998b). Sterilisation as an alternative strategy to control wildlife diseases: bovine tuberculosis in European badgers as a case study. *Biodiversity & Conservation* **7**, 705–723.

Tuyttens, F. A. M., Macdonald, D. W. & Cheeseman, C. L. (1997). The evolution and function of territorial scent-marking behaviour in the Eurasian badger (*Meles meles*). *Fourth Benelux Congress of Zoology*, 14–15 Nov, Utrecht.

Tyler, G. (1996). The ecology of the corncrake. Unpublished PhD thesis, University of East Anglia, UK.

Tyler, G. & Green, R. E. (1996). The incidence of nocturnal song by male corncrakes *Crex crex* is reduced during pairing. *Bird Study* **43**, 214–219.

Tyler, G., Smith, K. W. & Burgess, D. (1998). Reedbed management and breeding bitterns *Botaurus stellaris* in the UK. *Biological Conservation* **86**, 257–266.

Ulfstrand, S. (1996). Behavioural ecology as a tool in conservation biology: an introduction. *Oikos* **77**, 183

Underhill, L. G., Waltner, M. & Summers, R. W. (1989). Three-year breeding cycles in breeding productivity of knots *Calidris canutus* wintering in southern Africa suggest Taimyr Peninsula provenance. *Bird Study* **36**, 83–87.

Underhill, L. G., Prys-Jones, R. P., Syroechkovski, E. E., Jr, Groen, N. M., Karpov, V., Lappo, H. G., van Roomen, M. W. J., Rybkin, A., Schekkerman, H., Spiekman, H. & Summers, R. W. (1993). Breeding of waders (Charadrii) and brent geese *Branta bernicla bernicla* at Pronchishcheva Lake, northeastern Taimyr, Russia, in a peak and a decreasing lemming year. *Ibis* **135**, 277–292.

Uttley, J. D., Walton, P., Monaghan, P. & Austin, G. (1994). The effects of food abundance on breeding performance and adult time budgets of guillemots *Uria aalge*. *Ibis* **136**, 205–213.

Van der Meer, J. & Ens, B. J. (1997). Models of interference and their consequences for the spatial distribution of ideal and free predators. *Journal of Animal Ecology* **66**, 846–858.

Veitch, C. R. (1985). Methods of eradicating cats from offshore islands in New Zealand. *International Council for Bird Preservation Technical Bulletin* **3**, 125–141.

Venkataraman, A. B., Arumugam, R. & Sukumar, R. (1995). The foraging ecology of dhole (*Cuon alpinus*) in Mudumalai Sanctuary, Southern India. *Journal of Zoology* **237**, 543–561.

Vickers, W. T. (1994). From opportunism to nascent conservation: the case of the Siona-Secoya. *Human Nature* **5**, 307–337.

Vickery, J. A. & Summers, R. W. (1992). The cost-effectiveness of scaring brent geese *Branta b. bernicla* from fields of arable crops by a human bird scarer. *Crop Protection* **11**, 480–484.

Vining, D. R. (1986). Social versus reproductive success – the central theoretical problem of human sociobiology. *Behavioural & Brain Sciences* **9**, 167–260.

Vrijenhoek, R. C. (1985). Animal population genetics and disturbance: the effects of local extinctions and recolonisations on heterozygosity and fitness. In *The ecology of natural disturbance and patch dynamics* (Eds S. T. A. Pickett & P. S. White), pp. 265–285. Academic Press, Orlando.

Vucetich, J. A., Peterson, R. O. & Waite, T. A. (1997). Effects of social structure and prey dynamics on extinction risk in gray wolves. *Conservation Biology* **11**, 957–965.

Wallace, M. P. (1997). Carcasses, people, and power lines. In *ZooView (Spring)*. Los Angeles Zoo, Los Angeles, CA.

Wallace, M. P. & Temple, S. A. (1983). An evaluation of techniques for releasing hand-reared vultures to the wild. In *Vulture biology and management* (Eds S. R. Wilbury & J. A. Jackson), pp. 400–423. University of California Press, Berkeley.

Wallace, M. P. & Temple, S. A. (1985). A comparison between raptors and vulture hacking techniques. In *Proceedings of the international symposium on raptor reintroduction* (Eds D. K. Garcelon & G. W. Roemer). Institute for Wildlife Studies, California.

Wallace, M. P. & Temple, S. A. (1987a). Competitive interactions within and between species in a guild of avian scavengers. *Auk* **104**, 290–295.

Wallace, M. P. & Temple, S. A.(1987b). Releasing captive-reared Andean condors to the wild. *Journal of Wildlife Management* **51**, 541–550.

Wallace, M. R. (1994). Control of behavioral development in the context of reintroduction programs for birds. *Zoo Biology* **13**, 491–499.

Walters, C. J. (1978). Some dynamic programming applications in fisheries management. In *Dynamic programming and its applications* (Ed. M. L. Puterman), pp. 233–246. Academic Press, New York.

Walters, C. & Maguire, J.-J. (1996). Lessons for stock assessment from the northern cod collapse. *Reviews in Fish Biology and Fisheries* **6**, 125–137.

Walters, J. (1984). The evolution of parental behavior and clutch size in shorebirds. In *Shorebirds. Breeding behavior and populations* (Eds J. Burger & B.L. Olla), pp. 243–287. Plenum Press, New York.

Wandeler, A. I. (1991). The control of disease in wildlife. In *Applied animal behaviour: past, present and future. Society for Veterinary Ethology, proceedings of the international congress, Edinburgh* (Eds M. C. Appleby, R. I. Horrell, J. C. Petherick & S. M. Rutter), pp. 100–101. UFAW, Potters Bar, Herts.

Wanless, S., Harris, M. P., Calladine, J. & Rothery, P. (1996). Modelling responses of herring gull and lesser black-backed gull populations to reduction of reproductive output: implications for control measures. *Journal of Applied Ecology* **33**, 1420–1432.

Warburton, B. (1977). Ecology of the Australian brushtailed possum (*Trichosurus vulpecula* Kerr) in an exotic forest. MSc thesis, University of Canterbury, Christchurch.

Wardle, C. S. (1993). Fish behaviour and fishing gear. In *Behaviour of teleost fishes*, 2nd edn (Ed. T. J. Pitcher), pp. 609–643. Chapman & Hall, London.

Warner, R. R. (1988). Sex change and the size-advantage model. *Trends in Ecology and Evolution* **3**, 133–136.

Watkins, R. W., Gill, E. L., Bishop, J. D., Gurney, J. E., Scanlon, C. B., Feare, C. J. & Cowan, D. P. (1994). Cinnamamide: a vertebrate repellent. *Advances in the Biosciences* **93**, 473–478.

Watkins, R. W., Gurney, J. E. & Cowan, D. P. (1999). Taste-aversion conditioning of house mice (*Mus musculus*) using the non-lethal repellent, cinnamamide. *Applied Animal Behaviour Science* **57**, 171–177.

Watson, A., Moss, R., Parr, R., Moutford, M. D. & Rothery, P. (1994). Kin landownership, differential aggression between kin and nonkin, and population fluctuations in red grouse. *Journal of Animal Ecology* **63**, 39–50.

Weaver, L. T. & Beckerleg, S. (1993). Is health a sustainable state? A village study in The Gambia. *The Lancet* **341**, 1327–1331.

Weber, A. W. & Vedder, A. (1983). Population dynamics of the Virunga gorillas: 1959–1978. *Biological Conservation* **26**, 341–366.

Weber, T. P. & Piersma, T. (1996). Basal metabolic rate and the mass of tissues differing in metabolic scope: migration-related covariation between individual knots *Calidris canutus*. *Journal of Avian Biology* **27**, 215–224.

Weber, T. P., Ens, B. J. & Houston, A. I. (1997). Optimal avian migration: a dynamic model of fuel stores and site use.

Wehner, R. (1997). Sensory systems and behaviour. In *Behavioural ecology* (Eds J. R. Krebs & N. B. Davies), pp. 19–41. Blackwell Science, Oxford.

Weisenberger, M. E., Krausman, P. R., Wallace, M. C., Deyoung, D. W. & Maughan, O. E. (1996). Effects of simulated jet aircraft noise on heart-rate and behavior of desert ungulates. *Journal of Wildlife Management* **60**, 52–61.

Wenink, P. W., Baker, A.J. & Tilanus, M. G. J. (1994). Mitochondral control region sequences in two shorebird species, the turnstone and the dunlin, and their utility in population genetic studies. *Molecular Biology and Evolution* **11**, 22–31.

Werner, E. E., Gilliam, J. F., Hall, D. J. & Mittelbach, G. G. (1983). An experimental test of the effects of predation risk on habitat use in fish. *Ecology* **64**, 1540–1548.

West-Eberhard, M. J. (1983). Sexual selection, social competition, and speciation. *Quarterly Review of Biology* **58**, 155–183.

Western, D. (1975). Water availability and its influence on the structure and dynamics of a savannah mammal community. *East African Wildlife Journal* **13**, 265–286.

Western, D. & Henry, W. (1979). Economics and conservation in third world national parks. *Bioscience* **29**, 2764–2769.

White, P. C. L. & Harris, S. (1995). Bovine tuberculosis in badger (*Meles meles*) populations in southwest England: the use of a spatial stochastic model to understand the dynamics of the disease. *Philosophical Transactions of the Royal Society of London, Series B* **349**, 391–413.

White, P. C. L., Saunders, G. & Harris, S. (1996). Spatio-temporal patterns of home-range use by foxes (*Vulpes vulpes*) in urban environments. *Journal of Animal Ecology* **65**, 121–125.

Whitfield, D. P. (1985). Raptor predation on wintering waders in southeast Scotland. *Ibis* **127**, 544–558.

Whitfield, D. P. & Brade, J. J. (1991). The breeding behaviour of the knot *Calidris canutus*. *Ibis* **133**, 246–255.

Whitfield, D. P. & Tomkovich, P. S. (1996). Mating systems and timing of breeding in Holarctic waders. *Biological Journal of the Linnean Society* **57**, 277–290.

Whitmore, T. C. (1997). Tropical forest disturbance, disappearance and species loss. In *Tropical forest remnants: ecology, management and conservation of fragmented communities* (Eds W. F. Laurance & R. O. Bierregaard), pp. 3–12. The University of Chicago Press, Chicago.

Wickman, P.-O. (1988). Dynamics of mate-searching behaviour in a hilltopping butterfly, *Lasiommata megera* (L.): the effects of weather and male density. *Zoological Journal of the Linnaean Society* **93**, 357–377.

Wielebnowski, N. (1996). Reassessing the relationship between juvenile mortality and genetic monomorphism in captive cheetahs. *Zoo Biology* **15**, 353–369.

Wielgus, R. B. & Bunnell, F. L. (1994). Dynamics of a small, hunted brown bear *Ursus arctos* population in southwestern Alberta, Canada. *Biological Conservation* **67**, 161–166.

Wielgus, R. B., Bunnell, F. L., Wakkinen, W. L. & Zager, P. E. (1994). Population dynamics of Selkirk grizzly bears. *Journal of Wildlife Management* **58**, 266–272.

Wiens, J. A. (1985). Vertebrate responses to environmental patchiness in arid and semiarid ecosystems. In *The ecology of natural disturbance and patch dynamics* (Eds S. T. A. Pickett & P. S. White), pp. 169–193. Academic Press, Orlando.

Wiersma, P. & Piersma, T. (1994). Effects of microhabitat, flocking, climate and migratory goal on energy expenditure in the annual cycle of red knots. *Condor* **96**, 257–279.

Wilcoxon, H., Dragoin, W. & Kral, P. (1971). Illness-induced aversions in rat and quail: relative salience of visual and gustatory cues. *Science* **171**, 826–828.

Wild, M. I., Plosker, G. L. & Benfield, P. (1993). Fluvoxamine. an updated review of its pharmacology and therapeutic use in depressive illness. *Drugs* **46**, 895–924.

Wildt, D. E., Bush, M., Howard, J. G., O'Brien, S. J., Meltzer, D., van Dyk, A., Ebedes, H. & Brand, D. J. (1983). Unique seminal quality in the South African cheetah and a comparative evaluation in the domestic cat. *Biology of Reproduction* **29**, 1019–1025.

Wildt, D. E., O'Brien, S. J., Howard, J. G., Caro, T. M., Roelke, M. E., Brown, J. L. & Bush, M. (1987). Similarity in ejaculate-endocrine characteristics in captive versus free-living cheetahs of two subspecies. *Biology of Reproduction* **36**, 351–360.

Wildt, D. E., Brown, J. L., Bush, M., Barone, M. A., Cooper, K. A., Grisham, J. & Howard, J. G. (1993). The reproductive status of the cheetah (*Acinonyx jubatus*) in North American zoos: the benefits of physiological surveys for strategic planning. *Zoo Biology* **12**, 45–80.

Wiley, R. H. (1994). Errors, exaggeration and deception in animal communication. In *Behavioral mechanisms in evolutionary ecology* (Ed. L. Real), pp. 157–189. University of Chicago Press, Chicago.

Williams, C. K. & Twigg, L. E. (1996). Responses of wild rabbit populations to imposed sterility. In *Frontiers in population ecology* (Eds R. B. Floyd, A. W. Sheppard & P. J. De Barro), pp. 547–560. CSIRO, Melbourne.

Williams, G. C. (1966). *Adaptation and natural selection.* Princeton University Press, Princeton, NJ.

Williams, T. C. & Williams, J. M. (1990). The orientation of transoceanic migrants. In *Bird migration. Physiology and ecophysiology* (Ed. E. Gwinner), pp. 7–21. Springer-Verlag, Berlin.

Willis, E. O. (1967). The behaviour of bicolored antbirds. *University of California, Publications in Zoology* **79**, 1–127.

Willis, E. O. & Oniki, Y. (1978). Birds and army ants. *Annual Review of Ecology and Systematics* **9**, 243–263.

Wilson, J. D., Evans, J., Brown, S. J. & King, J. R. (1997). Territory distribution and breeding success of skylarks *Aluada arvensis* on organic and intensive farmland in southern England. *Journal of Applied Ecology* **34**, 1462–1478.

Winterhalder, B. P. (1997). Foraging strategy adaptations of the boreal forest Cree: an evaluation of theory and models from evolutionary ecology. PhD dissertation, Cornell University, USA.

Winterhalder, B. & Lu, F. (1997). A forager-resource population ecology model and implications for indigenous conservation. *Conservation Biology* **11**, 1354–1364.

Wittlin, W. A. & Brookshire, K. H. (1968). Apomorphine-induced conditioned taste aversion to a novel food. *Psychonomic Science* **12**, 217–218.

Wolff, J. O. (1994). More on juvenile dispersal in mammals. *Oikos* **71**, 349–352.

Wood, J. B. (1983). The conservation and management of animal populations. In *Conservation in perspective* (Eds A. Warren & F. B. Goldsmith), pp. 119–139. Wiley, Chichester.

Woodard, J. C., Forrester, D. J., White, F. H., Gaskin, J. M. & Thompson, N. P. (1977). An epizootic among knots (*Calidris canutus*) in Florida. I. Disease syndrome, histology and transmission studies. *Veterinary Pathology* **14**, 338–350.

Woodroffe, R. & Ginsberg, J. R. (1997). The role of captive breeding and reintroduction in wild dog conservation. In *The African wild dog: status survey and conservation action plan* (Eds R. Woodroffe, J. Ginsberg & D. Macdonald), pp. 100–111. IUCN/SSC Canid Specialist Group, IUCN, Cambridge.

Woodroffe, R., Macdonald, D. W. & da Silva, J. (1993). Dispersal and philopatry in the European badgers, *Meles meles*. *Journal of Zoology* **237**, 227–239.

Woodroffe, R., Ginsberg, J. R. & Macdonald, D. W. (Eds) (1997). *The African wild dog: status survey and conservation action plan*. IUCN/SSC Canid Specialist Group, IUCN, Cambridge.

Wright, P., Barrett, R. T., Greenstreet, S. P. R., Olsen, B. & Tasker, M. L. (1996). Effect of fisheries for small fish on seabirds in the eastern Atlantic. In *Seabird/fish interactions, with particular reference to seabirds in the North Sea* (Eds G. L. Hunt & R. W. Furness), pp. 44–55. ICES Cooperative Research Report No. 216. International Council for the Exploration of the Sea, Copenhagen.

Wright, S. (1931). Evolution in mendelian populations. *Genetics* **16**, 97–159.

Wuethrich, B. (1995). El-Niño goes critical. *New Scientist* **145**, 32–35.

Yanagimachi, Y. & Sato, A. (1968). Effects of a single oral administration of ethinyl estradiol on early pregnancy in the mouse. *Fertility and Sterility* **19**, 787–801.

Yasuda, Y., Kihara, T. & Nishimura, H. (1981). Effect of ethinylestradiol on development of mouse foetuses. *Teratology* **23**, 233–239.

Yerli, S., Canbolat, A. F., Brown, L. J. & Macdonald, D. W. (1997). Mesh grids protect loggerhead turtles *Caretta caretta* nests from red fox *Vulpes vulpes* predation. *Biological Conservation* **82**, 109–111.

Young, B. F. & Ruff, R. L. (1982). Population dynamics and movements of black bears in East Central Alberta. *Journal of Wildlife Management* **46**, 845–860.

Young, T. & Harcourt, A. H. (1997). !Viva Caughley! *Conservation Biology* **11**, 831–832.

Young, T. P. & Isbell, L. A. (1994). Minimum group size and other conservation lessons exemplified by a declining primate population. *Biological Conservation* **68**, 129–134.

Yuhki, N. & O'Brien, S. J. (1990). DNA variation of the mammalian major histocompatability complex reflects genomic diversity and population history. *Proceedings of the National Academy of Sciences USA* **87**, 836–840.

Zabel, C. J. & Taggeart, S. J. (1989). Shift in red fox, *Vulpes vulpes*, mating system associated with El Niño in the Bering Sea. *Animal Behaviour* **38**, 830–838.

Zahorik, D. M., Houpt, K. A. & Swartzman-Andert, J. (1990). Taste-aversion learning in three species of ruminants. *Applied Animal Behaviour Science* **26**, 27–39.

Zann, L. P. (1989). Traditional management and conservation of fisheries in Kiribati and Tuvalu Atolls. In *Traditional marine resource management in the Pacific Basin: an anthology* (Eds K. Ruddle & R. E. Johannes), pp. 77–102. UNESCO/ROSTSEA, Jakarta, Indonesia.

Zeigler, J. M., Gustavson, C. R., Holzer, G. A. & Gruber, D. (1983). Anthelmintic-based taste aversion in wolves. *Applied Animal Ethology* **9**, 373–377.

Zezulak, D. S. & Schwab, R. G. (1979). A comparison of density, home range and habitat utilization of bobcat populations of Lava Beds and Joshua Tree National Monuments, California. Paper presented at the Bobcat Research Conference, Oct 15–18, Front Royal, Virginia.

Zhou, S. J. & Shirley, T. C. (1997). Behavioural responses of red king crab to crab pots. *Fisheries Research* **30**, 177–189.

Zwarts, L. & Wanink, J. H. (1993). How the food supply harvestable by waders in the Wadden Sea depends on variation in energy content, body weight, biomass, burying depth and behaviour of tidal-flat invertebrates. *Netherlands Journal of Sea Research* **31**, 441–476.

Zwarts, L., Blomert, A.-M. & Wanink, J. H. (1992). Annual and seasonal variation in the food supply harvestable by knot *Calidris canutus* staging in the Wadden Sea in late summer. *Marine Ecology Progress Series* **83**, 129–139.

Zwarts, L., Ens, B. J., Goss-Custard, J. D., Hulscher, J. B. & Kersten, M. (1996). Why oystercatchers *Haematopus ostralegus* cannot meet their daily energy requirements in a single low water period. *Ardea* **84A**, 269–290.

Zwarts, L., van der Kamp, J., Overdijk, O., van Spanje, T. M., Veldkamp, R., West, R. & Wright, M. (1998). Wader count of the Banc d'Arguin, Mauritania, in January/February 1997. *Wader Study Group Bulletin* **86**, 53–69.

Index

Note: page numbers in *italics* refer to figures and tables

training 313
see also conditioned taste aversion (CTA)

badger, Eurasian 298, 318, 320–1
  disease spread 327–8
  dispersal 319
  encounters 327–8
Badger Removal Operations 327
bait digging, shorebirds 69–71
bait-shyness 281, 324
Balaenoptera musculus 240, *241*, 277
  see also whale, blue
Barabaig people (Tanzania)
  adjudication of disputes 46
  common land tenure 41
  conservation ethic 41
  dry season reserve 41, 42
  elders 45–6
  grassland preservation 41
  grazing
    movements 40
    patterns 40
    regulations controlling 41
    reserves 38–43, *44*, 45–7
    rotation 41
    strategy pay-offs 42
  herders 42
  poor herder cooperation 47
  punishment and regulation 46, 47
  wet-season pastures 40, 42
barrages, marine 250, 252
barriers, physical 337
Barro Colorado Island (Panama) 144, *145*, 157
bass, European 250, 252
bats, acoustic signals 262
bear
  black *128*, 133, 287
  brown/grizzly *128*, 133, 139, 323
behaviour
  incorporation into predictive models 198
  population viability 174–6
behaviour-based models, conservation studies
  218–19
behavioural diversity, population viability 325
behavioural ecology 3–4, 5, 6–7
behavioural incompatibility 5
benthic community, shellfishing 73
bias
  capture techniques 268–9
  sex from culling 317
biodiversity, destruction control 33
birds
  acoustic signals 262

edge effects for migratory 136–7
extravagant feather ornaments 170
human disturbance of populations 52
multiple clutching 306
sexually dichromatic species 166–7
song geographic variation 273
traffic noise 277
see also seabirds; shorebirds; wildfowl
birth interval 16
bittern 262, 267
  habitat quality 272
  quality of signal 271
  signalling costs 271
blue, silver-studded 8, *88*, 89, 92–5
bobcat 317, 320–1, 324
body size and extinctions 165–6
Boiga irregularis (brown tree snake) 307
Botarus stellaris 262
  see also bittern
boundary effects 98
brain stem 284, 289
Branta bernicla (brent goose) 215, 217
Branta leucopsis 201–2, 212
  see also goose, barnacle
breeding programmes for captive animals 335
breeding status and song variation 270–1
breeding system, human disturbance 321–2
brideprice 19
brood parasitism, interspecific 136–7
buffer areas, national parks 138
bunting, corn 274, *275*, 276
bustard, Houbara 309
Buteo jamaicensis (buzzard) 294–5
butterflies
  adult resources 97–8
  behaviour 95
  boundary effects 98
  breeding habitat 86
  corridors 99
  density 99–101
  dispersal 85
    behaviour 101–4
    density 99–100
    distance distribution 89
    estimate 88
    rates 92–5
  edges 98
  egg-laying sites 95, 96, 99, 102
  emigration 102
    density 100
    determinants 97, 98–9
    fraction 86–8, 90–1, 92
  female aggregation 101